# Resilient Control of Intelligent Connected Vehicle Platoon Systems

Chengwei Pan · Yong Chen

# Resilient Control of Intelligent Connected Vehicle Platoon Systems

Springer

Chengwei Pan
School of Automation Engineering
University of Electronic Science
and Technology of China
Chengdu, Sichuan, China

Yong Chen
School of Automation Engineering
University of Electronic Science
and Technology of China
Chengdu, Sichuan, China

ISBN 978-981-97-8849-1 ISBN 978-981-97-8850-7 (eBook)
https://doi.org/10.1007/978-981-97-8850-7



This Springer imprint is published by the registered company Springer Nature Singapore Pte Ltd.
The registered company address is: 152 Beach Road, #21-01/04 Gateway East, Singapore 189721, Singapore

# Acknowledgements

This work was supported by the National Natural Science Foundation of China under Grant no. 62403106, 61973331 and 61973257, the Postdoctoral Fellowship Program of CPSF under Grant no. GZC20240221, the National Key Research and Development Plan Programs of China no. 2022YFE0120700, the Sichuan Province high-tech field key research and development project under Grant no. 2023YFG0200, the Natural Science Foundation of Sichuan Province under Grant no. 2022NSFSC0867, and 2024NSFSC1491.

Meanwhile, I would also like to express my sincere gratitude to the Sichuan Province Engineering Technology Research Center for Electric Vehicle Driving System and Safety, and the Institute for Electric Vehicle Driving System and Safety Technology of University of Electronic Science and Technology of China. Their excellent research platform and extensive support have played an important role in the successful completion of this book. I am deeply thankful for the resources, guidance, and collaborative environment they have provided, which have greatly contributed to the development of this work.

# Contents

# Chapter 1
# Introduction

## 1.1 Research Background and Significance

Through highly integrated technology systems such as internet of vehicles communication, autonomous driving and cloud-edge computing, intelligent connected vehicles have become a new industrial form to realize the integrated development of human-vehicle-road-cloud ubiquitous perception, information fusion and collaborative control [1]. Driven by the transformation of the new four modernizations automobile industry such as electrification, intelligence, networking and sharing, as well as the policy guidance and promotion of the United States, China, the European Union and Japan, the technological innovation and system construction of the intelligent connected vehicles industry have also risen to become a development strategy at the national level.

As the key industry technology field of intelligent connected vehicles, intelligent connected vehicle platoon system (ICVPS) can organically integrate advanced control, mobile communication computing, and unmanned driving to realize complex environment perception, multilateral interactive mapping, distributed control and decision making [2]. From the perspective of the control field, ICVPS has a typical three-layer structure, namely, the physical entity layer, the network communication layer and the control layer, etc. Ensuring operation safety, improving traffic efficiency and reducing energy consumption are the basic characteristics [3]. The physical layer of ICVPS describes the dynamic characteristics of intelligent vehicles in complex road environments, while the network communication layer explicitly expresses the network behavior of ICVPS. With the new generation of cellular vehicle-to-everything (C-V2X) network communication technology, intelligent connected vehicles in the same lane can dynamically form vehicular ad hoc networks (VANETs) through the on-board unit (OBU). Vehicle-to-vehicle (V2V) and vehicle-to-infrastructure (V2I) communication interaction can be realized. Interactive mapping of multi-source data information of multiple intelligent vehicle physical

 
C. Pan and Y. Chen, *Resilient Control of Intelligent Connected Vehicle Platoon Systems*,
https://doi.org/10.1007/978-981-97-8850-7_1

entities, and complex road environment perception can also be achieved. Furthermore, As the core part of ICVPS, the control layer can realize the close interaction and closed-loop feedback between the physical layer and the network communication layer. The motion state of each intelligent vehicle can be adjusted through integrated calculation, data analysis, intelligent decision-making and advanced control, and the cooperative consistency of velocity and inter-vehicle distance are achieved. Thus, air resistance can be effectively reduced, energy consumption such as fuel can be saved, and road traffic flow passing efficiency can be improved [4].

## 1.2 Overview of Intelligent Connected Vehicle Platoon Systems

As one of the eight key technologies of intelligent transportation, ICVPS has attracted the attention of countries all over the world. In 1986, the University of California, Berkeley, and the California Department of Transportation jointly launched the PATH project, which first proposed the concept of ICVPS. In 2018, the U.S. Department of Transportation released "Preparing for the Future of Transportation: Automated Vehicles 3.0" and made strategic deployment for the development of ICVPS technology. The European Union launched the SARTRE project in 2009 and the ENSEMBLE project in 2017, and published "Connected, Cooperative and Automated Mobility Roadmap" in 2019 to advance the development of ICVPS driving technology. In 2021, Japanese officials announced the success of the automated truck platoon driving verification test, as shown in Fig. 1.1.

China also attaches great importance to the development of ICVPS. Led by the sub-Technical Committee of the Intelligent Connected Vehicle of the National Standards Committee, the first large-scale intelligent connected vehicle platoon tracking test was conducted in Tianjin in May 2019. In May 2020, Baidu also conducted a test verification of the intelligent connected vehicle platoon in the Cangzhou intelligent connected vehicle manned test network. In 2018, the Ministry of Industry and Information Technology of China issued "Action Plan of Development of the Internet of Vehicles (Intelligent Connected Vehicles) Industry", which clearly pointed out that the application of longitudinal platooning driving of trucks on highways should be promoted to improve the traffic efficiency. Subsequently, in February 2021, the Ministry of Industry and Information Technology of China, together with the Ministry of Transport and the National Standards Management Committee of China issued the "Guidance of the Construction of the National Internet of Vehicles Industry Standard System (Intelligent Connected Vehicles)" to plan the relevant standards for ICVPS transportation organizations.

The above analysis shows that ICVPS is a strong support for promoting the innovation and integration of the intelligent connected vehicles industry chain and the development of ecological standardization. ICVPS is also a strong driving force for promoting the transformation and upgrading of the automobile industry as well as

**Fig. 1.1** Connected vehicle platoon system testing and application in countries around the world

the development of intelligent, networked and shared development, and a basic guarantee for the integration and upgrading of new forms of intelligent transportation and new infrastructure.

From the perspective of system structure, ICVPS closely integrates multiple connected and unmanned vehicle physical entities, mobile wireless communication networks and control units. ICVPS is a large-scale complex connected cyber physical fusion system with multiple information sources with physical layer, network communication layer and control layer structure, as shown in Fig. 1.2. However, due to the openness of the wireless communication network, the complexity of the traffic environment and the vulnerability of the security of the ICVPS, random and unpredictable physical faults will affect the safe driving of intelligent connected vehicles. The cyber attacks induced by deliberate artificial network attack is also bound to cause the closed-loop stability and string stability of other subsystems and even the entire ICVPS control layer to decline, or even collapse. Finally, resulting in vehicle collisions and other traffic accidents.

### *1.2.1 Intelligent Connected Vehicle Platoon Systems with Physical Faults and Cyber Attacks*

On the one hand, there are many practical cases of physical faults affecting the safety performance of intelligent connected vehicles. In October 2021, Faw-volkswagen announced a recall of more than 104,000 Audi A3 and A4L models due to frequent

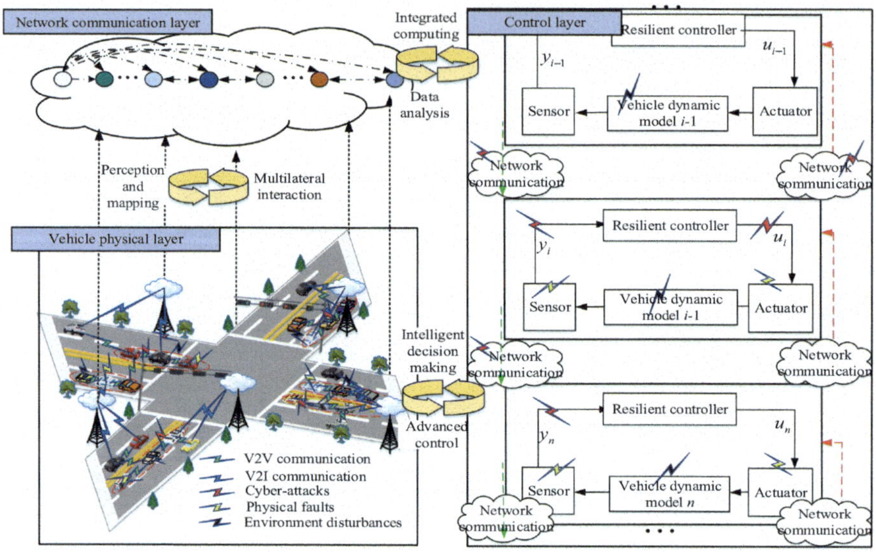


**Fig. 1.2** The structure diagram of ICVPS

faults of the electromechanical control unit of the gearbox, which can lead to sudden velocity increases. In April 2022, Tesla recalled more than 120,000 Model 3 electric vehicles after a small manufacturing difference in the power semiconductor components caused the rear inverter to malfunction, leading to the vehicle lose driving power. It can be seen from the above events that the physical fault is caused by the defect or failure of the vehicle subsystem component, and its generation is random and unpredictable. Physical fault can directly affect the vehicle control system, causing the control performance of the system to decay and destroy its stability.

Physical faults are usually caused by the faults of mechanical or electronic physical devices such as actuators, sensors and system components in the physical layer of the system. Among them, actuator faults pose the greatest threat to the stability of ICVPS due to their wide variety and direct effects on the transmission control system and motion control system of the vehicle [5]. On the other hand, executive devices such as hub motors, inverter rectifiers and hydraulic cylinders of intelligent unmanned vehicles are prone to sudden loss of efficiency, drift, random nonlinearity, periodicity and intermittency problems during long-term high-load operation [6, 7]. The types of physical faults are varied and can not be predicted accurately, which puts higher requirements on ICVPS driving safety and stability control.

On the other hand, deliberate cyber attacks can also exploit inherent "back doors" or vulnerabilities in network communication layer defense mechanisms and application programming interfaces to induce system communication faults and threaten ICVPS control security. In 2015, two researchers in the United States remotely hacked into a Jeep Cherokee, causing the vehicle's brakes, steering and acceleration controls to fail. In 2019, the automotive threat research team of 360 Company found that there were remote information control unit vulnerabilities in Mercedes-Benz cars, which seriously threatened the control security of the engine. In 2022, the hacker found that the remote communication system of Acura, Honda and other automobile brands had API endpoint vulnerabilities, which could easily cause engine start and stop faults. The above cases show that the cyber attack has the ability to enter the network communication environments of ICVPS and the control signal communication interaction network, induce communication faults, and destroy the system information interaction and transmission. Malicious attackers take advantage of cyber attacks to threaten the interaction process of information, which makes the system output information and control input information in the network communication layer incomplete, unavailable and untrustworthy.

Different from physical faults, the performance behaves of cyber attacks are usually generated based on the existing resource configuration of malicious attackers, with obvious intelligent characteristics and pertinence. cyber attacks usually can destroy the observability of the system and threaten the safety of ICVPS. Malicious attackers can use denial of service attacks to send massive useless data to the information exchange channel through network communication security defense vulnerabilities to occupy the normal network communication channel. It can result in system communication congestion or paralysis, and affect the availability of interactive information [8]. Compared with denial of service attack, deception attack mainly destroys the integrity and credibility of communication interaction information by tampering with it. Deception attack behavior is more hidden and targeted, and thus more destructive to system stability [9].

In July 2021, China's Ministry of Industry and Information Technology issued the "Opinions on Strengthening the Management of Intelligent Connected Vehicle Production Enterprises and Product Access", which requires accelerating the revision of safety standards related to automatic driving and strengthening the driving safety guarantee of intelligent connected vehicles. In November 2022, the Chinese government also issued the "Notice on the Pilot Work of Intelligent Connected Vehicles Access and road Access (Draft for Comments)" document, which also requires the protection of intelligent connected vehicles driving safety. Therefore, it is realistic, urgent and important to reduce the impact of internal physical faults and communication faults induced by malicious cyber attacks on ICVPS driving safety. Based on the practical requirements of ICVPS for system control performance such as high security, high stability, strong robustness and fast recovery, this book studies adaptive platoon resilient control technology under fault and cyber attack conditions from the perspective of control field. ICVPS driving safety can be ensured and the system's rapid recovery ability and resilient performance under fault and cyber attack conditions also can be improved. ICVPS' comprehensive security and stability assurance capabilities can also be strengthened. This work has important engineering application value and research significance.

## 1.3 Longitudinal Control of Intelligent Connected Vehicle Platoon Systems

Based on VANETs and the new generation of C-V2X omnidirectional network communication technology, the longitudinal control of ICVPS can realize that multiple intelligent connected vehicle subsystem units on the same lane can form a longitudinal moving platoon, and drive at the same speed and desired distance [10]. Therefore, longitudinal control is the basis of ICVPS control. However, due to the complexity of wireless communication network and traffic environment, as well as the vulnerability of ICVPS security under the constraint of physical equipment, the driving safety problem caused by faults and cyber attacks is inevitable. The security issues of ICVPS are very easy to cause the control instability of unmanned vehicles and the string instability of ICVPS, resulting in the threat of collision. Resilient control can effectively deal with the impact of faults or cyber attacks on the stability of ICVPS through feedback adjustment. It is one of the effective measures to improve the resilience and tolerance of ICVPS under fault or attack conditions. Therefore, it is of great significance to integrate resilient control technology into longitudinal secure control framework and study the longitudinal resilient control of ICVPS. Academician Li Keqiang of Tsinghua University and Han Qinglong of the European Academy of Sciences both noted: Compared to human-driven vehicles, longitudinal control of ICVPS can travel at higher speeds with shorter inter-vehicle spacing, which has advantages in reducing urban congestion, increasing traffic capacity, ensuring driving safety, and reducing fuel consumption [11, 12]. The longitudinal control issue of ICVPS was first proposed by Levine from the University of Maryland, College Park, USA [13]. The key objective of longitudinal control of the ICVPS is to regulate a platoon of following intelligent connected vehicles to track the leader vehicle with the desirable velocity, and acceleration simultaneously, keep a safe and the required inter-vehicle spacing. The vast majority of existing methods about the distributed longitudinal control of the vehicle platoons are based on five correlative parts, which are vehicle longitudinal dynamics modelling, the information flow topology, the inter-vehicle spacing policy, the distributed longitudinal controller of the ICVPS, stability performances [14, 15], as shown in Fig. 1.3. On the basis of the above, the controller is designed to guarantee the closed-loop control stability of the entire vehicle platoon system, and to ensure that faults and disturbances will not cause large transient spacing errors during the layer-by-layer propagation of the platoons [16]. So the string instability and collisions can be avoided. The vehicle platoon system can finally travels safely at the same speed and desired inter-vehicle distance.

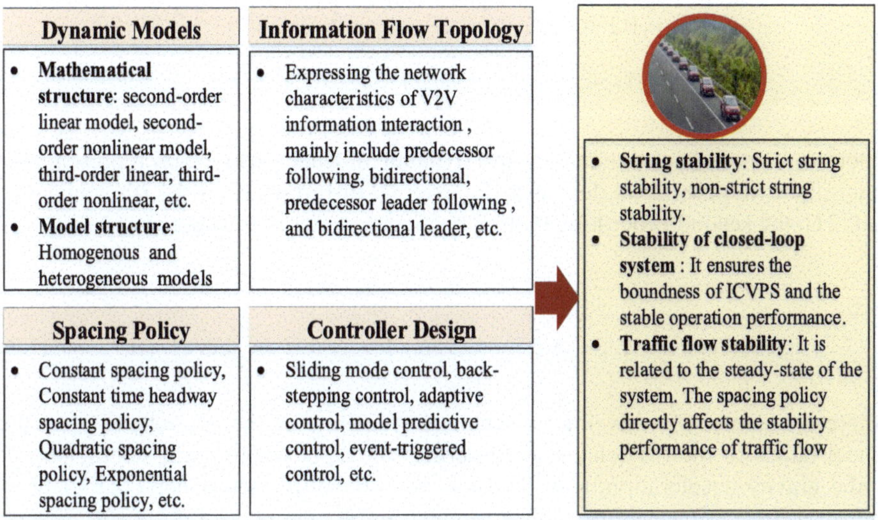


**Fig. 1.3** Longitudinal control structure of ICVPS

## *1.3.1 Model of Intelligent Connected Vehicle Longitudinal Dynamics*

Linear and nonlinear dynamical models are widely used to describe the longitudinal dynamical behavior of ICVPS. For one thing, according to the different mathematical forms of the modeling, the longitudinal dynamics model can be divided into second-order and third-order models. In this section, the review over several popular linear and nonlinear vehicle models is given. For example, in [17, 18], considering $n + 1$ intelligent connected vehicles driving on longitude movements within a straight lane by forming a platoon, in which one leader vehicle and $n$ follower vehicles are included. The second-order longitudinal linear dynamic model is considered as,

$$\begin{aligned} \dot{\bar{p}}_i(t) &= \bar{v}_i(t) \\ \dot{\bar{v}}_i(t) &= u_i(t), i = 0, 1, 2, \ldots, n \end{aligned} \tag{1.1}$$

where $\bar{p}_i(t), \bar{v}_i(t) \in \mathbb{R}$ denote the position and velocity of the $i$th vehicle, $u_i(t) \in \mathbb{R}$ is the control input of $i$th vehicle.

Due to the small external disturbances in the vehicle platoon system can cause large spacing errors between vehicles down the platoon. On this basis, in [19–21], the external disturbances or parameter uncertainties are further considered in the model, and the vehicle platoon control design of second-order nonlinear model is studied,

$$\begin{aligned}\dot{\bar{p}}_i(t) &= \bar{v}_i(t)\\ \dot{\bar{v}}_i(t) &= \frac{1}{m_i}u_i(t) + \omega_i(t), i = 1, 2, \ldots, n\end{aligned} \tag{1.2}$$

where $\dot{\bar{p}}_0(t) = \bar{v}_0(t)$ is the longitudinal dynamic of leader vehicle, $m_i$ represents the mass of $i$th vehicle. $\omega_i(t)$ describes the external disturbances of the $i$th vehicle. In [20, 21], the second-order longitudinal nonlinear model is considered as

$$\begin{aligned}\dot{\bar{p}}_i(t) &= \bar{v}_i(t)\\ \dot{\bar{v}}_i(t) &= \frac{1}{m_i}\left(\bar{\eta}_i\frac{1}{r_i}u_i(t) - \bar{C}_i\bar{v}_i^2(t)\right) - g_i\bar{f}_i, i = 1, 2, \ldots, n\end{aligned} \tag{1.3}$$

where $\bar{\eta}_i$ denotes the driveline mechanical efficiency of the $i$th vehicle, $r_i$ is the wheel radius of the $i$th vehicle, $\bar{C}_i$ describes the aerodynamic drag coefficient, $g_i$ is the gravity acceleration, and $\bar{f}_i$ denote the tire rolling resistance coefficient of the $i$th vehicle. However, the above research in platoon control design is an ideal second-order model based on vehicle kinematics, which takes acceleration as the control input of the system and only considers the position and velocity state of the vehicle, ignoring the internal dynamics and nonlinear characteristics of the vehicle such as throttle and brake torque, mechanical resistance, engine and transmission system and so on.

For getting closer to the real system, and improving the applicability of the model, striking the balance between control accuracy and conciseness. The third-order linear dynamics model [22–24] is considered as

$$\begin{aligned}\dot{\bar{p}}_i(t) &= \bar{v}_i(t)\\ \dot{\bar{v}}_i(t) &= \bar{a}_i(t)\\ \dot{\bar{a}}_i(t) &= \frac{1}{\tau_i}u_i(t) + \frac{1}{\tau_i}\bar{a}_i(t), i = 1, 2, \ldots, n\end{aligned} \tag{1.4}$$

where $\bar{a}_i(t)$ is the acceleration of the $i$th vehicle, $u_i(t)$ denote the control input, which can represent the desired driving/braking force. $\tau_i$ is the engine time constant. The third-order linear dynamics model not only inherits the simplification of the second-order model, but also covers the dynamics of the drive system. However, the third-order linear model is obtained by the exact feedback linearization of the third-order nonlinear system on the basis of the exact known system parameters. In the actual system, the complete parameters and other prior knowledge are difficult to directly obtain.

The third-order nonlinear model considers the inherent nonlinear characteristics of the vehicle and the characteristics of the transmission system, which is closer to the real system. The existing third-order nonlinear dynamic model [25] has the form as

$$\begin{aligned}
\dot{\bar{p}}_i(t) &= \bar{v}_i(t) \\
\dot{\bar{v}}_i(t) &= \frac{1}{m_i}\left(\eta_i \frac{1}{r_i}\bar{F}_i(t) - \bar{C}_i \bar{v}_i^2(t) - m_i g_i \bar{f}_i\right) \\
\dot{\bar{F}}_i(t) &= \frac{1}{\tau_i}u_i(t) - \frac{1}{\tau_i}\bar{F}_i(t), i = 1, 2, \ldots, n
\end{aligned} \tag{1.5}$$

where $\bar{F}_i(t)$ is the driving/braking force of the $i$th vehicle's engine. In [26], the third-order longitudinal nonlinear dynamic model is

$$\begin{aligned}
\dot{\bar{p}}_i(t) &= \bar{v}_i(t) \\
\dot{\bar{v}}_i(t) &= \bar{a}_i(t) \\
\dot{\bar{a}}_i(t) &= \frac{1}{m_i \tau_i}u_i(t) - \frac{2\bar{C}_i}{m_i}\bar{v}_i(t)\bar{a}_i(t) \\
&- \frac{1}{\tau_i}\left(a_i(t) + \frac{\bar{C}_i}{m_i}\bar{v}_i^2(t) + \frac{D_{mi}}{m_i}\right), i = 1, 2, \ldots, n
\end{aligned} \tag{1.6}$$

where $D_{mi}$ is the mechanical drag. However, due to the rate of acceleration change of the third-order nonlinear model is sensitivity to environmental disturbances and faults, the influence on the closed-loop control stability of ICVPS is increased.

For another, from the perspective of vehicle physical model structure, ICVPS can be divided into two types: homogeneous [23, 27] and heterogeneous [28, 29], according to the vehicle dynamics. Although homogeneous ICVPS can simplify the longitudinal control design and theoretical analysis, heterogeneous ICVPS is undoubtedly more in line with real traffic scenarios. How to ensure the safety and stability of heterogeneous ICVPS longitudinal control is one of the difficult problems to be solved due to the difference in sensitivity to external disturbances and faults and in structural of vehicle dynamics. Based on the above analysis, it is more challenging and practical engineering significance to study the resilient control problem of heterogeneous ICVPS based on third-order nonlinear longitudinal dynamics model to improve the safety robustness and fault resilient recovery ability of the system.

### 1.3.2 Information Flow Topology

The information flow topology shows the network characteristics of information interaction between vehicles in the network communication layer, which directly affects the control performance and string stability of ICVPS. In the early days, due to the functional limitations of sensor devices such as on-board radars, V2V communication only allowed adjacent vehicles to communicate with each other. During this period, the topology of information flow only included predecessor following (PF) and bi-directional (BD) [30, 31]. The PF type information flow topology is the simplest form of topological communication, and its control design often only needs the interaction information of neighboring predecessor vehicle. Based on the PF information flow topology, Su Hongye team of Zhejiang University studied the cooperative control problem of third-order nonlinear homogeneous vehicle platoon under the influence of disturbances and uncertainties in [16]. In [32], the influence of random noise and delay on collision of vehicle platoon systems under PF information flow topology is studied. Because of the small amount of interactive information in the PF type topology, it is difficult to ensure the stability of the ICVPS by using only the state information of the predecessor vehicle. In [33], based on the BD information flow topology, the disturbances amplification problem of the vehicle platoon system is analyzed. With the emergence of VANETs and C-V2X, the communication range of intelligent connected vehicles has been expanded, and more types of information flow topologies have been studied, such as predecessor leader following (PLF), Bi-directional leader (BDL), two predecessor following, two predecessor-following leader, and r-predecessor following, etc. [14, 15, 34–37], as shown in Fig. 1.4.

With the increase of interactive information in the information flow communication topology, the stability performance of the vehicle platoon system will be enhanced. However, on the one hand, in the actual traffic environment, due to the high-speed movement of the ICVPS and the limitations of the vehicle unit communication range, media access times in VANETs are usually short. On the other hand, unpredictable physical faults, deliberate artificial cyber attacks and complex environmental disturbances can also lead to the instability of wireless transmission channels. Therefore, on the premise of ensuring the closed-loop control stability and string stability of the ICVPS, how to further reduce the transmission of interactive

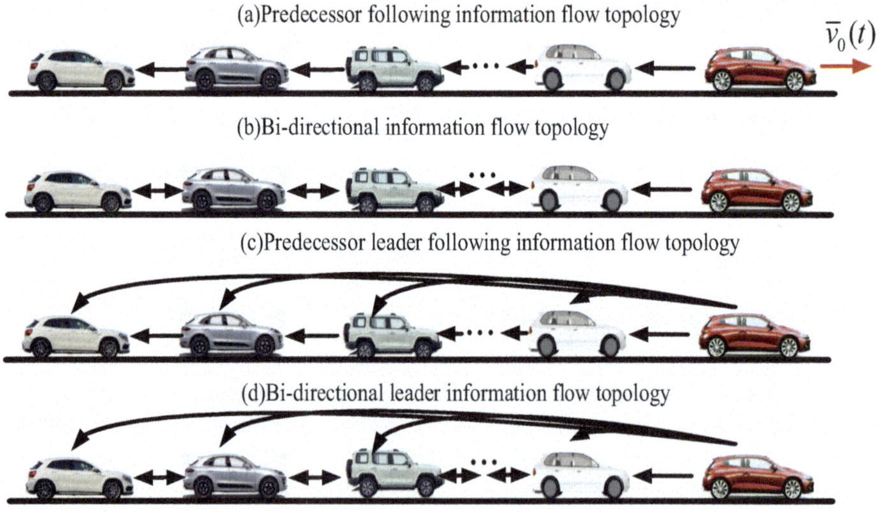


**Fig. 1.4** Typical information flow topology

information, save communication resources, and improve the resilient performance of the vehicle platoon system is the key direction of the information flow topology design.

### *1.3.3 String Stability and Traffic Flow Stability Performance*

IEEE Fellow and professor Darbha in the United States pointed out that the vehicle platoon spacing policy and controller design are favorable guarantees for the realization of closed-loop control stability of each vehicle system, string stability and traffic flow stability [38]. Closed-loop control stability and string stability depend heavily on vehicle spacing policy and controller design. Different from the stability control of a single vehicle system, all vehicle inter-vehicle spacing errors, velocity errors and acceleration errors in ICVPS are required to be bounded through the spacing strategy and controller design for achieving the closed-loop control stability of ICVPS. The string stability of the vehicle platoon ensures that the disturbance and fault affecting the spacing error and velocity error of the upstream vehicles will not be amplified and propagated among the downstream neighboring vehicles. In hence, the risk of collision can be avoided and the safety of the longitudinal driving of the ICVPS is realized.

In existing studies, string stability is mainly categorized into strict string stability [23, 38] and non-strict string stability [22, 24]. Strict string stability ensures that the spacing error decreases layer by layer along the length of the ICVPS. For example, by the definition of strict string stability in [38], the transient spacing errors $\bar{e}_i(t)$ not amplifying with the length of vehicle platoon for zero or nonzero initial spacing error, if the spacing error propagation transfer function $\bar{G}_i(s) := \frac{\bar{E}_{i+1}(s)}{\bar{E}_i(s)}$ satisfies the condition that $\left\|\bar{G}_i(s)\right\| \leq 1, \forall i = 1, \ldots, n$, which means $\|\bar{e}_n(t)\| \leq \|\bar{e}_i(t)\| \leq \cdots \leq \|\bar{e}_1(t)\|$, where $\bar{E}_i(s)$ is the Laplace transform of $\bar{e}_i(t)$. As for the non-strict string stability, which guarantees that the spacing error has an asymptotic convergence property during the propagation of disturbances and faults in the ICVPS. By the definition of non-strict string stability in [22], the origins of the ICVPS with longitudinal dynamics is non-strict string stability, if given any $\bar{\varepsilon} > 0$, there exists a constant $\bar{\mu} > 0$ such that $\|\bar{e}_i(0)\|_\infty < \bar{\mu} \Rightarrow \sup_i \|\bar{e}_i(\cdot)\|_\infty < \bar{\varepsilon}$. In practice, the non-strict form of string stability has fewer limitations and is more favorable to implement. In practice, the non-strict form of string stability has fewer limitations and is more favorable to implement.

For the traffic flow stability in ICVPS, which is stable when the gradient of traffic flow along the traffic density is positive (i.e., when the traffic density is lower than the critical characteristic density) [39]. In the definition of traffic flow stability, a traffic flow is stable if the gradient $\frac{\partial \bar{Q}}{\partial \bar{\rho}} > 0$ holds, where $\bar{Q}$ is the traffic flow rate, $\bar{\rho}$ is the traffic flow density, $\frac{\partial \bar{Q}}{\partial \bar{\rho}} = 0$ is the critical characteristic density. During the steady state of the vehicle platoon system, the traffic flow density disturbances to the vehicle platoon system dissipate in the form of forward traveling waves and eventually decay to the end of the vehicle flow.

Traffic flow stability avoids the decrease of traffic flow rate due to the increase of traffic flow density under the influence of perturbation, and its guarantees the exemption from the threat of traffic congestion. From the control point of view, since the traffic flow stability characterization is related to the steady state process of the ICVPS. The spacing policy will directly affect the traffic flow stability of the vehicle platoon system.

### 1.3.4 Inter-Vehicle Spacing Policy

Based on the above stability requirements of the ICVPS, scholars around the world have studied the vehicle platoon spacing policies such as Constant spacing (CS) policy, Constant Time Headway (CTH) policy, and nonlinear spacing policies such as Quadratic spacing policy and Exponential spacing (ES) [40–42]. The CS policy is only related to the desired distance constant of the vehicle platoon. In [40], the CS policy is

$$\bar{e}_i(t) = \bar{p}_{i-1}(t) - \bar{p}_i - L_{i-1} - \bar{d}_{des,i}, i = 0, 1, 2, \ldots, n \tag{1.7}$$

where $\bar{d}_{des,i} = \bar{d}_i$ in the CS policy, $\bar{e}_i(t)$ is the inter-vehicle spacing errors of rear-to-bumper distance between two neighboring vehicles, $L_{i-1}$ is the length of $i-1$th vehicle, $\bar{d}_{des,i}$ is the desired inter-vehicle spacing, and $\bar{d}_i$ is the positive static spacing when the vehicle platoon is in steady state. Kwon et al. [43] from Yonsei University, Korea, combined with the CS policy, proposed a coupled sliding mode control method for the first time, which not only enhances the string stability, but also improves the traffic flow capacity during the steady state process of the vehicle platoon. In [44], event-triggered cooperative control based on CS policy was studied to deal with the actuator delay caused by braking of the vehicle platoon system. However, the above CS policy is based on the assumption that the initial spacing error and velocity error are zero. And under the condition that the velocity and acceleration of the predecessor vehicle are known and the inter-vehicle distance is constant, the CS policy that the inter-vehicle distance is independent of the vehicle velocity cannot effectively guarantee the string stability of ICVPS in a complex environment.

For removing the assumption that the acceleration of the predecessor vehicle is known in the CS policy, the CTH policy was studied in the [45, 46]. The CTH policy is

$$\begin{aligned} \bar{e}_i(t) &= \bar{p}_{i-1}(t) - \bar{p}_i - L_{i-1} - \bar{d}_{des,i} \\ \bar{d}_{des,i} &= \bar{h}_i \bar{v}_i(t) + \bar{d}_i, i = 0, 1, 2, \ldots, n \end{aligned} \tag{1.8}$$

where $\bar{h}_i$ is a positive headway time constant. The CTH policy indicates that the desired inter-vehicle spacing can be adjusted based on the velocity $\bar{v}_i(t)$. At the same time, in order to reduce vehicle spacing and remove the assumption that the initial spacing error, velocity error and acceleration error are zero, Guo Xianggui et al. [47–49] from University of Science and Technology Beijing also proposed an improved CTH policy. However, when the ICVPS is in a stable state, the inter-vehicle distance based on CTH policy will continue to increase at high velocity, which will result in lower traffic flow density or instability of traffic flow [50, 51]. Specifically, the traffic flow gradient based on the CTH will always be negative, that is, the traffic flow will continue to be unstable, leading to the collision threat. In order to achieve better string stability and traffic flow stability, the ES policy was first proposed in [52], the form is

$$\begin{aligned}\bar{e}_i(t) &= \bar{p}_{i-1}(t) - \bar{p}_i - L_{i-1} - \bar{d}_{des,i}\\ \bar{d}_{des,i} &= \frac{\phi\bar{v}_i(t)}{2A_{\max}} + \alpha(1 - e^{(-\frac{\bar{v}_i(t)}{\beta})}), i = 0, 1, 2, \ldots, n\end{aligned} \tag{1.9}$$

where $\phi$ denotes the safety coefficient, $A_{\max}$ is the maximum deceleration, $\alpha$ and $\beta$ are the designed constants. However, it should be pointed out that this spacing policy does not take into account the effects of non-zero initial spacing errors and physical actuator faults, which may lead to string instability. At the same time, when the intelligent connected vehicle suffers from actuator faults, the maximum acceleration of the vehicle may be abrupt. The ES policy cannot guarantee the string stability and traffic flow in this case. Guo Ge et al. [53–55] of Northeastern University proposed a QS policy, which introduced fault upper bound factor to ensure string stability and traffic flow stability under actuator fault. The QS policy is

$$\begin{aligned}\bar{e}_i(t) &= \bar{p}_{i-1}(t) - \bar{p}_i - L_{i-1} - \bar{d}_{des,i}\\ \bar{d}_{des,i} &= \bar{h}_i\bar{v}_i(t) + \frac{\phi\bar{v}_i^2(t)}{2A_{\max}(1-\bar{\lambda}_i)}, i = 0, 1, 2, \ldots, n\end{aligned} \tag{1.10}$$

where $\bar{\lambda}_i$ is the physical fault factor of the vehicle. However, the QS policy is designed based on the headway time constant of the control system. When the system is in the steady state, the platoon may be unstable if the velocity is too low and the headway time constant is too small. The traffic flow rate will also be reduced if the headway time constant is too large, causing the traffic flow instability.

The vehicle spacing policy is a kind of platoon forming strategy in essence, which realizes the coordination and unification of the vehicle platoon forming and control system design of multiple vehicle subsystems. On the premise of meeting the closed-loop control stability of the system, the string stability is maintained to achieve a smaller spacing error. In the steady-state process of vehicle platoon system, it is an important goal to ensure the stability of traffic flow and obtain higher traffic flow density. Therefore, how to realize the high performance requirements of vehicle platoon stability and traffic flow stability, and how to realize the unity of opposites between traffic flow stability and string stability are the key research directions of vehicle spacing policy design. On the other hand, in the actual complex traffic environment, disturbances, physical fault and cyber attacks are important reasons that affect the stability of ICVPS and traffic flow. How to incorporate fault, attack and disturbance factors into the design of inter-vehicle spacing error policy is also the key content of ICVPS control and needs further research.

## 1.4 Resilient Control of Intelligent Connected Vehicle Platoon Systems

ICVPS is a kind of large-scale complex connected cyber physical fusion system with multi-information source. It is highly coupled by multiple vehicle physical subsystems through the network communication layer and the control layer. Due to the complex physical structure, the variable coupling of topology and control, and the massive and multi-source information exchange, the probability of fault and cyber attack is large. Therefore, how to achieve the goals of the safety, stability and reliability under fault and cyber attack conditions has become the primary requirement and task of ICVPS longitudinal control. Experts in the field of resilient control, such as professor Zhou Donghua of Tsinghua University and professor Ju H Park of Yeungnam University, all pointed out that the technical requirements of the resilient control scheme that the influence of faults and cyber attacks on the closed-loop system can be effectively compensated, and the safety, reliability and accuracy of the system under long-term and high-intensity operation conditions can be ensured [56, 57]. At present, resilient control method has become one of the most comprehensive cross-research directions of engineering significance in the field of control, and it is also an important part of the engineering design process. By combining adaptive control technology, adaptive resilient control technology is incorporated into ICVPS longitudinal control framework, which can effectively improve the security, stability performance and the recovery ability under fault and cyber attack conditions of ICVPS.

### *1.4.1 Physical Fault Issues of the Intelligent Vehicle Platoon Resilient Control*

For a large and complex network cyber physical fusion system such as ICVPS, the fault issues can be mainly divided into physical faults [58–60] and communication faults induced by cyber attacks [61–63]. Physical faults mainly refers to the fault of actuators such as wheel hub motors and hydraulic cylinders in the vehicle subsystem of ICVPS, the fault of position sensors such as throttle and crankshaft, the fault of speed and acceleration sensors, and the fault of air intake sensors and oxygen intake

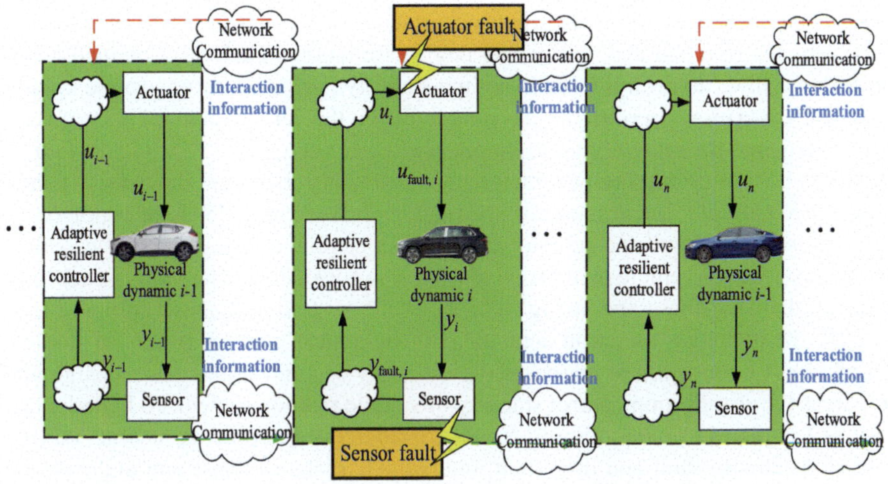


**Fig. 1.5** Longitudinal control block diagram of ICVPS under physical fault

sensors, etc. Physical faults also include the fault of electronic or mechanical unit components such as power amplifiers. Physical faults mainly occur at the physical layer of ICVPS, as shown in Fig. 1.5. It can also be summarized as loss of effectiveness fault, stuck fault, and additive fault such as sudden change, interruption, migration, random nonlinear, periodic and intermittent [64, 65]. In [58, 60], the models of actuator faults and sensor faults are

$$u_{fault,i}(t) = \bar{\lambda}_i u_i(t) + \bar{\beta}_i \widehat{u_i}(t) \tag{1.11}$$

$$y_{fault,i}(t) = \bar{\varsigma}_i(t) y_i(t) + (1 - \bar{\varsigma}_i(t)) \bar{U}_i y_i(t) \tag{1.12}$$

where $u_{fault,i}(t)$ denotes the actuator fault function, $u_i(t)$ is the healthy control input of the $i$th vehicle, $\bar{\lambda}_i$ is the loss of effectiveness fault factor, and $\bar{\beta}_i = \{0, 1\}$, $\bar{\beta}_i = 1$ means that there exists the bias faults of the $i$th vehicle, and $\widehat{u_i}(t)$ denote the bias fault function. $y_{fault,i}(t)$ denotes the sensor fault function, and $y_i(t)$ denotes the healthy system output of the $i$th vehicle. $\bar{\varsigma}_i(t) = \{0, 1\}$, $\bar{\varsigma}_i(t) = 1$ means that the sensor faults occur, and $\bar{U}_i$ is the fault parameter. Physical faults are usually based on internal changes of the vehicle dynamic system, with obvious physical characteristics and relatively fixed forms. Physical faults do not affect the observable performance of the ICVPS. In [66], the problem of motor control gain reduction caused by motor winding fault and inverter phase fault in four-wheel independent drive electric vehicles is studied, which causes motor control signal failure and affects the longitudinal control stability of vehicles. In [67], the loss of effectiveness fault, stuck fault and abrupt fault of a single wheel hub motor of a four-wheel independent drive vehicle are considered. The involved faults are easy to damage the availability and integrity of the control input signal and destroy the stability of vehicle lateral tracking. The stuck fault, basis fault and noise disturbance problems of the wheel angle of steer-by-wire vehicles were studied, and wheel angle estimation signal was obtained based on Kalman filter algorithm to improve the stability performance of vehicle system in [68]. Furthermore, in [69], the physical faults on the vehicle platoon system were considered, and the sensor fault problems of a single vehicle in a ICVPS are studied. Based on CS policy and fault characteristic matrix, an active fault diagnosis method is designed to reduce the influence of the fault on the system. In [54], the problems of actuator fault, drift fault and actuator saturation constraint existing in a single vehicle in a heterogeneous vehicle platoon system are also considered. The faults issues have threaten the robustness and string stability of the system and cause collision risk. In [59], the unknown direction actuator fault in vehicle platoon system is studied. This kind of fault problem increases the difficulty of control algorithm design and threatens the closed-loop control stability of vehicle platoon system. Ju H Park et al. [70] of Lingnan University, South Korea considered the problems of time-varying direction actuator fault and drift fault, intervehicle distance constraint and external disturbances in homogeneous vehicle platoon system. The fault mode made the conventional adaptive fault compensation method difficult to work.

### *1.4.2 Cyber-Attack Issues of the Intelligent Connected Vehicle Platoon Resilient Control*

The operation and control performance of ICVPS have high requirements for the quality of interactive information in the network communication layer. In practical application scenarios, due to the openness of vehicle wireless communication network and the complexity of traffic environment, the credibility, availability and integrity of the interactive data of ICVPS wireless network communication layer face serious threats. After penetrating the wireless network defense mechanism, malicious cyber attacks can destroy the interactive data at the network communication layer through feedback loop communication channels, such as control-actuator, sensor-controller, and V2V communication channels, as shown in Fig. 1.6. Thus, the cyber attacks can cause communication fault, partially or completely destroy the control system, and achieve the purpose of reducing the control performance of ICVPS or even completely crash the system. Therefore, the malicious cyber attack is one of the main reasons for the communication faults of the networked cyber physical system [71–73]. The communication fault caused by malicious cyber attack mainly occurs in the network communication layer. Different from physical faults, the performance of deliberate artificial cyber attacks is usually based on the existing resource allocation of malicious attackers [74]. Malicious cyber attack has obvious intelligence characteristics and can usually destroy the observable performance of the states of vehicle systems [75, 76].

Typical cyber attacks that cause communication faults mainly include denial of service attacks and deception attacks. Denial-of-service attacks have been studied

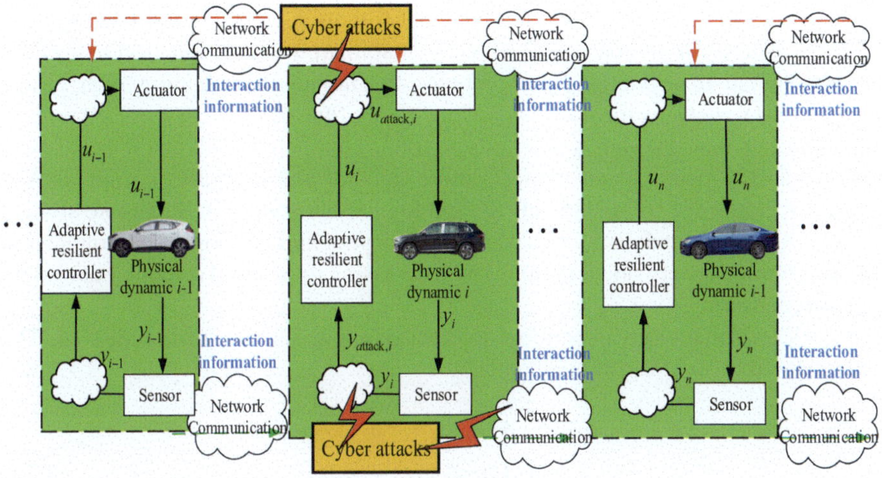


Fig. 1.6 Longitudinal control block diagram of ICVPS under cyber attacks

extensively because of their ease of launch and low cost. Denial of service attacks mainly include probabilistic model based on Bernoulli random distribution and Markov chain [77, 78], periodic model based on dwell time [79], and non-periodic model based on attack frequency and duration [80, 81]. In [82], the problem of periodic denial of service attack existed in the platoon system was studied, which extended the service duration of the platoon network communication and damaged the control performance of the system. In [83], a probabilistic denial-of-service attack model based on Bernoulli random distribution is studied for heterogeneous first-order vehicle platoon systems, which destroys real-time V2V information exchange, increases network communication service time, and leads to vehicle collision threat.

Different from denial of service attacks, deception attacks include false data injection attacks, which directly tamper with system communication interaction data by monitoring system communication transmission. By using the obtained system information to destroy the integrity and credibility of communication interaction data, thus deception attacks can cause communication faults and having stronger destructive power to ICVPS control systems [84, 85]. In [86], an intermittent communication fault based on deception attack is constructed to apply to the multi-sensor transmission channel of the networked cyber physical system, which is intermittent and random, so as to avoid the detection of the fault detector, make the state of the control end unusable and untrustworthy, and destroy the control performance of the system. In [87], a state-dependent probabilistic random injection attack is proposed to create network communication faults. The form of the probabilistic random injection attack is

$$u_{attack,i}(t) = u_i(t) + \alpha_i(t)G_i(t)\varphi_i(\bar{p}_i(t), \bar{v}_i(t), \bar{a}_i(t)) \qquad (1.13)$$

where $u_{attack,i}(t)$ denotes the actuator attack function, $u_i(t)$ is the healthy control input of the $i$th vehicle, $\alpha_i(t)$ is the probabilistic random function and obeys the exponential distribution, $G_i(t)$ denotes the unknown weighting matrix, and $\varphi_i(\bar{p}_i(t), \bar{v}_i(t), \bar{a}_i(t))$ is the state-dependent attack function, which means the malicious attacker can use the system information. Compared with traditional injection attacks [88], this attack mode is more general and difficult to control. Wen Changyun et al. [89] from Nanyang Technological University in Singapore studied the data injection attack, which leads to time-varying and unknown communication link weights of the networked system, thus increasing the control difficulty of system for fault recovery.

In [90, 91], malicious injection nonlinear dynamic attacks are considered. The mathematical form of the dynamic attacks is

$$\begin{aligned} u_{attack,i}(t) &= u_i(t) + \bar{\varphi}_i(t) \\ y_{attack,i}(t) &= y_i(t) + \bar{\varphi}_i(t) \end{aligned} \tag{1.14}$$

$$\begin{aligned} \dot{J}_i(t) &= \bar{H}_i(J_i(t), \bar{p}_i(t), \bar{v}_i(t), \bar{a}_i(t)) \\ \bar{\varphi}_i(t) &= \bar{k}_i(J_i(t)) \end{aligned} \tag{1.15}$$

where $u_{attack,i}(t)$ and $y_{attack,i}(t)$ are the actuator attack function and the sensor attack function, respectively. $\bar{\varphi}_i(t)$ is the malicious injection dynamic attack function. $\bar{H}_i(J_i(t), \bar{p}_i(t), \bar{v}_i(t), \bar{a}_i(t))$, $J_i(t)$, and $\bar{k}_i(J_i(t))$ are unknown bounded functions. The attack data is generated by unknown nonlinear functions dependent on the system state, which has strong generality and can be used to represent additive actuator faults, such as abrupt faults and intermittent faults [92]. It can also be used to represent uniformly bounded injection attacks [93, 94] and state-dependent attacks [95] to trigger the controller-actuator communication faults of the system. In [96], the injection attack problem of homogeneous platoon system is studied, which leads to random loss or delay of intervehicle communication information and damages the control stability of system. In [97], the location sensor attacks on local vehicles in the platoon system are considered, which damages the availability of state information at the controller end and damages the longitudinal dynamic performance of the vehicle platoon system. In [98], an intermittent false data injection attack for vehicle platoon systems is constructed. The attack switches through multiple sampling time series to enhance the concealment and intelligence of the attack and destroy the information transmission. Ge Xiaohua et al. [99] of Swinburne University of Technology in Australia considered the problem of unknown forged control signal injection attacks on the vehicle platoon system, which would destroy the integrity of the control transmission data and cause the vehicle to brake.

From the above analysis and Table 1.1, it can be seen that the physical fault originates from the defects or loss of effectiveness changes of the internal components of the vehicle subsystem. It has obvious physical characteristics, its form is relatively fixed, and its generation is random and unpredictable. However, the malicious cyber attacks are mainly based on the existing resource allocation of malicious attackers,

**Table 1.1** Main characteristics of physical faults and cyber attacks

| | Type | Location occurred | Forms | Observability |
|---|---|---|---|---|
| Physical faults | Loss of effectiveness faults, stuck faults, and additive faults such as abrupt change, bias, periodicity, etc. | Physical components in the physical layer, actuators such as hub motors, hydraulic cylinders, and sensors such as position, speed, and intake, etc. | It is usually based on the internal changes of the vehicle system, with obvious physical characteristics, and its form is relatively fixed | Do not affect |
| Cyber attacks | Deception attacks, denial of service attacks, etc. | Communication layer, such as feedback closed-loop communication channel, V2V communication channel, etc. | It is usually generated based on the resource configuration handled by malicious attackers, and has obvious intelligent characteristics and pertinence | Usually can destroy system observability |

and the damage to the system has obvious intelligence and pertinence. Based on the above characteristic analysis, for improving the safety, stability and fast recovery performance of the ICVPS, combined with adaptive control technology, resilient control technology for physical fault and cyber attack conditions has become a research hotspot of scholars, and its typical application methods mainly include sliding mode adaptive control method [43, 48, 100], backstepping adaptive control method [14, 16], adaptive fuzzy control [101] and neural network control method [51, 61, 72], model predictive adaptive control methods [18, 102], $H_\infty$ adaptive control methods [73, 94], etc. On the other hand, because of its great advantages in saving communication resources and controlling energy, Event-triggered control is also an important research direction [36, 44, 79, 80, 88]. However, on the basis of ensuring the stability of the system under fault and attack conditions, how to avoid the "Zeno" or singular phenomenon triggered by mistake, reduce the conservatism of the event triggered control, prolong the triggering time interval, and better save the control energy and communication resources of the vehicle platoon system is still worthy of in-depth study.

From the perspective of control objectives, how to ensure closed-loop control stability and string stability of the system under fault and attack conditions, avoid collision risk, and improve the robustness of the system is the focus of ICVPS adaptive resilient control research. For the problems of heterogeneous ICVPS suffering from actuator faults and bias faults, in [53, 54], based on QS policy, an adaptive coupled sliding mode control strategy is constructed to ensure uniform bounded stability and strict string stability of the vehicle platoon system. A homogeneous platoon system with actuator faults and cyber attacks is also studied. Based on the fault observer, a robust adaptive fault-tolerant tracking control scheme is constructed to achieve the exponential stability of the system in [73]. In [103], for a ICVPS with abnormal signal actuator faults, measurement noise and external disturbances, a state observer is designed to estimate unmeasurable state information and abnormal actuator signals. Meanwhile, an adaptive linear feedback vehicle platoon fault-tolerant control scheme is designed combined with CS policy to ensure the uniform bounded stability, string stability and resilience performance of the system. In [104], for the sensor fault problem of homogeneous platoon system, firstly, dynamic linearization is used to transform the nonlinear platoon system model into a linear data model, and the output data is introduced into the control design. Secondly, an event-triggered adaptive model-free resilient control scheme is designed to ensure the bounded stability and robustness of the system. In [105], the problems of time-varying actuator fault, actuator input saturation and disturbances suffered by the ICVPS in two-dimensional space were studied. The problem of unknown time-varying actuator fault was solved by constructing Nussbaum function, and then combining CS policy, an adaptive sliding mode resilient control scheme based on disturbance observer is designed to ensure the semi-global uniform bounded stability and safety of the vehicle platoon system. In [106], the problem of bounded injection attack in V2V communication is studied, and $H_\infty$ adaptive resilient control scheme is constructed to achieve uniform bounded stability of the system. From the above analysis, it can be seen that the research on the design of adaptive resilient control scheme to ensure the asymptotic stability of the closed-loop system of ICVPS with physical faults and cyber attack has achieved fruitful results.

Based on the performance requirements of the system, how to further maintain the stability of the closed-loop system under fault and attack conditions, shorten the recovery time of the system under fault and attack conditions and enhance the resilient capability of the system has more practical engineering significance. It is also a hot research direction of ICVPS adaptive resilient control in recent years. Compared with the asymptotic stability performance, the finite-time stability control method can shorten the convergence time of ICVPS closed-loop under fault and attack conditions, and has stronger robustness. Professor Li Yongfu et al. [107] from Chongqing University of Posts and Telecommunications studied the finite time uniformity and stability control of homogeneous vehicle platoon systems under different communication topologies. Professor Ren Wei et al. [108] from the University of California, Riverside, United States, combined with the finite time observer, realized the distributed finite time feedback tracking control of the heterogeneous networked vehicle platoon system to ensure the safety of control. For the fault problem of heterogeneous platoon systems suffering communication interruption and delay, in [109], combined with CTH policy, adaptive finite time control based on sliding mode is studied to reduce spacing error fluctuation, maintain the string stability, and improve system robustness and security. In [110], the homogeneous vehicle platoon system with sensor bias fault, and loss of effectiveness actuator fault is considered. Distributed finite time observer is designed to realize sensor fault detection, and finite time parameter estimation law is used to compensate the influence of actuator fault. Finally, a terminal sliding mode adaptive resilient controller is constructed to realize the global finite time stability, which ensures the safety of the platoon system under fault conditions.

However, the above work based on the finite-time adaptive resilient control method must depend on the initial conditions and system parameters, and the convergence time cannot be estimated in advance. When the initial state of ICVPS is far from the equilibrium point, the convergence speed of the system will be greatly reduced.

As an extension of finite time stability control, the fixed-time stability control method has stronger resilience performance to deal with the physical faults and communication cyber attacks. The fixed-time control method can estimate the stability time of the system in advance without initial state information. Recently, fixed time stability control has become a hot topic in control engineering field. In [111], state variables with uncertainty and measurement noise are used as inputs to construct an augmented sliding mode observer to estimate uncertainty and filter measurement noise. Then, a fixed-time back-stepping resilient controller is established based on observation errors to ensure fixed-time convergence of the closed-loop system. In [112], a fixed-time adaptive law is designed to compensate for the saturation and uncertainty of mechanical system. On this basis, a distributed back-stepping adaptive fixed-time resilient control scheme based on neural network is constructed to achieve the actual fixed-time stability of the system. In [113], a practical fixed-time resilient scheme is proposed to solve the fixed-time consistency tracking stability problem of higher-order multi-agent systems by combining fuzzy adaptive control and back-stepping control methods. In recent years, some scholars have also applied the fixed-time control method to ICVPS to accelerate the convergence speed of the closed-loop system of the intelligent connected vehicle platoon and improve the robustness of the system. For example, in [114], distributed trajectory optimization algorithm is introduced to shorten intervehicle distance error, and a terminal sliding mode fixed time control strategy is constructed based on CS policy to achieve global fixed time convergence within the closed-loop vehicle platoon system and ensure the string stability of the system. In [115], robust $H_\infty$ adaptive control is used to deal with the external disturbance problems existing in heterogeneous platoon systems with third-order nonlinear, and then CS policy and back-stepping control are used to construct a distributed nonlinear fixed-time adaptive tracking controller to achieve practical fixed-time stability of ICVPS.

Further, in order to ensure the desired transient/steady state performance of the closed-loop system under fault and attack conditions and improve the control accuracy of the system, scholars began to introduce the prescribed performance [116] constraint into the fixed time control, so as to achieve the fixed time stability of the system under the predefined constraints. In [117], for the problems of actuator faults such as steering mechanism fault and motor traction fault of unmanned ground vehicles, the hyperbolic tangent function is used as the prescribed performance function to conduct predefined constraints on path tracking error and speed control. An adaptive resilient control scheme with fixed time prescribed performance is constructed to achieve the fixed time stability of unmanned vehicle closed-loop system under prescribed conditions. In [118], for the fault issues of output contactors and actuators in microgrid systems, a fixed-time performance decay function is designed to constrain the convergence region of voltage recovery synchronization errors. Based on the event-triggered mechanism, a fixed-time adaptive control algorithm based on backstepping is constructed to achieve the fixed-time stability of the system. In view of the problems of multiplicative and additive actuator faults existing in connected systems, a fault observer was proposed in [119] to estimate the system state and fault information within a fixed time, and a prescribed performance decay function was introduced to constrain the tracking errors. Then, a sliding mode resilient controller was constructed to realize the fixed-time stability of the closed-loop system under the meaning of prescribed performance. Professor Jiang Bin et al. [120] of Nanjing University of Aeronautics and Astronautics, for the problems of actuator loss of effectiveness fault and external disturbances on the multiple heterogeneous multi-agents of UAVs and unmanned vehicles, constructed distributed and decentralized fixed-time adaptive formation sliding mode resilient control algorithms on the x-y and z axes, respectively. Then, the formation error can be guaranteed to converge to the prescribed constraint domain within a fixed time by introducing a finite time performance decay function.

The research of prescribed constraint performance and fixed time control is of great significance to satisfy the transient and steady state performance requirements of ICVPS. For ICVPS, due to the physical structure of ICVPS and the communication characteristics of wireless mobile network, the closed-loop control stability and string stability of the system are vulnerable to sudden physical fault and cyber attacks, and are also susceptible to physical environment constraints. For example, V2V wireless communication distance exist limitations. When the inter-vehicle distance is too large, it is easy to cause instability and fragmentation of the vehicle platoon system. When the inter-vehicle distance is too small, vehicle collision is also easy to occur, resulting in system collapse. In [55], the saturation function based on Gaussian error is used to deal with the actuator saturation problem of non-smooth in the vehicle platoon system. The spacing error is pre-defined by the fixed-time performance function, and the fixed-time prescribed performance sliding mode controller is constructed based on QS policy to ensure the closed-loop control stability and the string stability of the vehicle platoon system.

However, the existing fixed-time adaptive resilient control methods for ICVPS with prescribed performance are limited. How to maintain the desired inter-vehicle safety distance under fault and attack conditions, ensure the desired ICVPS transient and steady-state performance, reduce the convergence time for the system to enter the steady-state is meaningful issues. Furthermore, how to improve the system's rapid recovery and resilience ability under fault and attack conditions, and ensure the closed-loop system control stability and the string stability also are still open research directions.

**Table 1.2** Summary of Longitudinal resilient control of ICVPS

| | Category |
|---|---|
| Longitudinal dynamic model | • **Mathematical structure**: second-order linear model [17, 18], second-order nonlinear model [19–21], third-order linear dynamics model [22–24], third-order nonlinear model [14, 25, 26]<br>• **Model structure**: homogeneous ICVPS [27] and heterogeneous ICVPS [28, 29] |
| Information flow topology | • Predecessor following (PF) and Bi-directional (BD) [30, 31];<br>• Predecessor leader following (PLF), Bi-directional leader (BDL), two predecessor following, two predecessor-following leader, r-predecessor following, etc. [15, 16, 34–37] |
| Spacing error policy | • Constant Spacing (CS) [30, 44], Constant Time Headway (CTH) [45, 46];<br>• Nonlinear spacing policy: Quadratic Spacing (QS), Exponential Spacing (ES), etc. [40–42] |
| Resilient control | • Resilient control of ICVPS in the sense of asymptotic stability [53, 54, 73, 103–106],<br>• Finite time resilient control of ICVPS [107–110],<br>• Fixed time resilient control of ICVPS [111–115],<br>• Resilient control of cyber physical system in the sense of fixed time with prescribed performance [117–120], etc. |

Based on the above analysis and Table 1.2, it can be seen that the scholars for all over the world have achieved a lot of achievements in the field of longitudinal resilient control of ICVPSs. However, with the complexity of intelligent transportation environment, the opening of wireless communication network and the intelligence of unmanned vehicles, there is also a higher realistic demand for the high safety, strong robustness and fast recovery performance of the longitudinal control of ICVPS.

## 1.5 Key Technical Problems of the Book

As the key technology field of intelligent connected vehicles, ICVPS deeply integrates advanced control, communication computing and unmanned vehicles. It has significant advantages in improving traffic efficiency, driving safety and reducing energy consumption. However, the external disturbances and physical fault in the physical layer seriously affect the closed loop platoon system's control performance and the string stability. The malicious cyber attacks always threaten the driving safety of ICVPS. There are still a lot of technical problems in the study of adaptive resilient control of ICVPS under fault and cyber attack conditions. This paper will focus on the following technical problems:

- For the resilient control design of ICVPS with actuator saturation and complex environment disturbances, how to improve the fault recovery response speed of the system based on the existing control methods, improve the robustness and safety of ICVPS resilient control, and ensure the string stability under actuator fault and saturated conditions is the basic problem to be solved.
- How to improve the inter-vehicle spacing policy, reduce the conservative of the existing spacing error policy, and improve the stability of traffic flow while ensuring the string stability of ICVPS under fault and attack conditions is an important research problem to be solved.
- For the resilient control of ICVPS with multiple cyber attacks and external disturbances, how to guarantee the transient/steady state performance of the system with multiple cyber attacks and external disturbances under the condition of predefined convergence domain of asymmetric spacing error, and achieve the internal closed-loop fast finite time stability and string stability of ICVPS is an innovative research problem.
- How to design a distributed fixed-time resilient controller in the sense of saving communication resources and energy to eliminate the influence of cyber attacks and disturbances on the basis of guaranteeing the closed-loop fixed-time stability and string stability in ICVPS. How to design a fixed time observer to quickly estimate the external disturbances of the vehicle platoon system to improve the limitation of the traditional disturbances observer is a research problem of practical significance.
- Considering the unknown dead zone of ICVPS and the faults of multi-form actuators, how to design a distributed resilient control framework in the sense of fixed time Lyapunov stability with prescribed performance on the basis of existing methods to ensure the transient/steady state performance of ICVPS in the predefined asymmetric constraint domain? It is a challenging research problem.

## 1.6 Organizational Structure and Logic of the Book

In view of the above key technical problems, this book studies the platoon distributed resilient control of heterogeneous intelligent connected vehicles based on the longitudinal control performance requirements of ICVPSs under complex conditions such as physical faults, deliberate cyber attacks and external disturbances. Starting from multiple technical dimensions of "longitudinal control + distributed resilient control + finite time control + fixed time stability of prescribed performance", and the comprehensive security and stability of the system is improved.

In this chapter, the background and significance of the research on distributed longitudinal resilient control of ICVPS under physical fault and cyber attack conditions are introduced, and the research status and the basic structures of longitudinal control of ICVPS, physical faults and cyber attacks issues of ICVPS and the overview of resilient control methods are analyzed and summarized. Finally, the key technical problems in distributed resilient control of vehicle platoon system are presented, and the organization structure of this book are claimed.

In Chap. 2, the design processes and the basic structures of distributed longitudinal resilient control of the intelligent connected vehicle platoon system are introduced and summarized. Some simulation cases also have been provided.

In Chap. 3 of this book, a distributed adaptive resilient control strategy is proposed to study the global finite-time stability and strict string stability of the closed-loop ICVPS, for the problems of single vehicle with actuator fault, input saturation and complex road environment disturbances.

In order to further improve the control accuracy under physical fault conditions, based on the Chap. 3 and the improved exponential spacing policy, the observer-based distributed finite-time adaptive resilient control strategy is proposed in Chap. 4. The global finite-time stability, string stability and traffic flow stability under fault conditions are achieved.

In Chap. 5, a distributed adaptive prescribed performance finite time resilient control method is proposed based on the exponential spacing strategy and control architecture in Chap. 4 to study the fast finite time stability and string stability of the system under the predefined asymmetric constraint domain.

In order to further improve the convergence speed of the closed-loop system and save control energy and communication resources on the basis of Chap. 5, in Chap. 6, a distributed adaptive event-triggered fixed-time resilient control method is proposed to study the practical fixed-time stability and string stability of the closed-loop ICVPS, by considering the unknown external disturbances and the unknown communication faults induced by the cyber attacks.

On the basis of Chap. 6, in Chap. 7, the practical fixed time stability and string stability of closed-loop system under unknown dead zone and faults are further studied. A distributed adaptive resilient control method under fixed time frame of prescribed performance is proposed to enhance the convergence speed and control accuracy of the system and ensure the transient/steady state performance of the system.

In Chap. 8, the research work of this book is summarized, and the future research direction of adaptive resilient control of ICVPS is clarified.

From the perspective of control objectives, Chaps. 2–7 of this book are studied the longitudinal control problems of ICVPS under complex conditions such as physical faults or cyber attacks and external disturbances. And the distributed adaptive platoon resilient control methods of ICVPS are designed to achieve safe driving of multiple vehicle subsystems at the same velocity and desired inter-vehicle distance under fault and attack conditions. On the premise of meeting the closed-loop control stability of the system, the string stability of the platoon is also maintained, and the smaller spacing error of the platoon is achieved.

From the perspective of string stability control of ICVPS, the strict string stability of the system is studied in Chap. 3, which ensures that the spacing error decreases layer by layer along the length of the vehicle platoon. In order to reduce the limitations of the control method in practical application, the non-strict string stability of the system is studied in Chaps. 4, 5, 6, and 7, which can effectively ensure the bounded convergence of the spacing error during the propagation process of the disturbances and faults of the platoon system. It can be guaranteed that the disturbances and faults affecting the spacing error and velocity error of the upstream vehicles of the platoon will not be amplified and propagated among the downstream vehicles. Thus, the risk of collision can be avoided.

From the perspective of system convergence speed, distributed adaptive resilient control strategies are designed in Chaps. 3, 4, and 5 to ensure the stability of the closed-loop system of ICVPS under fault and cyber attack conditions in the sense of finite time. On the basis of the above, Chap. 6 ensures the stability of the system under cyber attack conditions in the sense of practical fixed time. In Chap. 7, the stability of the prescribed performance under the faults and disturbances conditions is realized in the sense of fixed time. The control accuracy and robustness of the system are enhanced, and the transient/steady state performance of ICVPS in the predefined constraint domain is also ensured.

## References


1. Shen X, Fantacci R and Chen S(2020) Internet of vehicles. *Proceedings of the IEEE*, 108(2): 242-245.
2. Jia D, Lu K, Wang J, et al(2016) A survey on platoon-based vehicular cyber physical systems. *IEEE Communications Surveys and Tutorials*, 18(1): 263-284.
3. Lesch V, Breitbach M, Segata M, et al(2022) An overview on approaches for coordination of platoons. *IEEE Transactions on Intelligent Transportation Systems*, 23(8): 10049-10065.
4. Besselink B, Johansson K H(2017) String stability and a delay-based spacing policy for vehicle platoons subject to disturbances. *IEEE Transactions on Automatic Control*, 62(9): 4376-4391.
5. Wang R, Wang J(2014) Actuator-redundancy-based fault diagnosis for four-wheel independently actuated electric vehicles. *IEEE Transactions on Intelligent Transportation Systems*, 15(1): 239-249.
6. Yu M, Wang D(2014) Model-based health monitoring for a vehicle steering system with multiple faults of unknown type. *IEEE Transactions on Industrial Electronics*, 61(7): 3574-3586.
7. Guo B, Chen Y(2020) Robust adaptive fault-tolerant control of four-wheel independently actuated electric vehicles. *IEEE Transactions on Industrial Informatics*, 18(5): 2882-2894.
8. Yuan Y, Yuan H, Guo L, et al(2016) Resilient control of networked control system under DOS attacks: a unified game approach. *IEEE Transactions on Industrial Informatics*, 12(5): 1786-1794.
9. Han Z, Zhang S, Jin Z, et al(2022) Secure state estimation for event-triggered cyber-physical systems against deception attacks. *Journal of the Franklin Institute*, 359(18): 11155-11185.
10. Dressler F, Klingler F, Segata M, et al(2019) Cooperative driving and the tactile internet. *Proceedings of the IEEE*, 107(2): 436-446.
11. Li K Q, Wang J, and Zheng Y(2022) Cooperative formation of autonomous vehicles in mixed traffic flow: beyond platooning. *IEEE Transactions on Intelligent Transportation Systems*, 23(9): 15951-15966.
12. Xiao S, Ge X, Han Q L, et al(2022) Resource-efficient platooning control of connected automated vehicles over VANETs. *IEEE Transactions on Intelligent Vehicles*, 7(3): 579-589.
13. Levine W, Athans M(1966) On the optimal error regulation of a string of moving vehicles. *IEEE Transactions on Automatic Control*, 11(3), 355-361.

14. Zhu Y, Zhu F(2018) Distributed adaptive longitudinal control for uncertain third-order vehicle platoon in a networked environment. *IEEE Transactions on Vehicular Technology*, 67(10): 9183-9197.
15. Feng S, Zhang Y, Li S E, et al(2019) String stability for vehicular platoon control: Definitions and analysis methods. *Annual Reviews in Control*, 47: 81-97.
16. Zhu Y, Wu J, Su H(2022) V2V-based cooperative control of uncertain, disturbed and constrained nonlinear CAVs platoon. *IEEE Transactions on Intelligent Transportation Systems*, 23(3): 1796-1806.
17. Liu Y, Xu B, and Ding Y(2017) Convergence analysis of cooperative braking control for interconnected vehicle systems. *IEEE Transactions on Intelligent Transportation Systems*, 18(7): 1894-1906.
18. Wang P, Deng H, Zhang J, et al(2022) Model predictive control for connected vehicle platoon under switching communication topology. *IEEE Transactions on Intelligent Transportation Systems*, 23(7): 7817-7830.
19. Darbha S, Hedrick J K, Choi S B(2001) Direct adaptive longitudinal control of vehicle platoons. *IEEE Transactions on Vehicular Technology*, 50(1): 150-161.
20. Naus, G J, Vugts R P, Ploeg J, et al(2010) String-stable CACC design and experimental validation: A frequency-domain approach. *IEEE Transactions on Vehicular Technology*, 59(9), 4268-4279.
21. Li Y, Tang C, Peeta S, et al(2019) Integral-sliding-mode braking control for a connected vehicle platoon: theory and application. *IEEE Transactions on Industrial Electronics*, 66(6): 4618-4628.
22. Ploeg J, Wouw N, and Nijmeijer H(2014) LP string stability of cascaded systems: application to vehicle platooning. *IEEE Transactions on Control Systems Technology*, 22(2): 786-793.
23. Li S E, Qin X, Li K, et al(2017) Robustness analysis and controller synthesis of homogeneous vehicular platoons with bounded parameter uncertainty. *IEEE/ASME Trans. Mechatronics*, 22(2): 1014-1025.
24. Liu A, Li T, Gu Y, et al(2021) Cooperative extended state observer based control of vehicle platoons with arbitrarily small time headway. *Automatica*, 129: 109678.
25. Guo G, Yue W(2012) Autonomous platoon control allowing range limited sensors. *IEEE Transactions on Vehicular Technology*, 61(7): 2901-2912.
26. Hu J, Bhowmick P, Arvin F, et al(2020) Cooperative control of heterogeneous connected vehicle platoons: an adaptive leader-following approach. *IEEE Robotics and Automation Letters*, 5(2): 977-984.
27. Peters A A, Middleton R H, Mason O(2014) Leader tracking in homogeneous vehicle platoons with broadcast delays. *Automatica*, 50(1): 64-74.
28. Monteil J, Russo G, Shorten R(2019) On $L_\infty$ string stability of nonlinear bidirectional asymmetric heterogeneous platoon systems. *Automatica*, 105: 198-205.
29. Zhao C, Cai L, and Cheng P(2021) Stability analysis of vehicle platooning with limited communication range and random packet losses. *IEEE Internet of Things Journal*, 8(1): 262-277.
30. Zheng Y, Li S E, Wang J, et al(2016) Stability and scalability of homogeneous vehicular platoon: study on the influence of information flow topologies. *IEEE Transactions on Intelligent Transportation Systems*, 17(1): 14-26.
31. Li S E, Zheng Y, Li K, et al(2017) Dynamical modeling and distributed control of connected and automated vehicles: challenges and opportunities. *IEEE Intelligent Transportation Systems Magazine*, 9(3): 46-58.
32. Somarakis C, Ghaedsharaf Y, and Motee N(2020) Risk of collision and detachment in vehicle platooning: Timedelay induced limitations and tradeoffs. *IEEE Transactions on Automatic Control*, 65(8): 3544-3559.
33. Herman I, Knorn S, Ahlén A(2017) Disturbance scaling in bidirectional vehicle platoons with different asymmetry in position and velocity coupling. *Automatica*, 82: 13-20.
34. Gao F, Hu X, Li S E, et al(2018) Distributed adaptive sliding mode control of vehicular platoon with uncertain interaction topology. *IEEE Transactions on Industrial Electronics*, 65(8): 6352-6361.

35. Wu Y, Li S E, Cortés J, et al(2020) Distributed sliding mode control for nonlinear heterogeneous platoon systems with positive definite topologies. *IEEE Transactions on Control Systems Technology*, 28(4): 1272-1283.
36. Xiao S, Ge X, Han Q L, et al(2022) Dynamic event-triggered platooning control of automated vehicles under random communication topologies and various spacing policies. *IEEE Transactions on Cybernetics*, 52(11): 11477-11490.
37. Wang J, Deng X, Guo J, et al(2023) A fully distributed antiwindup control protocol for intelligent-connected electric vehicles platooning with switching topologies and input saturation. *IEEE/ASME Transactions on Mechatronics*, 28 (3): 1683-1694.
38. Darbha S, Rajagopal K R(1999) Intelligent cruise control systems and traffic flow stability. *Transportation research. Part C, Emerging technologies*, 7 (6): 329-352.
39. Zhou J, Peng H(2005) Range policy of adaptive cruise control vehicles for improved flow stability and string stability. *IEEE Transactions on Intelligent Transportation Systems*, 6(2): 229-237.
40. Rödönyi G(2018) An adaptive spacing policy guaranteeing string stability in multi-brand ad hoc platoons. *IEEE Transactions on Intelligent Transportation Systems*, 19(6): 1902-1912.
41. Horn J, Seeland F, Miekautsch F, et al(2020) Control of heterogeneous platoons using a delay-based spacing policy, *IFAC-Papers OnLine*, 53(2): 15281-15287.
42. Wijnbergen P, Jeeninga M, Besselink B(2021) Nonlinear spacing policies for vehicle platoons: A geometric approach to decentralized control. *Systems and Control Letters*, 153: 104954.
43. Kwon J, Chwa D(2014) Adaptive bidirectional platoon control using a coupled sliding mode control method. *IEEE Transactions on Intelligent Transportation Systems*, 15(5): 2040-2048.
44. Wen S, Guo G, Chen B, et al(2018) Event-triggered cooperative control of vehicle platoons in vehicular ad hoc networks. *Information Sciences*, 459: 341-353.
45. Ali A, Garcia G, Martinet P(2015) The flatbed platoon towing model for safe and dense platooning on highways. *IEEE Intelligent Transportation Systems Magazine*, 7(1): 58-68.
46. Vite L, Juárez L, Gomez M A(2022) Dynamic predictor-based adaptive cruise control. *Journal of the Franklin Institute*, 359(12): 6123-6141.
47. Guo X, Wang J, Liao F, et al(2017) Distributed adaptive sliding mode control strategy for vehicle-following systems with nonlinear acceleration uncertainties. *IEEE Transactions on Vehicular Technology*, 66(2): 981-991.
48. Guo X, Wang J, Liao F, et al(2017) Distributed adaptive control for vehicular platoon with unknown dead-zone inputs and velocity/acceleration disturbances. *International Journal of Robust and Nonlinear Control*, 27(16): 2961-2981.
49. Guo X, Wang J, Liao F, et al(2019) Neuroadaptive quantized PID sliding-mode control for heterogeneous vehicular platoon with unknown actuator deadzone. *International Journal of Robust and Nonlinear Control*, 29(1): 188-208.
50. Baskar L D, Schutter B D, Hellendoorn H(2012) Traffic management for automated highway systems using model-based predictive control. *IEEE Transactions on Intelligent Transportation Systems*, 13(2): 838-847.
51. Liu Y, Gao H, Xu B, et al(2014) Autonomous coordinated control of a platoon of vehicles with multiple disturbances. *IET Control Theory and Applications*. 8(18): 2325-2335.
52. Sungu H, Inoue M, Imura J(2015) Nonlinear spacing policy based vehicle platoon control for local string stability and global traffic flow stability. *2015 European Control Conference, Linz, Austria*, 3396-3401.
53. Guo G, Li P, Hao L Y(2020) Adaptive fault-tolerant control of platoons with guaranteed traffic flow stability. *IEEE Transactions on Vehicular Technology*, 69(7): 6916-6927.
54. Guo G, Li P, Hao L Y(2022) A new quadratic spacing policy and adaptive fault-tolerant platooning with actuator saturation. *IEEE Transactions on Intelligent Transportation Systems*, 23(2): 1200-1212.
55. Gao Z, Zhang Y, Guo G(2022) Fixed-time prescribed performance adaptive fixed-time sliding mode control for vehicular platoons with actuator saturation. *IEEE Transactions on Intelligent Transportation Systems*, 23(12): 24176-24189.

56. Liu Y, Wang Z, and Zhou D H(2021) Resilient actuator fault estimation for discrete-time complex networks: A distributed approach. *IEEE Transactions on Automatic Control*, 66(9): 4214-4221.
57. Song S, Park J H, Zhang B, et al(2022) Adaptive NN finite-time resilient control for nonlinear time-delay systems with unknown false data injection and actuator faults. *IEEE Transactions on Neural Networks and Learning Systems*, 33(10): 5416-5428.
58. Zhang D, Liu G, Zhou H, et al(2018) Adaptive sliding mode fault-tolerant coordination control for four-wheel independently driven electric vehicles. *IEEE Transactions on Industrial Electronics*, 65(11): 9090-9100.
59. Guo X, Xu W, Wang J, et al(2021) Distributed neuroadaptive fault-tolerant sliding-mode control for 2-D plane vehicular platoon systems with spacing constraints and unknown direction faults. *Automatica*, 129: 109675.
60. Ji Y, Liu Y C(2022) Force allocation control for distributed drive electric vehicles under split-friction regions and actuator faults. *ISA Transactions*, 129: 398-412.
61. Izadi H A, Gordon B W, Zhang Y M(2013) Hierarchical decentralized receding horizon control of multiple vehicles with communication failures. *IEEE Transactions on Aerospace and Electronic Systems*, 49(2): 744-759.
62. Saraoglu M, Hart F, Morozov A, et al(2018) Fault-tolerant path planning in networked vehicle systems in presence of communication failures. *IFAC-PapersOnLine*, 51(23): 82-87.
63. Zhao W, Liu H, Valavanis K P, et al(2022) Fault-tolerant formation control for heterogeneous vehicles via reinforcement learning. *IEEE Transactions on Aerospace and Electronic Systems*, 58(4): 2796-2806.
64. Zhang G, Zhang H, Huang X, et al(2016) Active fault-tolerant control for electric vehicles with independently driven rear in-wheel motors against certain actuator faults. *IEEE Transactions on Control Systems Technology*, 24(5): 1557-1572.
65. Witczak M, Mrugalski M, Pazera M, et al(2020) Fault diagnosis of an automated guided vehicle with torque and motion forces estimation: A case study. *ISA Transactions*, 104: 370-381.
66. Wang R, Wang J(2011) Fault-tolerant control with active fault diagnosis for four-wheel independently driven electric ground vehicles. *IEEE Transactions on Vehicular Technology*, 60(9): 4276-4287.
67. Zhang B, Lu S(2020) Fault-tolerant control for four-wheel independent actuated electric vehicle using feedback linearization and cooperative game theory. *Control Engineering Practice*, 101: 104510.
68. Zou S, Zhao W, Liang W, et al(2022) Fault diagnosis and fault-tolerant compensation strategy for wheel angle sensor of steer-by-wire vehicle via extended kalman filter. *IEEE Sensors Journal*, 22(2): 1756-1766.
69. Lopes A, Araújo R E(2020) Active fault diagnosis method for vehicles in platoon formation. *IEEE Transactions on Vehicular Technology*, 69(4): 3590-3603.
70. Guo X, Xu W, Wang J, et al(2022) BLF-Based neuroadaptive fault-tolerant control for nonlinear vehicular platoon with time-varying fault directions and distance restrictions. *IEEE Transactions on Intelligent Transportation Systems*, 23(8): 12388-12398.
71. Janaideh M, Hammad E, Farraj A, et al(2019) Mitigating attacks with nonlinear dynamics on actuators in cyber-physical mechatronic systems. *IEEE Transactions on Industrial Informatics*, 15(9): 4845-4856.
72. Huang X, Dong J(2019) On modeling and secure control of cyber physical systems with attacks/faults changing system dynamics: an average dwell-time approach. *International Journal of Robust and Nonlinear Control*, 29(16): 5481-5498.
73. Liu C, Zhao J, and Patton R J(2023) Distributed anti-attack fault-tolerant tracking control for vehicle platoon systems under cyber-physical threats. *IEEE Transactions on Industrial Informatics*, 19(6): 7825-7834.
74. Zhou C, Hu B, Shi Y, et al(2021) A unified architectural approach for cyberattack-resilient industrial control systems. *Proceedings of the IEEE*, 109(4): 517-541.

75. Teixeira A, Sou K C, Sandberg H(2015) Secure control systems: a quantitative risk management approach. *IEEE Control Systems Magazine*, 35(1): 24-45.
76. Wu C, Li X, Pan W, et al(2021) Zero-sum game-based optimal secure control under actuator attacks. *IEEE Transactions on Automatic Control*, 66(8): 3773-3780.
77. Amin S, Schwartz G A, Saatry S S(2013) Security of interdependent and identical networked control systems. *Automatica*, 49: 186-192.
78. He H, Chen Y, Qi W, et al(2022) Observer-based resilient control of positive systems with heterogeneous DoS attacks: A Markov model approach. *Journal of the Franklin Institute*, 359(1): 272-293.
79. Jiao S, Xu S, Yuan D, et al(2023) Sliding mode control for networked control systems under DoS attacks via an event-triggered scheme. *Journal of the Franklin Institute*, 360(4): 2931-2954.
80. Liu M, Yu J, Liu Yu(2022) Dynamic event-triggered asynchronous fault detection for Markov jump systems with partially accessible hidden information and subject to aperiodic DoS attacks. *Applied Mathematics and Computation*, 431: 127317.
81. Wang Y W, Zeng Z H, Liu X K, et al(2022) Input-to-state stability of switched linear systems with unstabilizable modes under DoS attacks. *Automatica*, 146: 110607.
82. Biron Z A, Dey S, Pisu P(2018) Real-time detection and estimation of denial of service attack in connected vehicle systems. *IEEE Transactions on Intelligent Transportation Systems*, 19(12): 3893-3902.
83. Petrillo A, Pescapé A, Santini S, et al(2021) A secure adaptive control for cooperative driving of autonomous connected vehicles in the presence of heterogeneous communication delays and cyberattacks. *IEEE Transactions on Cybernetics*, 51(3): 1134-1149.
84. Li Y, Yang Y, Chai T, et al(2022) Stochastic detection against deception attacks in CPS: Performance evaluation and game-theoretic analysis. *Automatica*, 144: 110461.
85. Góes R, Marchand H, Lafortune S, et al(2023) Dealing with sensor and actuator deception attacks in supervisory control. *Automatica*, 147: 110736.
86. Huang X, Dong J(2020) Learning-based switched reliable control of cyber-physical systems with intermittent communication faults. *IEEE/CAA Journal of Automatica Sinica*, 7(3): 711-724.
87. Cao Z, Niu Y, Song J(2020) Finite-time sliding-mode control of markovian jump cyber-physical systems against randomly occurring injection attacks. *IEEE Transactions on Automatic Control*, 65(3): 1264-1271.
88. Shoukry Y, Tabuada P(2016) Event-triggered state observers for sparse sensor noise/attacks. *IEEE Transactions on Automatic Control*, 61(8): 2079-2091.
89. Li X, Wen C, Chen C, et al(2022) Adaptive resilient secondary control for microgrids with communication faults. *IEEE Transactions on Cybernetics*, 52(8): 8493-8503.
90. Huang X, Dong J(2021) ADP-Based robust resilient control of partially unknown nonlinear systems via cooperative interaction design. *IEEE Transactions on Systems, Man, and Cybernetics: Systems*, 51(12): 7466-7474.
91. Huang X, Dong J(2022) Robust Dynamic actuator failure compensation control of nonlinear systems via cooperative interaction design. *IEEE Transactions on Cybernetics*, 52(7): 6013-6023.
92. Lee T H, Lim C P, Nahavandi S, et al(2019) Observer-based H$\infty$ fault-tolerant control for linear systems with sensor and actuator faults. *IEEE Systems Journal*, 13(2): 1981-1990.
93. Liu M, Cao X, and Shi P(2012) Fuzzy-model-based fault-tolerant design for nonlinear stochastic systems against simultaneous sensor and actuator faults. *IEEE Transactions on Fuzzy Systems*, 21(5): 789-799.
94. Gusrialdi A, Qu Z, Simaan M A(2018) Competitive interaction design of cooperative systems against attacks. *IEEE Transactions on Automatic Control*, 63(9): 3159-3166.
95. Jin X, Haddad W M, Yucelen T(2017) An adaptive control architecture for mitigating sensor and actuator attacks in cyber-physical systems. *IEEE Transactions on Automatic Control*, 62(11): 6058-6064.

96. Rezaee H, Parisini T, Polycarpou M P(2020) Control of vehicular platoons: Stochastic robustness against jamming attacks. *IFAC-PapersOnLine*, 53(2): 17041-17046.
97. Ju Z, Zhang H, Tan Y(2020) Deception attack detection and estimation for a local vehicle in vehicle platooning based on a modified UFIR estimator. *IEEE Internet of Things Journal*, 7(5): 3693-3705.
98. Yang F, Gu Z, Hua L, et al(2022) A resource-aware control approach to vehicle platoons under false data injection attacks. *ISA Transactions*, 131: 367-376.
99. Xiao S, Ge X, Han Q L, et al(2022) Secure and collision-free multi-platoon control of automated vehicles under data falsification attacks. *Automatica*, 145: 110531.
100. Liu X, Zhao M, Sun D, et al(2023) Adaptive fault-tolerant controller and operation strategies designed for mixed vehicular platoons in the vicinity of a signalized intersection. *Physica A: Statistical Mechanics and its Applications*, 611: 128438.
101. Ma Y, Li Z, Malekian R, et al(2019) Hierarchical fuzzy logic-based variable structure control for vehicles platooning. *IEEE Transactions on Intelligent Transportation Systems*, 20(4): 1329-1340.
102. Guo H, Liu J, Dai Q, et al(2020) A distributed adaptive triple-step nonlinear control for a connected automated vehicle platoon with dynamic uncertainty. *IEEE Internet of Things Journal*, 7(5): 3861-3871.
103. Ge X, Han Q L, Wang J, et al(2022) Scalable and resilient platooning control of cooperative automated vehicles. *IEEE Transactions on Vehicular Technology*, 71(4): 3595-3608.
104. Yue B F, and Che W W(2023) Data-driven dynamic event-triggered fault-tolerant platooning control. *IEEE Transactions on Industrial Informatics*, 19(7): 8418-8426.
105. Xu W D, Guo X G, Wang J L, et al(2022) Nonlinear disturbance observer-based fault-tolerant sliding-mode control for 2-D plane vehicular platoon with UTVFD and ANAS. *IEEE Transactions on Cybernetics*, https://doi.org/10.1109/TCYB.2022.3222496.
106. Mousavinejad E, Vlacic L(2022) Secure platooning control of automated vehicles under cyber attacks. *ISA Transactions*, 127: 229-238.
107. Li Y, Tang C, Li K, et al(2018) Nonlinear finite-time consensus-based connected vehicle platoon control under fixed and switching communication topologies. *Transportation Research Part C: Emerging Technologies*, 93: 525-543.
108. Du C, Bian Y, Liu H, et al(2021) Cooperative startup control for heterogeneous vehicle platoons: a finite-time output tracking-based approach. *IEEE Transactions on Control of Network Systems*, 8(4): 1767-1777.
109. Guo G, Kang J, Lei H, et al(2022) Finite-time stabilization of a collection of connected vehicles subject to communication interruptions. *IEEE Transactions on Intelligent Transportation Systems*, 23(8): 10627-10635.
110. Han J, Zhang J, He C, et al(2023) Distributed finite-time safety consensus control of vehicle platoon with senor and actuator failures. *IEEE Transactions on Vehicular Technology*, 2023, 72(1): 162-175.
111. Sun J, Pu Z, Yi J, et al(2020) Fixed-time control with uncertainty and measurement noise suppression for hypersonic vehicles via augmented sliding mode observers. *IEEE Transactions on Industrial Informatics*, 16(2): 1192-1203.
112. Liu D, Liu Z, Chen C L P, et al(2022) Distributed adaptive neural fixed-time tracking control of multiple uncertain mechanical systems with actuation dead zones. *IEEE Transactions on Systems, Man, and Cybernetics: Systems*, 52(6): 3859-3872.
113. Zhang L, Chen B, Lin C, et al(2022) Fuzzy adaptive fixed-time consensus tracking control of high-order multi-agent systems. *IEEE Transactions on Fuzzy Systems*, 30(2): 567-578.
114. Guo G, Zhao Z, and Zhang R(2023) Distributed trajectory optimization and fixed-time tracking control of a group of connected vehicles. *IEEE Transactions on Vehicular Technology*, 72 (2): 1478-1487.
115. Liu Y, Yao D, Wang L, et al(2023) Distributed adaptive fixed-time robust platoon control for fully heterogeneous vehicles. *IEEE Transactions on Systems, Man, and Cybernetics: Systems*, 53(1): 264-274.

116. Stamouli C J, Bechlioulis C P, Kyriakopoulos K J(2022) Robust dynamic average consensus with prescribed transient and steady state performance. *Automatica*, 144: 110503.
117. Liang Z, Wang Z, Zhao J, et al(2023) Fixed-time and fault-tolerant path-following control for autonomous vehicles with unknown parameters subject to prescribed performance. *IEEE Transactions on Systems, Man, and Cybernetics: Systems*, 53 (4): 2363-2373.
118. Liu G, Sun Q, Wang R, et al(2022) Event-based fuzzy adaptive consensus FTC for microgrids with nonlinear item via prescribed fixed-time performance. *IEEE Transactions on Circuits and Systems I: Regular Papers*, 69(7): 2982-2993.
119. Liu Q, Zhang K, Jiang B(2022) Fixed-time fault estimation and prescribed performance fault-tolerant control for interconnected systems. *IEEE Transactions on Cybernetics*, https://doi.org/10.1109/TCYB.2022.3188812.
120. Cheng W, Zhang K, Jiang B(2023) Fixed-time fault-tolerant formation control for a cooperative heterogeneous multi-agent system with prescribed performance. *IEEE Transactions on Systems, Man, and Cybernetics: Systems*, 53(1): 462-474.

# Chapter 2
# Resilient Control Design of Intelligent Connected Vehicle Platoon Systems

To facilitate the research, in this Chapter, we introduce several resilient control methods of the intelligent connected vehicle platoon systems, such as coupled sliding mode control, command filtered back-stepping control. We discuss and analysis the basic structures and design processes of existing distributed longitudinal resilient control methods of coupled sliding mode control, command filtered back-stepping control. Some simulation cases also have been provided. All of the abovementioned resilient control methods will be used in the design of distributed longitudinal resilient control of the intelligent connected vehicle platoon systems.

## 2.1 Coupled Sliding Mode Based Distributed Resilient Control

The sliding mode control theory is one of the important methods in the design of longitudinal resilient control of the intelligent connected vehicle platoon system, and has been widely used. The core of the sliding mode control algorithm is as follows: First, the sliding mode surface is designed for each vehicle based on the inter-vehicle spacing error, and the coupled sliding mode surface is designed to connect the stability convergence relation of each vehicle subsystem in the vehicle platoon system. On this basis, then, the control laws of the vehicle platoon system are designed. By this way, the trajectory of the inter-vehicle spacing error of each vehicle subsystem can approach the sliding mode surface in a finite time [1, 2]. The sliding mode control method has good robustness to bounded faults and attacks. Specifically, through the following cases, the controller design process and stability performances analysis process of the sliding mode control method during the longitudinal resilient



C. Pan and Y. Chen, *Resilient Control of Intelligent Connected Vehicle Platoon Systems*,
https://doi.org/10.1007/978-981-97-8850-7_2

control design processing of the vehicle platoon are introduced. Several sets of simple simulation cases are also given to verify the effect of the control method.

### 2.1.1 Model Description of Longitudinal Dynamic with Physical Faults

In this case, considering that the intelligent connected vehicle platoon system consists of $n$ following vehicles and one leader vehicle. We take the following third-order nonlinear longitudinal dynamics [3, 4] of the $i$th following vehicles as example,

$$\begin{aligned}\dot{\bar{p}}_i(t) &= \bar{v}_i(t)\\ \dot{\bar{v}}_i(t) &= \bar{a}_i(t)\\ \dot{\bar{a}}_i(t) &= \frac{\bar{\eta}_i}{m_i r_i \tau_i}\bar{u}_i(t) - \frac{1}{\tau_i}\bar{a}_i(t) - \frac{1}{m_i\tau_i}[\bar{C}_{Ai}(\bar{v}_i^2 + 2\tau_i\bar{v}_i\bar{a}_i(t))] + \bar{d}_i \end{aligned} \tag{2.1}$$

where $i = 1, 2, \ldots, n$, $\bar{p}_i(t)$, $\bar{v}_i(t)$ and $\bar{a}_i(t)$ are the position, velocity, and acceleration of the $i$th following vehicle, respectively. $\tau_i$ is the engine time lag, $u_i(t)$ describes the engine's control input. $m_i$ denotes the vehicle's mass, $\bar{C}_{Ai}$ describe the aerodynamic drag coefficient. $\bar{\eta}_i$ represents the vehicle's driveline mechanical efficiency. $r_i$ is the vehicle's tire radius, $\bar{d}_i$ is the external disturbance functions, which is bounded and satisfied that $\left|(m_i r_i \tau_i \bar{d}_i)/\bar{\eta}_i\right| \leq \widehat{d}_i$, $\widehat{d}_i$ is the unknown positive constant. The leader vehicle's longitudinal dynamic model is

$$\begin{aligned}\dot{\bar{p}}_0(t) &= \bar{v}_0(t)\\ \dot{\bar{v}}_0(t) &= \bar{a}_0(t)\end{aligned} \tag{2.2}$$

where the states information of the leader vehicle is available here.

We take the actuator fault model of the $i$th following vehicle as

$$\bar{u}_{fi}(t) = (1 - \bar{\rho}_i)\bar{u}_i(t) + \bar{\Delta}_i \tag{2.3}$$

where $\bar{\rho}_i$ denote the partial loss of effectiveness fault factor, $\bar{\rho}_i$ satisfy the condition that $0 < \bar{\rho}_{i1} \leq \bar{\rho}_i \leq \bar{\rho}_{i2} < 1$. $\bar{\Delta}_i$ is the bias fault, $\bar{\Delta}_i \leq \bar{\bar{\Delta}}_i$. $\bar{\rho}_{i1}$, $\bar{\rho}_{i2}$ and $\bar{\bar{\Delta}}_i$ are the unknown positive constants. $\bar{\rho}_{i1} = \bar{\rho}_{i2} = 0$, and $\bar{\bar{\Delta}}_i = 0$ mean that there is no faults occur.

Based on the formulas (2.1) and (2.3), we have

$$
\begin{aligned}
\dot{\bar{p}}_i(t) &= \bar{v}_i(t) \\
\dot{\bar{v}}_i(t) &= \bar{a}_i(t) \\
\dot{\bar{a}}_i(t) &= \frac{\bar{\eta}_i}{m_i r_i \tau_i}(1-\bar{\rho}_i)\bar{u}_i(t) - \frac{1}{\tau_i}\bar{a}_i(t) \\
&\quad - \frac{1}{m_i \tau_i}[\bar{C}_{Ai}(\bar{v}_i^2 + 2\tau_i \bar{v}_i \bar{a}_i(t))] - \frac{\bar{\eta}_i}{m_i r_i \tau_i}\bar{\varpi}_i
\end{aligned}
\tag{2.4}
$$

where $\bar{\varpi}_i = -\bar{\Delta}_i - (m_i r_i \tau_i \bar{d}_i)/\bar{\eta}_i$, based on the abovementioned conditions, $\bar{\Delta}_i \leq \bar{\bar{\Delta}}_i$, and $\left|(m_i r_i \tau_i \bar{d}_i)/\bar{\eta}_i\right| \leq \overset{\frown}{d}_i$, it can derives that there exists a positive parameter $\bar{\bar{\varpi}}_i$ that satisfied that $\left|\dot{\bar{\varpi}}_i\right| \leq \bar{\bar{\varpi}}_i$.

For achieving the vehicle platoon control, and guaranteeing the string stability and the closed loop stability of each vehicles, we take the QS policy as example,

$$
\begin{aligned}
\bar{e}_i(t) &= \overset{\frown}{\bar{e}}_i(t) - \bar{\Gamma}_i \\
\overset{\frown}{\bar{e}}_i(t) &= \bar{\delta}_i(t) - \bar{\delta}_d - \bar{h}\bar{v}_i(t) - \frac{\bar{\phi}\bar{v}_i^2}{2\bar{\varphi}_{\max}(1-\bar{\rho}_{i2})} \\
\bar{\Gamma}_i &= (\overset{\frown}{\bar{e}}_i(0) + [\bar{\varsigma}_i \overset{\frown}{\bar{e}}_i(0) + \dot{\overset{\frown}{\bar{e}}}_i(0)]t \\
&\quad + 0.5[\bar{\varsigma}_i^2 \overset{\frown}{\bar{e}}_i(0) + 2\bar{\varsigma}_i \dot{\overset{\frown}{\bar{e}}}_i(0) + \ddot{\overset{\frown}{\bar{e}}}_{ii}(0)]t^2)e^{-\bar{\varsigma}_i t}
\end{aligned}
\tag{2.5}
$$

where $\bar{\delta}_i(t) = \bar{p}_{i-1}(t) - \bar{p}_i(t)$ denotes the actual measured spacing between the $i-1$th vehicle and $i$th vehicle, $\bar{e}_i(t)$ is the inter-vehicle spacing errors, $\bar{\delta}_d$ is the desired inter-vehicle distance at vehicle platoon standstill, $\bar{\phi}$ is the environmental safety factor, $\bar{\varphi}_{\max}$ denote the maximum deceleration of the vehicle, $\bar{h}$ is the headway time constant, $\bar{\varsigma}_i$ is the strictly positive constant. Furthermore, through (2.5), $\overset{\frown}{\bar{e}}_i(0) = \overset{\frown}{\bar{e}}_i(t)|_{t=0}$, $\dot{\overset{\frown}{\bar{e}}}_i(0) = \dot{\overset{\frown}{\bar{e}}}_i(t)|_{t=0}$, $\ddot{\overset{\frown}{\bar{e}}}_i(0) = \ddot{\overset{\frown}{\bar{e}}}_i(t)|_{t=0}$. Then, we can derived that $\bar{e}_i(t)|_{t=0} = 0$, $\dot{\bar{e}}_i(t)|_{t=0} = 0$, $\ddot{\bar{e}}_i(t)|_{t=0} = 0$, which means that $\bar{e}_i(t)$ can always be zero for arbitrary initial spacing errors $\bar{e}_i(0)$.

**Control objective:** The control objectives are shown as below:

- The closed-loop stability of each vehicle is achieved. The acceleration, velocity of each vehicle in the vehicle platoon can track the leader vehicles', and the connected vehicle platoon system can maintain the desired safety inter-vehicle spacing errors, $\lim_{t\to\infty} |\bar{v}_i(t) - \bar{v}_0(t)| = 0$, $\lim_{t\to\infty} |\bar{a}_i(t) - \bar{a}_0(t)| = 0$, $\lim_{t\to\infty} |\bar{e}_i(t)| = 0$.
- The strict string stability is guaranteed for the vehicle platoon system with the actuator fault and external disturbances.

### 2.1.2 Super Twisting Based Distributed Resilient Control Laws

Firstly, constructing the PID based sliding mode surface of the $i$th following vehicle as

$$j_i(t) = \dot{\bar{e}}_i(t) + \bar{\beta}_i \bar{e}_i(t) + \int_0^t \bar{\beta}_{i1} \bar{e}_i(\tau) d\tau \tag{2.6}$$

where $\bar{\beta}_i$, $\bar{\beta}_{i1}$ are designed positive constants.

Then, the coupled sliding mode surface is designed to connect the stability convergence relation of each following vehicle subsystem,

$$J_i(t) = \bar{\mu}_1 j_i(t) - \bar{\mu}_2 j_{i+1}(t) \tag{2.7}$$

where $\bar{\mu}_1$ and $\bar{\mu}_2$ are positive parameters and satisfy that $0 < \frac{\bar{\mu}_1}{\bar{\mu}_2} \leq 1$. Let $J_{n+1} = 0$ and $j_{n+1} = \dot{j}_{n+1} = 0$ for the vehicle $i = n$. Then, through formula (2.7), we can derive the relation between sliding mode surface $j_i(t)$ and the coupled sliding mode surface $J_i(t)$ as

$$J = \bar{Q} j \tag{2.8}$$

where $J = \begin{bmatrix} J_1(t) & J_2(t) & \ldots & J_n(t) \end{bmatrix}^T$, $j = \begin{bmatrix} j_1(t) & j_2(t) & \ldots & j_N(t) \end{bmatrix}^T$, and $\bar{Q} = \begin{bmatrix} \bar{\mu}_1 & -\bar{\mu}_2 & \ldots & 0 & 0 \\ 0 & \bar{\mu}_1 & -\bar{\mu}_2 & \ldots & 0 \\ & & \vdots & & \\ 0 & 0 & \ldots & \bar{\mu}_1 & -\bar{\mu}_2 \\ 0 & 0 & \ldots & 0 & \bar{\mu}_1 \end{bmatrix}$ is invertible matrix. Thus, it can concluded that it's equivalence between $J = 0$ and $j = 0$. If $J_i(t)$ can converges to zero it means that $j_i(t)$ will also converges to zero at the same time.

Combining with QS policy in (2.5), and taking the time derivative of $J_i(t)$ in (2.7),

$$\begin{aligned} \dot{J}_i(t) &= \bar{\mu}_1(\bar{\beta}_i \dot{\bar{e}}_i(t) + \ddot{\bar{e}}_i(t) + \bar{\beta}_{i1} \bar{e}_i(t)) - \bar{\mu}_2 \dot{j}_{i+1} \\ &= \bar{\mu}_1 \left( \dot{\bar{v}}_{i-1} - \dot{\bar{v}}_i - \frac{\bar{\phi} a_i^2}{\bar{\varphi}_{\max}(1 - \bar{\rho}_{i2})} - \bar{h} \dot{\bar{a}}_i - \frac{\bar{\phi} \bar{v}_i \dot{\bar{a}}}{\bar{\varphi}_{\max}(1 - \bar{\rho}_{i2})} - \bar{\Gamma}_i \right) \\ &+ \bar{\mu}_1(\beta_i \dot{\bar{e}}_i + \bar{\beta}_{i1} e_i) - \bar{\mu}_2 \dot{j}_{i+1} \\ &= -\bar{\mu}_1 \left( \frac{\bar{\phi} \bar{v}_i}{\bar{\varphi}_{\max}(1 - \bar{\rho}_{i2})} + \bar{h} \right) \dot{\bar{a}}_i + \bar{\Pi}_i \end{aligned} \tag{2.9}$$

where $\Pi_i(t) = \bar{\mu}_1(\dot{\bar{v}}_{i-1} - \dot{\bar{v}}_i - \frac{\bar{\phi} a_i^2}{\bar{\varphi}_{\max}(1 - \bar{\rho}_{i2})} - \bar{\Gamma}_i) + \bar{\mu}_1(\beta_i \dot{\bar{e}}_i + \bar{\beta}_{i1} e_i) - \bar{\mu}_2 \dot{j}_{i+1}$, $i = 1, 2, \ldots, n-1$.

For the vehicle $i = n$,

$$\dot{J}_n(t) = -\bar{\mu}_1\left(\frac{\bar{\phi}\bar{v}_n}{\bar{\varphi}_{\max}(1-\bar{\rho}_{n2})}+\bar{h}\right)\dot{\bar{a}}_n + \bar{\Pi}_n \tag{2.10}$$

where $\bar{\Pi}_n(t) = \bar{\mu}_1(\dot{\bar{v}}_{n-1} - \dot{\bar{v}}_n - \frac{\bar{\phi}a_n^2}{\bar{\varphi}_{\max}(1-\bar{\rho}_{n2})} - \bar{\Gamma}_n) + \bar{\mu}_1(\beta_n\dot{\bar{e}}_n + \bar{\beta}_{n1}e_n)$.

Then, based on the (2.9) and (2.10), we have

$$\begin{aligned}
\dot{J}_i(t)J_i(t) &= -\bar{\mu}_1\left(\frac{\bar{\phi}\bar{v}_i}{\bar{\varphi}_{\max}(1-\bar{\rho}_{i2})}+\bar{h}\right)\left[\frac{\bar{\eta}_i(1-\bar{\rho}_i)\bar{u}_i(t)}{m_i r_i\tau_i} - \frac{\bar{a}_i(t)}{\tau_i}\right]J_i(t) + \bar{\Pi}_i J_i\\
&\quad - \bar{\mu}_1\left(\frac{\bar{\phi}\bar{v}_i}{\bar{\varphi}_{\max}(1-\bar{\rho}_{i2})}+\bar{h}\right)\left[-\frac{[\bar{C}_{Ai}(\bar{v}_i^2+2\tau_i\bar{v}_i\bar{a}_i(t))]}{m_i\tau_i} - \frac{\bar{\eta}_i}{m_i r_i\tau_i}\bar{\varpi}_i\right]J_i(t)\\
&\le -\bar{\mu}_1\left(\frac{\bar{\phi}\bar{v}_i}{\bar{\varphi}_{\max}(1-\bar{\rho}_{i2})}+\bar{h}\right)\left[\frac{\bar{\eta}_i(1-\bar{\rho}_i)\bar{u}_i(t)}{m_i r_i\tau_i} - \frac{\bar{a}_i(t)}{\tau_i}\right]J_i(t)\\
&\quad - \bar{\mu}_1\left(\frac{\bar{\phi}\bar{v}_i}{\bar{\varphi}_{\max}(1-\bar{\rho}_{i2})}+\bar{h}\right)\left[-\frac{[\bar{C}_{Ai}(\bar{v}_i^2+2\tau_i\bar{v}_i\bar{a}_i(t))]}{m_i\tau_i} - \frac{\bar{\eta}_i}{m_i r_i\tau_i}\bar{\varpi}_i\right]J_i(t)\\
&\quad + (1-\bar{\rho}_{i2})\bar{\Theta}_i\left|\bar{\Pi}_i J_i\right|
\end{aligned} \tag{2.11}$$

By the coupled sliding mode control method, the spacing errors can converge to zero and guarantee $\dot{J}_i(t)J_i(t) < 0$ under the Lyapunov theory. But the control input of each vehicle will switch between $J_i(t) < 0$ and $J_i(t) > 0$.

For eliminating the chattering caused by the switch, different traditional reaching laws have been used, such as the terms including the sign function $\dot{J}_i(t) = -(\alpha^* + sign(J_i(t)))$ or $\dot{J}_i(t) = -(\alpha^* + \frac{J_i(t)}{|J_i(t)|+\varepsilon^*})$. Here, we take the super twisting law as the sliding mode surface reaching law for eliminating the chattering and improving the approach speed of sliding mode motion. The super twisting based distributed platoon resilient control law is designed as

$$\begin{aligned}
\bar{u}_i(t) &= \frac{k_{1i}}{\bar{\wp}_i}\sqrt{|J_i|}sign(J_i) + \frac{k_{2i}}{\bar{\wp}_i}\int sign(J_i)dt + \frac{\bar{\Im}_i}{\bar{\wp}_i}\frac{\bar{a}_i(t)}{\tau_i}\\
&\quad + \frac{(1-\bar{\rho}_{i2})\bar{\Theta}_i}{\bar{\wp}_i}\frac{\left|\bar{\Pi}_i J_i\right|}{J_i(t)} + \frac{\bar{\Im}_i}{\bar{\wp}_i}\frac{1}{m_i\tau_i}[\bar{C}_{Ai}(\bar{v}_i^2+2\tau_i\bar{v}_i\bar{a}_i)]
\end{aligned} \tag{2.12}$$

where $i = 1, 2, \ldots, n$, $\bar{\Im}_i{=}\mu_1\left(\frac{\bar{\phi}\bar{v}_i}{\bar{\varphi}_{\max}(1-\bar{\rho}_{i2})}+\bar{h}\right)$ $\bar{\wp}_i = \bar{\Im}_i\frac{\bar{\eta}_i}{m_i r_i\tau_i}(1-\bar{\rho}_{i2})$, $1 \le 1/(1-\bar{\rho}_{i2}) \le \bar{\Theta}_i$, $\bar{\Theta}_i > 0$ is the positive parameter, and $k_{1i} > 0$, $k_{2i} > 0$, $k_{2i} > \bar{\Im}_i\frac{\bar{\eta}_i}{m_i r_i\tau_i}\bar{\bar{\varpi}}_i$.

### 2.1.3 Vehicle Platoon Stability Performances Analysis and Proof

In this section, firstly, we will prove the closed loop stability of each vehicle subsystem. Then, the strict string stability of the connected vehicle platoon system will be proved later.

***Proof*** First, selecting the Lyapunov function candidate as $V(t) = (\sum_{i=1}^{n-1} \frac{1}{2} J_i^2(t)) + V_n(t)$, $V_n = \frac{1}{2} J_n^2(t)$, then, taking the time derivative of $V(t)$ for the vehicle $i = 1, 2, \ldots, n$, and based on (2.11), one has,

$$\begin{aligned}\dot{V}(t) \leq &\sum_{i=1}^{n-1} \left\{ -\bar{\Im}_i \left[ \frac{\bar{\eta}_i (1 - \bar{\rho}_i) \bar{u}_i(t)}{m_i r_i \tau_i} - \frac{\bar{a}_i(t)}{\tau_i} \right] J_i(t) + (1 - \bar{\rho}_{i2}) \bar{\Theta}_i \left| \bar{\Pi}_i J_i \right| \right\} \\ &+ \sum_{i=1}^{n-1} \{ -\bar{\Im}_i \left[ -\frac{[\bar{C}_{Ai}(\bar{v}_i^2 + 2\tau_i \bar{v}_i \bar{a}_i(t))]}{m_i \tau_i} - \frac{\bar{\eta}_i}{m_i r_i \tau_i} \bar{\varpi}_i \right] J_i(t) \} + \dot{V}_n(t) \end{aligned} \tag{2.13}$$

Substituting the distributed bi-directional platoon resilient control law (2.12) into (2.13),

$$\begin{aligned}\dot{V}(t) \leq &\sum_{i=1}^{n} \left\{ -\bar{\wp}_i J_i(t) \left\{ \frac{(1 - \bar{\rho}_{i2}) \Theta_i}{\bar{\wp}_i} \frac{\left| \bar{\Pi}_i J_i \right|}{J_i} + \bar{\Im}_i \frac{\bar{C}_{Ai}(\bar{v}_i^2 + 2\tau_i \bar{v}_i \bar{a}_i(t))}{\bar{\wp}_i m_i \tau_i} \right\} \right\} \\ &+ \sum_{i=1}^{n} \left\{ -\bar{\wp}_i J_i(t) \left\{ \frac{k_{2i}}{\bar{\wp}_i} \int sign(J_i) dt + \frac{k_{1i}}{\bar{\wp}_i} \sqrt{|J_i|} sign(J_i) + \frac{\bar{\Im}_i a_i(t)}{\bar{\wp}_i \tau_i} \right\} \right\} \\ &+ \sum_{i=1}^{n} \left\{ \bar{\Im}_i \frac{a_i(t) J_i(t)}{\tau_i} + \bar{\Im}_i \frac{\bar{\eta}_i \bar{\varpi}_i J_i(t)}{m_i r_i \tau_i} + \bar{\Im}_i \frac{\bar{C}_{Ai}(\bar{v}_i^2 + 2\tau_i \bar{v}_i \bar{a}_i(t))}{m_i \tau_i} J_i(t) \right\} \\ &+ \sum_{i=1}^{n} \{ (1 - \bar{\rho}_{i2}) \bar{\Theta}_i \left| \bar{\Pi}_i J_i \right| \} \end{aligned} \tag{2.14}$$

By combining terms of the same kind with (2.14), we get

$$\begin{aligned}\dot{V}(t) &\leq \sum_{i=1}^{n} \left\{ -J_i(t) \{ k_{2i} \int sign(J_i) dt + k_{1i} \sqrt{|J_i|} sign(J_i) \} \right\} + \sum_{i=1}^{n} \left\{ \bar{\Im}_i \frac{\bar{\eta}_i \bar{\varpi}_i J_i(t)}{m_i r_i \tau_i} \right\} \\ &\leq \sum_{i=1}^{n} \left\{ - |J_i(t)| \int k_{2i} dt - k_{1i} \sqrt{|J_i|} \, |J_i(t)| + \left| \bar{\Im}_i \frac{\bar{\eta}_i \bar{\varpi}_i J_i(t)}{m_i r_i \tau_i} \right| \right\} \end{aligned} \tag{2.15}$$

Furthermore, according to the fact that $\left| \dot{\bar{\varpi}}_i \right| \leq \bar{\bar{\varpi}}_i$ in (2.4), we derive (2.15) and get

$$
\begin{aligned}
\dot{V} &\le \sum_{i=1}^{n}\left\{-|J_i(t)|\int k_{2i}dt - k_{1i}\sqrt{|J_i|}\,|J_i(t)|\}+\sum_{i=1}^{n}\{\bar{\Im}_i\frac{\bar{\eta}_i}{m_i r_i \tau_i}|J_i(t)|\int\left|\dot{\bar{\varpi}}_i\right|dt\right\}\\
&\le \sum_{i=1}^{n}\left\{-|J_i(t)|\int k_{2i}dt - k_{1i}\sqrt{|J_i|}\,|J_i(t)|\}+\sum_{i=1}^{n}\{\bar{\Im}_i\frac{\bar{\eta}_i}{m_i r_i \tau_i}|J_i(t)|\int\bar{\bar{\varpi}}_i dt\right\}
\end{aligned}
\tag{2.16}
$$

Finally, we can have

$$
\begin{aligned}
\dot{V} &\le \sum_{i=1}^{n}\{-k_{1i}\sqrt{|J_i|}\,|J_i(t)|\}\\
&\le 0
\end{aligned}
\tag{2.17}
$$

where $k_{2i} > \bar{\Im}_i\frac{\bar{\eta}_i}{m_i r_i \tau_i}\bar{\bar{\varpi}}_i$.

It can concluded that $\dot{V}$ is the negative semi-definite due to $\dot{V} \le 0$. Then, based on (2.6) to (2.8), the coupled sliding mode surface $J_i(t)$ and the sliding mode surface $j_i(t)$ can converge to zero at the same time. The closed loop stability of each vehicle subsystem is proved and the connected vehicle platoon system is also stable.

Secondly, the connected vehicle platoon system may lose control, and the transient/steady spacing errors may be enlarged along with the whole system and lead to the accidents since the string instability of the vehicle platoon. Based on the last part of the system closed loop stability, we have that $J_i(t)$ will converge to zero after a finite time, then, $J_i(t) = \bar{\mu}_1 j_i(t) - \bar{\mu}_2 j_{i+1}(t) = 0$, one gets

$$
\begin{aligned}
&\bar{\mu}_1(\dot{\bar{e}}_i(t) + \bar{\beta}_i\bar{e}_i(t) + \int_0^t \bar{\beta}_{i1}\bar{e}_i(\tau)d\tau)\\
&= \bar{\mu}_2(\dot{\bar{e}}_{i+1}(t) + \bar{\beta}_{i+1}\bar{e}_{i+1}(t) + \int_0^t \bar{\beta}_{i+11}\bar{e}_{i+1}(\tau)d\tau)
\end{aligned}
\tag{2.18}
$$

Then, taking the Laplace transform of (2.18),

$$
\begin{aligned}
&\bar{\mu}_1\left(\bar{\beta}_i\bar{E}_i(s) + s\bar{E}_i(s) + \bar{\beta}_{i1}\frac{\bar{E}_i(s)}{s}\right)\\
&= \bar{\mu}_2\left(\bar{\beta}_{i+1}\bar{E}_{i+1}(s) + s\bar{E}_{i+1}(s) + \bar{\beta}_{(i+1)1}\frac{\bar{E}_{i+1}(s)}{s}\right)
\end{aligned}
\tag{2.19}
$$

Then, we have the following result that

$$
\bar{G}(s) = \frac{\bar{E}_{i+1}(s)}{\bar{E}_i(s)} = \frac{\bar{\beta}_i s + s^2 + \bar{\beta}_{i1}}{\bar{\beta}_{(i+1)}s + s^2 + \bar{\beta}_{(i+1)1}} = \frac{\bar{\mu}_1}{\bar{\mu}_2}
\tag{2.20}
$$

Based on the definition of strict string stability in Chap. 1, the considered connected vehicle platoon system can guarantee the strict string stability when $\left|\bar{G}(s)\right| \leq 1$, that is $0 < \frac{\bar{\mu}_1}{\bar{\mu}_2} \leq 1$. When the condition $0 < \frac{\bar{\mu}_1}{\bar{\mu}_2} \leq 1$ is met, the strict string stability of the vehicle platoon system also can be ensured.

Next, we will carry out a set of simulation experiments on the connected vehicle platoon system with 6 following vehicles and one leader vehicle. In this process, we assume that the third vehicle and the fifth vehicle have the sudden actuator faults within a certain period of time. For the case of vehicle 3, the actuator loss of effectiveness fault function is $\bar{\rho}_3 = \begin{cases} 0.2, & 15s \leq t < 17s \\ 0, & otherwise \end{cases}$, and the bias fault is $\Delta_i = 0.05\cos(5t)$. For the case of vehicle 5, the actuator loss of effectiveness fault function is $\bar{\rho}_5 = \begin{cases} 0.32, & 20s \leq t < 22s \\ 0, & otherwise \end{cases}$, and the bias fault is $\Delta_i = 0.045\cos(3t)$. The external disturbance function is $\bar{d}_i = 0.01\sin(2t+1)$. The simulation results are shown in Figs. 2.1, 2.2, 2.3, 2.4, 2.5, 2.6 and 2.7.

Figure 2.1a and b show the actual measured inter-vehicle distances in the connected vehicle platoon system. Figure 2.2a and b show the inter-vehicle spacing error of the connected vehicle platoon system. Longitudinal position, velocity, acceleration, coupled sliding mode surface and the control input are shown in Figs. 2.3, 2.4, 2.5, 2.6 and 2.7, respectively. It is obvious that the designed distributed resilient control scheme has good convergence performances through the coupled sliding mode control method.

## 2.2 Command Filtered Back-Stepping Based Distributed Resilient Control

Back-stepping control theory is also an important technology for the design of longitudinal distributed resilient control of the connected vehicle platoon system [4, 5]. Different from the sliding mode control method, back-stepping control method has not chattering inherent problem. The essence of back-stepping method is to use Lyapunov s second method to design the state feedback controller of the system. Through the iterative design of each state differential equation of the dynamic system, a complete set of control scheme is finally formed. Back-stepping control method mainly includes the traditional back-stepping control, dynamic surface back-stepping control, and command filtered back-stepping control. Dynamic surface back-stepping control method was designed by Swaroop et al. [6]. Dynamic surface control method was used for solving the problem of computational complexity inherent in the traditional back-stepping control by introducing first-order filters. However, the issue of the errors caused by the filters has not been taken into account. Therefore, if the error wants to be as small as possible, only the filters parameters be selected as small as possible, which increases the difficulty of control implementation. Farrell et al.

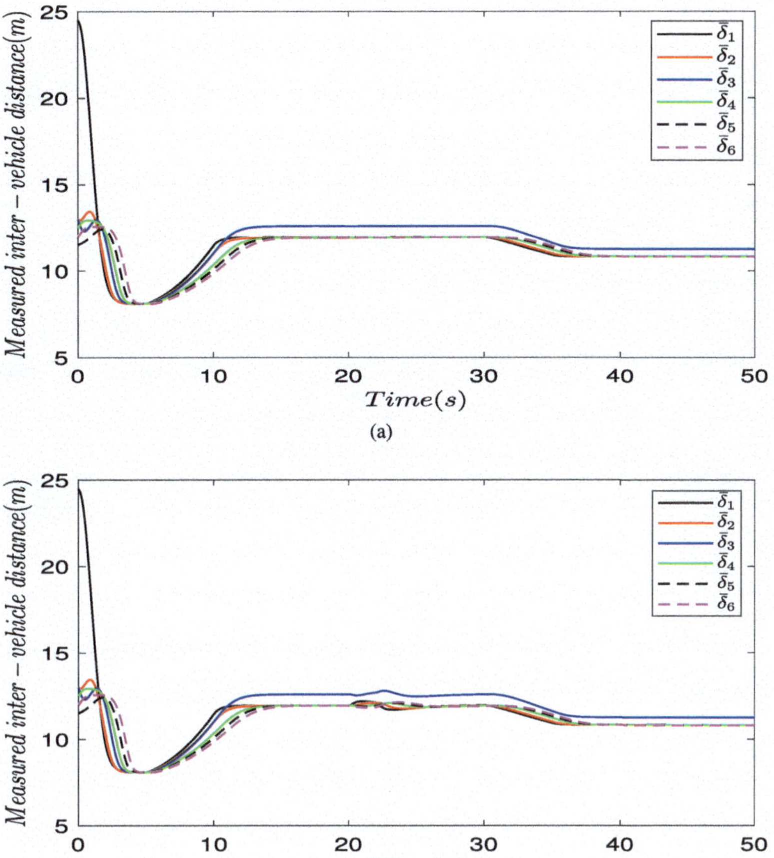


(a)

(b)

**Fig. 2.1** Actual measured inter-vehicle distance $\bar{\delta}_i(t)$ in the connected vehicle platoon system. **a** Vehicle $i = 3$ with physical faults, **b** Vehicle $i = 5$ with physical faults

[7, 8] developed the command filtered back-stepping control for the issues of the compensation of filters error and the computational complexity of terms.

The design of command filtered back-stepping distributed adaptive resilient control of the connected vehicle platoon system is mainly divided into five parts, such as inter-vehicle spacing error coordinate transformation, command filter design, error compensation mechanism design, virtual control laws design, and Lyapunov stability proof. Firstly, the high-order nonlinear dynamic model of the vehicle platoon system is decomposed into several first-order dynamic system models, and a reasonable inter-vehicle spacing policy is designed, and the coordinate transformation is carried out for the inter-vehicle spacing error of the vehicle platoon system. Secondly, the virtual control laws, the derivative of the virtual control laws and the actual control input are obtained by using a suitable command filter, which can avoid the com-

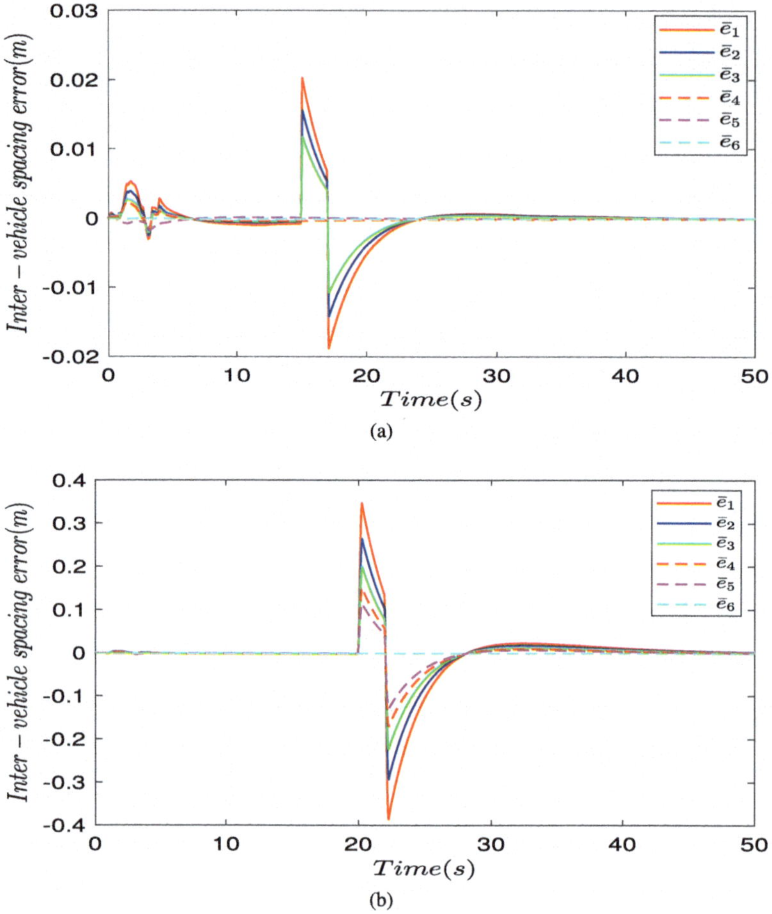


**Fig. 2.2** Inter-vehicle spacing error of the connected vehicle platoon system. **a** Vehicle $i = 3$ with physical faults, **b** Vehicle $i = 5$ with physical faults

putational complexity problem caused by the multi-order derivation of the virtual control function. Further, the filtering error compensation mechanism is designed to eliminate the errors caused by introducing the command filter. Finally, based on Lyapunov stability theory, it is proved that the designed control scheme and filtering error compensation mechanism can guarantee the closed-loop stability of the connected vehicle platoon system.

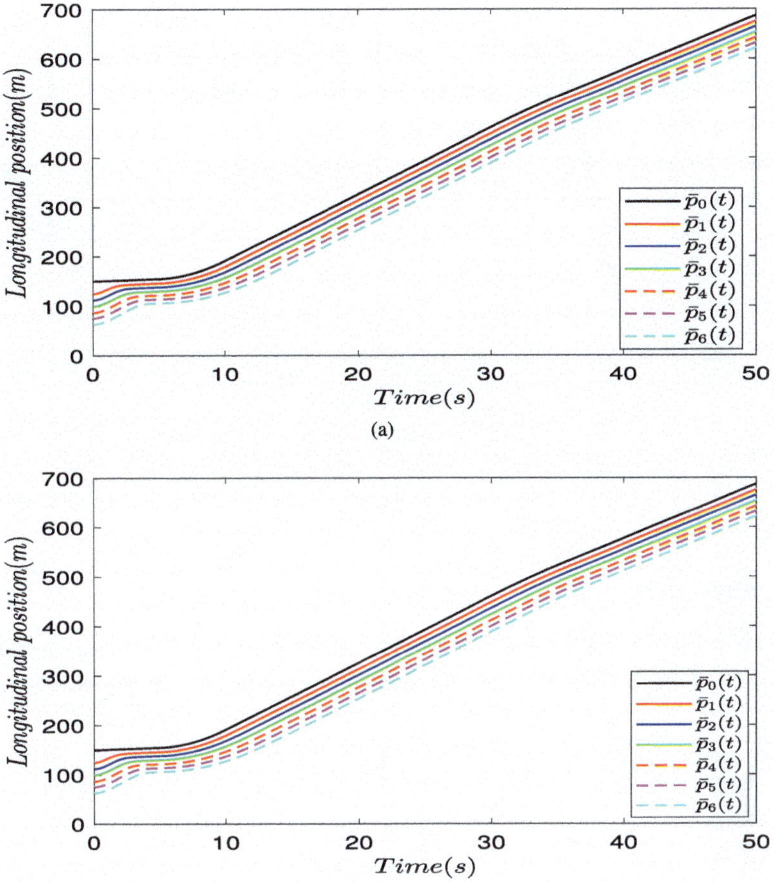


(a)

(b)

**Fig. 2.3** Longitudinal position of the connected vehicle platoon system. **a** Vehicle $i = 3$ with physical faults, **b** Vehicle $i = 5$ with physical faults

## *2.2.1 Model Description of Longitudinal Dynamic with Cyber Attacks*

Firstly, considering a connected vehicle platoon system consists of $n + 1$ connected vehicles driving on a straight road. The leader vehicle is labeled as vehicle 0, and the following vehicles are labeled as vehicle $i(i = 1, 2, \ldots, n)$. The longitudinal dynamic model of following vehicles in the vehicular platoon [9, 10] with the cyber attacks is presented as

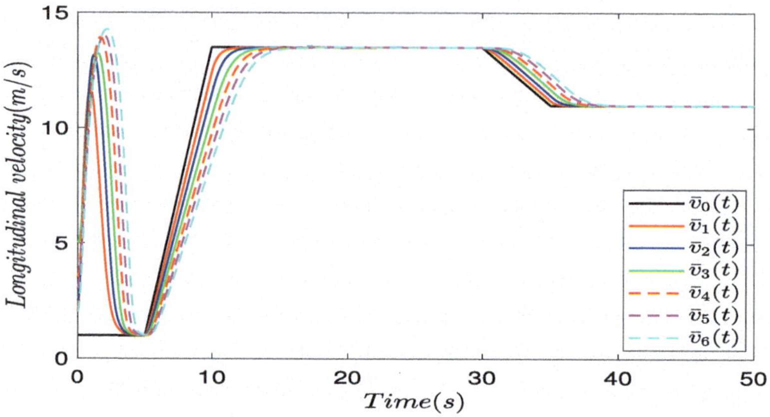


(a)

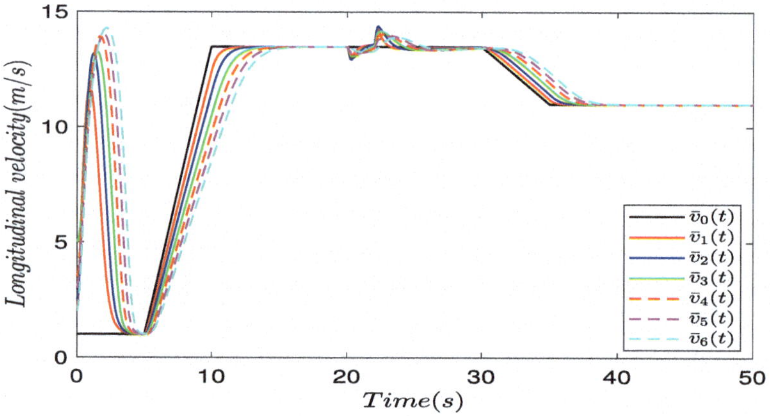


(b)

**Fig. 2.4** Longitudinal velocity of the connected vehicle platoon system. **a** Vehicle $i = 3$ with physical faults, **b** Vehicle $i = 5$ with physical faults

$$\begin{aligned}\dot{\bar{p}}_i(t) &= \bar{v}_i(t)\\ \dot{\bar{v}}_i(t) &= \bar{a}_i(t)\\ \dot{\bar{a}}_i(t) &= \bar{\vartheta}_i^T \bar{\mathrm{B}}_i(\bar{X}_i) + \frac{1}{\tau_i} u_i^A(t) + \bar{\varpi}_i(t)\end{aligned} \tag{2.21}$$

where $\bar{\mathrm{B}}_i(\bar{X}_i) = [\,-\bar{v}_i\bar{a}_i \;\; -\bar{a}_i \;\; -\bar{v}_i^2 \;\; -1\,]^T$, $\bar{\vartheta}_i^T = \left[\, \frac{\varphi^* \bar{A}_i \bar{k}_{ai}}{\bar{M}_i} \;\; \frac{1}{\tau_i} \;\; \frac{\varphi^* \bar{A}_i \bar{k}_{ai}}{2\tau_i \bar{M}_i} \;\; \frac{\bar{D}_{mi}}{\tau_i \bar{M}_i} \,\right]$. The position, velocity, and acceleration of the vehicle $i$ are repressed by $\bar{p}_i(t)$, $\bar{v}_i(t)$ and $\bar{a}_i(t)$, respectively. $\tau_i$ can be used to denote the time constant of engine. $u_i^A(t)$ describes the control input under the cyber attacks. $\bar{\vartheta}_i^T \bar{\mathrm{B}}_i(\bar{X}_i)$ are used to express the unknown nonlinear function with respect to the system states, $\bar{X}_i = [\bar{v}_i, \bar{a}_i]^T$. The air's specific mass is denoted by $\varphi^*$. The mass of vehicle, the cross-sectional area of vehicle, the

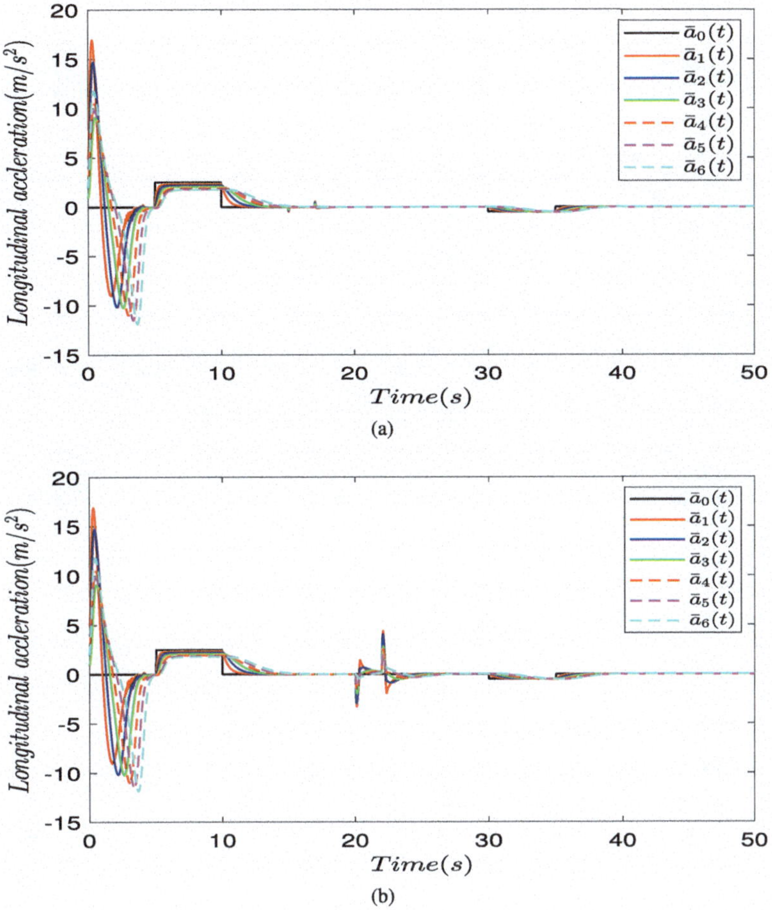


**Fig. 2.5** Longitudinal acceleration of the connected vehicle platoon system. **a** Vehicle $i = 3$ with physical faults, **b** Vehicle $i = 5$ with physical faults

mechanical drag, and air drag coefficient are denoted by $\bar{M}_i$, $\bar{A}_i$, $\bar{D}_{mi}$, and $\bar{k}_{ai}$, respectively. The unknown external disturbance is described by $\bar{\varpi}_i(t)$, which satisfies that $|\bar{\varpi}_i(t)| < \bar{\varpi}_{i1}$, $\bar{\varpi}_{i1} \in \mathrm{R}^+$.

Here, we consider that the adversaries can be easier to change the control signal in the closed-loop of the vehicles control system through the cyber attacks. Hence, the false data injection attack model used by adversaries on the controller-actuator channel can be built as

$$u_i^A(t) = u_i(t) + \Delta_i(\bar{X}_i) \tag{2.22}$$

where the control input of the engine can be described by $u_i(t) = \bar{c}_i \bar{M}_i^{-1}$, and $\bar{c}_i$ means the engine's input torque. $\Delta_i(\bar{X}_i)$ is the unknown nonlinear false data injection attack function.

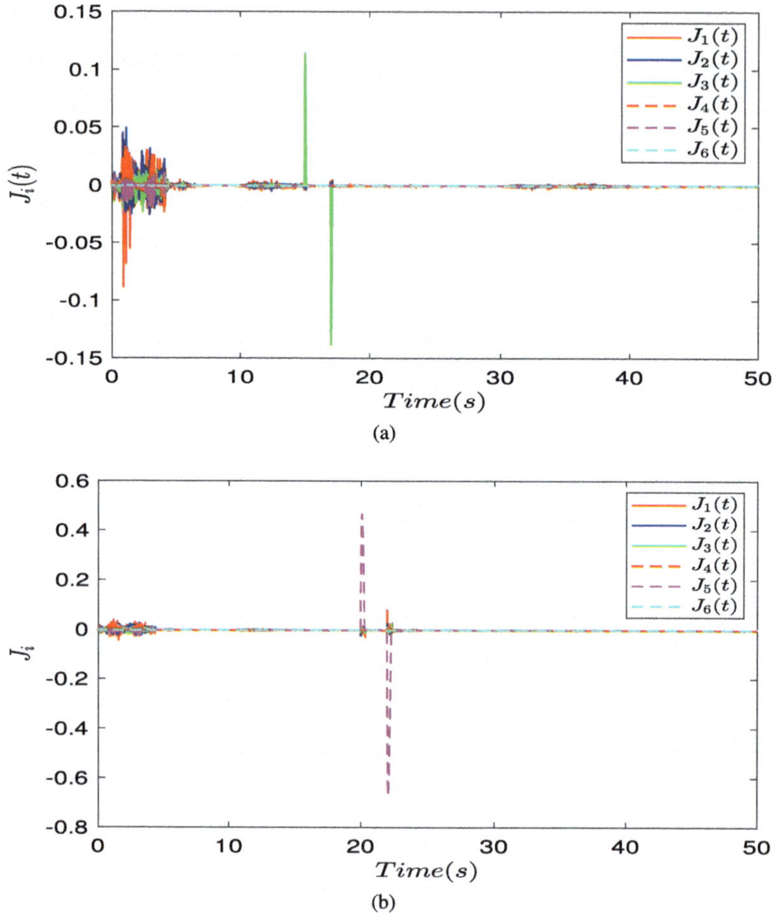


**Fig. 2.6** Coupled sliding mode surface of the connected vehicle platoon system. **a** Vehicle $i = 3$ with physical faults, **b** Vehicle $i = 5$ with physical faults

Assuming that the dynamic model of the leader vehicle can be given as

$$\begin{aligned}
\dot{\bar{p}}_0(t) &= \bar{v}_0(t) \\
\dot{\bar{v}}_0(t) &= \bar{a}_0(t) \\
\dot{\bar{a}}_0(t) &= u_0(t)
\end{aligned} \tag{2.23}$$

where $u_0(t)$ denotes the reference control input of the leader vehicle.

For the spacing policy of the connected vehicle platoon system, we use the constant time headway spacing policy in [2] to ensure the string stability performance and achieve the platoon formation regulation. The mathematical expression of the spacing policy is shown as

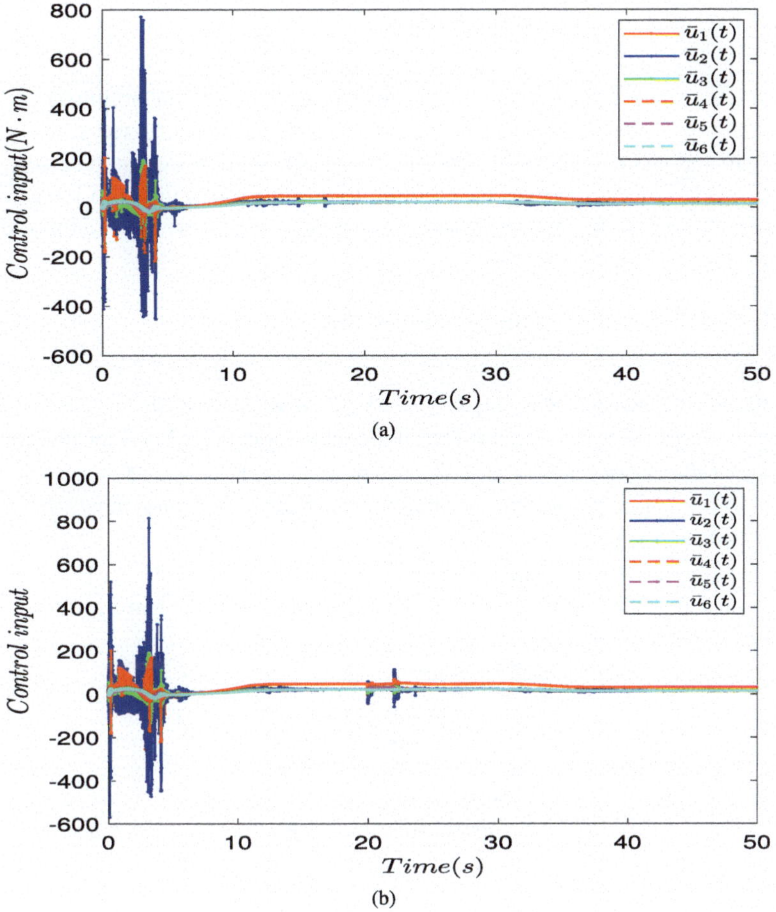


**Fig. 2.7** Control input of the connected vehicle platoon system. **a** Vehicle $i = 3$ with physical faults, **b** Vehicle $i = 5$ with physical fault

$$
\begin{aligned}
\bar{e}_i(t) &= \bar{p}_{i-1}(t) - \bar{p}_i(t) - L_{i-1} - \bar{y}_{r,i}(t) - \bar{\Lambda}_i(t) \\
\bar{y}_{r,i} &= \bar{d}_{i-1,i} + h_{1i}(\bar{v}_i - \bar{v}_0) \\
\bar{\Lambda}_i(t) &= \{\bar{e}_i^*(0) + \bar{\Lambda}_i^a t + 0.5\bar{\Lambda}_i^b t^2\} e^{-\varsigma_i t}
\end{aligned} \tag{2.24}
$$

where $\bar{e}_i^*(t) = \bar{p}_{i-1}(t) - \bar{p}_i(t) - L_{i-1} - y_{r,i}(t)$, $\bar{\Lambda}_i^a = \varsigma_i \bar{e}_i^*(0) + \dot{\bar{e}}_i^*(0)$, $\bar{\Lambda}_i^b = \varsigma_i^2 \bar{e}_i^*(0) + 2\varsigma_i \dot{\bar{e}}_i^*(0) + \ddot{\bar{e}}_i^*(0)$, $h_{1i}$ is the time headway positive constant. The vehicle's length and the required minimum inter-vehicle distance are indicated by $L_{i-1}$ and $\bar{d}_{i-1,i}$, respectively. $\varsigma_i$ is the positive parameter. By (2.24), and through the properties of exponential decay function $\bar{\Lambda}_i(t)$, we know that $\bar{e}_i(0) = 0$, $\dot{\bar{e}}_i(0) = 0$ and $\ddot{\bar{e}}_i(0) = 0$. It indicates that the initial $\bar{e}_i(t)$ can always be zero.

**Control Objective:** In this section, the control objective is same with that in Sect. 2.1.

### 2.2.2 *Fuzzy Logic System Theory*

Fuzzy logic system theory is one of the important methods for nonlinear function approximation in nonlinear system control design. By using fuzzy rules, fuzzy logic system can approximate the nonlinear function of the system well in the given precision range [11, 12]. Considering the continuous nonlinear function $\bar{f}(X)$ that defined on a compact set $\bar{\Gamma} \in \mathrm{R}^N$ and any positive constant $\bar{\chi} > 0$, there exists a Fuzzy Logic Systems (FLSs) $\bar{\theta}^T \bar{\delta}(X)$ such as

$$\sup_{X \in \Gamma} \left| \bar{f}(X) - \bar{\theta}^T \bar{\delta}(X) \right| \leq \chi \tag{2.25}$$

where the FLSs input vector $X = [X_1, \ldots, X_N]^T \in \mathrm{R}^N, \bar{\theta}^T = [\bar{\theta}_1, \ldots, \bar{\theta}_M] \in \mathrm{R}^M$ is the weight vector; $M$ is the total number of the fuzzy rules. $\bar{\kappa}_{F_i^l}(X_i)$ is the membership functions of the fuzzy set $F_i^l$, $i = 1, 2, \ldots, N$ and $l = 1, 2, \ldots, M$.

$\bar{\delta}(X) = [\bar{\delta}_1(X), \ldots, \bar{\delta}_M(X)]^T$ is the fuzzy basis function vector, in which $\bar{\delta}_l(X) = (\prod_{i=1}^N \bar{\kappa}_{F_i^l}(X_i)) \sum_{l=1}^M (\prod_{i=1}^N \bar{\kappa}_{F_i^l}(X_i))^{-1}$.

For the unknown nonlinear function $\bar{f}(X)X \in \mathrm{R}^N$, by (2.25), a FLSs and the optimal ideal fuzzy weight vector $\bar{\theta}^*$ of FLSs for approximation are given as

$$\begin{aligned} \bar{f}(X) &= \bar{\theta}^{*T} \bar{\delta}(X) + \bar{\chi}(X) \\ \bar{\theta}^* &= \arg \min_{\bar{\theta} \in \Gamma_\theta} [\sup_{X \in \Gamma} \left| \bar{f}(X) - \bar{\theta}^T \bar{\delta}(X) \right|] \end{aligned} \tag{2.26}$$

where $\bar{\theta} \in \Gamma_\theta$ and $X \in \bar{\Gamma}$, $\Gamma_\theta$ presents the compact set for $\bar{\theta}$, $\bar{\Gamma}$ is the compact set for $X$. $|\bar{\chi}(X)| \leq \bar{\delta}$, $\bar{\chi}(X)$ is the approximation error and $\bar{\delta} > 0$ is a constant.

### 2.2.3 *Distributed Adaptive Fuzzy Resilient Control Laws*

Based on the command filtered back-stepping control method, the design procedure of distributed adaptive fuzzy resilient control scheme will be shown in the following parts. Firstly, selecting the coordinate changes as

$$E_{i,1} = \bar{e}_i(t) \tag{2.27}$$

$$E_{i,2} = \bar{v}_{i-1}(t) - \bar{v}_i(t) - h_{1i}\bar{a}_i + h_{1i}\bar{a}_0 - \dot{\bar{\Lambda}}_i(t) - \bar{w}_{2,i} \tag{2.28}$$

where $i = 1, 2, \ldots, n$, the filtered output is denoted by $\bar{w}_{2,i}$, which is also the command filter control signal. The command filter is used to filter the virtual control signal $\bar{\alpha}_{1,i}(t)$ for avoiding the problem of repeated derivation of virtual control signal in the traditional back-stepping control technique. The considered command filter is

$$\begin{aligned}\dot{\bar{\gamma}}_{2,i,1} &= \bar{\lambda}_{2,i,1}\bar{\gamma}_{2,i,2} \\ \dot{\bar{\gamma}}_{2,i,2} &= -2\bar{\bar{\lambda}}\bar{\lambda}_{2,i,1}\bar{\gamma}_{2,i,2} - \bar{\lambda}_{2,i,1}\bar{\Im}_{1,i}\end{aligned} \tag{2.29}$$

where $\bar{\Im}_{1,i} = \bar{\gamma}_{2,i,1} - \bar{\alpha}_{1,i}$, $\bar{\alpha}_{1,i}(t)$ can be the command filter input, which will be designed later. $\bar{w}_{2,i} = \bar{\gamma}_{2,i,1}$, $\dot{\bar{w}}_{2,i} = \bar{\gamma}_{2,i,2}$ are the command filter outputs. $\bar{\lambda}_{2,i,1}$ and $\bar{\bar{\lambda}} \in (0, 1)$ are the positive parameters. There exists that $\bar{w}_{2,i}(0) = \bar{\gamma}_{2,i,1}(0) = \bar{\alpha}_{1,i}(0)$, and $\dot{\bar{w}}_{2,i}(0) = \bar{\gamma}_{2,i,2}(0) = 0$.

The error compensation scheme will be designed for reducing errors caused by introducing the above command filters (2.29). We defined that $\bar{\upsilon}_{i,1} = E_{i,1} - \bar{\xi}_{i,1}$, and $\bar{\upsilon}_{i,2} = E_{i,2} - \bar{\xi}_{i,2}$ are the compensated error signals, $\bar{\xi}_{i,1}$ and $\bar{\xi}_{i,2}$ are the compensating signals. The error compensation scheme is developed as

$$\dot{\bar{\xi}}_{i,1} = -k_{i,1}\bar{\xi}_{i,1} + \bar{\xi}_{i,2} + \bar{w}_{2,i} - \bar{\alpha}_{1,i} \tag{2.30}$$

$$\dot{\bar{\xi}}_{i,2} = -k_{i,2}\bar{\xi}_{i,2} - \bar{\xi}_{i,1} - 1.5\bar{\xi}_{i,2} \tag{2.31}$$

where $k_{i,1}, k_{i,2} > 0$ are the positive designed constants, $\bar{\xi}_{i,1}(0) = \bar{\xi}_{i,2}(0) = 0$.

The virtual control laws based on the command filters (2.29) are defined as

$$\bar{\alpha}_{1,i} = -k_{i,1}E_{i,1} \tag{2.32}$$

$$\begin{aligned}\bar{\alpha}_{2,i} = &-k_{i,2}E_{i,2} - E_{i,1} - \bar{C}_i + \dot{\bar{w}}_{2,i} - 1.5E_{i,2} \\ &- \frac{\bar{\upsilon}_{i,2}}{2\bar{m}_{2i}^2}\hat{\vartheta}_{i,1}\delta_i^T\delta_i - \frac{\bar{\upsilon}_{i,2}}{2\bar{m}_{1i}^2}\hat{\bar{\theta}}_{\Delta,i}\delta_{\Delta,i}^T\delta_{\Delta,i}\end{aligned} \tag{2.33}$$

$$u_i(t) = -b_i^{-1}\bar{\alpha}_{2,i} \tag{2.34}$$

where $b_i = h_{1i}\tau_i^{-1} > 0$, $\bar{C}_i = \bar{a}_{i-1}(t) - \bar{a}_i(t) + h_{1i}u_0(t) - \ddot{\bar{\Lambda}}_i(t)$, $\bar{m}_{1i}$, $\bar{m}_{2i}$ are positive parameters. Define $\hat{\vartheta}_{i,1}$, and $\hat{\bar{\theta}}_{\Delta,i}$ as the estimation of $\vartheta_{i,1}$, and $\bar{\theta}_{\Delta,i}$, respectively. Define that $\vartheta_{i,1} = \|\vartheta_i^*\|^2$, $\bar{\theta}_{\Delta,i} = \|\theta_{\Delta,i}^*\|^2$, in which $\theta_{\Delta,i}^*$ with respect to the false data injection attack function and will be introduced later. The adaptive updated laws of $\hat{\bar{\theta}}_{\Delta,i}$, and $\hat{\vartheta}_{i,1}$ are developed as

$$\dot{\hat{\bar{\theta}}}_{\Delta,i} = \frac{\varsigma_{\theta i}}{2\bar{m}_{1i}^2}\bar{\upsilon}_{i,2}^2\delta_{\Delta,i}^T\delta_{\Delta,i} - \bar{\varsigma}_{\theta i}\hat{\bar{\theta}}_{\Delta,i} \tag{2.35}$$

$$\dot{\hat{\vartheta}}_{i,1} = \frac{\varsigma_{\vartheta i}}{2\bar{m}_{2i}^2}\bar{\upsilon}_{i,2}^2\delta_i^T\delta_i - \bar{\varsigma}_{\vartheta i}\hat{\vartheta}_{i,1} \tag{2.36}$$

where $\varsigma_{\vartheta i}$, $\varsigma_{\theta i}$, $\bar{\varsigma}_{\vartheta i}$, $\bar{\varsigma}_{\theta i}$ are small positive constants. Here, $\delta_i(\bar{X}_i)$ and $\delta_{\Delta,i}(\bar{X}_i)$ are Gaussian functions of the FLSs and can be denoted by $\delta_i$ and $\delta_{\Delta,i}$ for simplicity.

The following part will show the design process of the command filter backstepping based distributed adaptive fuzzy resilient control method. We select the $i$th vehicle as the example to illustrate the control principle.

**Step 1:** Firstly, we need to prove that $E_{i,1}$ in (2.27) and the spacing error $\bar{e}_i(t)$ can converge to zero. We select the Lyapunov function candidate as $\bar{V}_{i,1} = 0.5\bar{\upsilon}_{i,1}^2$. By taking the time derivative of the above Lyapunov function, and with (2.28), we have

$$\begin{aligned}\dot{\bar{V}}_{i,1} &= \bar{\upsilon}_{i,1}(\bar{v}_{i-1} - \bar{v}_i - h_{1i}\bar{a}_i + h_{1i}\bar{a}_0 - \dot{\bar{\Lambda}}_i - \dot{\bar{\xi}}_{i,1}) \\ &= \bar{\upsilon}_{i,1}(E_{i,2} + \bar{w}_{2,i} - \dot{\bar{\xi}}_{i,1})\end{aligned} \tag{2.37}$$

Taking the error compensation scheme (2.30) into account,

$$\dot{\bar{V}}_{i,1} = \bar{\upsilon}_{i,1}(E_{i,2} - \bar{\xi}_{i,2} + \kappa_{i,1}\bar{\xi}_{i,1} + \bar{\alpha}_{1,i}) \tag{2.38}$$

Next, by using the virtual control law $\bar{\alpha}_{1,i}$ in (2.32), (2.38) becomes

$$\begin{aligned}\dot{\bar{V}}_{i,1} &= \bar{\upsilon}_{i,1}(\bar{\upsilon}_{i,2} - k_{i,1}\bar{\upsilon}_{i,1}) \\ &= \bar{\upsilon}_{i,1}\bar{\upsilon}_{i,2} - k_{i,1}\bar{\upsilon}_{i,1}^2\end{aligned} \tag{2.39}$$

**Step 2:** Similar to Step 1, by taking the time derivative of $\bar{\upsilon}_{i,2} = E_{i,2} - \bar{\xi}_{i,2}$, and combining that with formula (2.28), yields,

$$\dot{\bar{\upsilon}}_{i,2} = \bar{a}_{i-1}(t) - \bar{a}_i(t) - h_{1i}\dot{\bar{a}}_i + h_{1i}\dot{\bar{a}}_0 - \ddot{\bar{\Lambda}}_i(t) - \dot{\bar{w}}_{2,i} - \dot{\bar{\xi}}_{i,2} \tag{2.40}$$

Then, by the help of (2.21)–(2.23) into (2.40), it produces that

$$\dot{\bar{\upsilon}}_{i,2} = \bar{C}_i + \bar{\Delta}_i(\bar{X}_i) - b_i u_i(t) - \bar{F}_i(X_i) - h_{1i}\bar{\varpi}_i(t) - \dot{\bar{w}}_{2,i} - \dot{\bar{\xi}}_{i,2} \tag{2.41}$$

where $\bar{F}_i(\bar{X}_i) = h_{1i}\bar{\vartheta}_i^T \mathrm{B}_i(\bar{X}_i)$, and $\bar{\Delta}_i(\bar{X}_i) = -h_{1i}\Delta_i(\bar{X}_i)\tau_i^{-1}$. Then, we use the FLSs $\vartheta_i^{*T}\delta_i$ and $\theta_{\Delta,i}^{*T}\delta_{\Delta,i}$ to approximate the unknown nonlinear function $\bar{F}_i(\bar{X}_i)$ and the cyber attack function $\bar{\Delta}_i(\bar{X}_i)$.

$$\bar{F}_i(\bar{X}_i) = \vartheta_i^{*T}\delta_i + \bar{\chi}_i(\bar{X}_i) \tag{2.42}$$

$$\bar{\Delta}_i(\bar{X}_i) = \theta_{\Delta,i}^{*T}\delta_{\Delta,i} + \bar{\chi}_{\Delta,i}(\bar{X}_i) \tag{2.43}$$

where the approximation errors of FLSs are denoted by $\bar{\chi}_i(\bar{X}_i)$ and $\bar{\chi}_{\Delta,i}(\bar{X}_i)$, and $\left|\bar{\chi}_i(\bar{X}_i)\right| \le \bar{\delta}_i^*$, and $\left|\bar{\chi}_{\Delta,i}(\bar{X}_i)\right| \le \bar{\delta}_{\Delta,i}^*$. $\bar{\delta}_i^* > 0$ and $\bar{\delta}_{\Delta,i}^* > 0$ are the upper bound of the FLSs approximation error. Then, through (2.42) and (2.43), we derive (2.41) as

$$\bar{\upsilon}_{i,2}\dot{\bar{\upsilon}}_{i,2} = \bar{\upsilon}_{i,2}(\bar{C}_i + \theta^{*T}_{\Delta,i}\delta_{\Delta,i} + \bar{\chi}_{\Delta,i}(\bar{X}_i) - \vartheta^{*T}_i\delta_i - \bar{\chi}_i(\bar{X}_i)) + \bar{\upsilon}_{i,2}(-b_i u_i(t) - h_{1i}\varpi_i(t) - \dot{\bar{w}}_{2,i} - \dot{\xi}_{i,2}) \tag{2.44}$$

With Young's equality, we have the facts that

$$\bar{\upsilon}_{i,2}(\theta^{*T}_{\Delta,i}\delta_{\Delta,i} - \vartheta^{*T}_i\delta_i) \leq \frac{\bar{\upsilon}^2_{i,2}}{2\bar{m}^2_{1i}}\bar{\theta}_{\Delta,i}\delta^T_{\Delta,i}\delta_{\Delta,i} + \frac{\bar{\upsilon}^2_{i,2}}{2\bar{m}^2_{2i}}\vartheta_{i,1}\delta^T_i\delta_i + \frac{\bar{m}^2_{1i}}{2} + \frac{\bar{m}^2_{2i}}{2} \tag{2.45}$$

$$\bar{\upsilon}_{i,2}(\bar{\chi}_{\Delta,i}(\bar{X}_i) - \bar{\chi}_i(\bar{X}_i)) \leq \bar{\upsilon}^2_{i,2} + \frac{1}{2}\bar{\delta}^{*2}_{\Delta,i} + \frac{1}{2}\bar{\delta}^{*2}_i \tag{2.46}$$

$$-\bar{\upsilon}_{i,2}h_{1i}\bar{\varpi}_i \leq 0.5\bar{\upsilon}^2_{i,2} + 0.5h^2_{1i}\bar{\varpi}^2_{i1} \tag{2.47}$$

We select the Lyapunov function $\bar{V}_{i,2} = \bar{V}_{i,1} + 0.5\bar{\upsilon}^2_{i,2}$, and take the time derivative of it. by (2.45)–(2.47), one has

$$\dot{\bar{V}}_{i,2} \leq \dot{\bar{V}}_{i,1} + \bar{\upsilon}_{i,2}\left(\bar{C}_i + \frac{\bar{\upsilon}_{i,2}}{2\bar{m}^2_{1i}}\bar{\theta}_{\Delta,i}\delta^T_{\Delta,i}\delta_{\Delta,i} + \frac{\bar{\upsilon}_{i,2}}{2\bar{m}^2_{2i}}\vartheta_{i,1}\delta^T_i\delta_i\right) + \bar{\upsilon}_{i,2}(-b_i u_i(t) + 1.5\bar{\upsilon}_{i,2} - \dot{\bar{w}}_{2,i} - \dot{\xi}_{i,2}) + \bar{M}^*_i \tag{2.48}$$

where $\bar{M}^*_i = 0.5(\bar{m}^2_{1i} + \bar{m}^2_{2i} + \bar{\delta}^{*2}_{\Delta,i} + \bar{\delta}^{*2}_i + h^2_{1i}\bar{\varpi}^2_{i1})$.

By the error compensating scheme in (2.31), (2.48) can be

$$\dot{\bar{V}}_{i,2} \leq -k_{i,1}\bar{\upsilon}^2_{i,1} + \bar{\upsilon}_{i,2}(E_{i,1} + \bar{C}_i - b_i u_i(t) + 1.5E_{i,2}) + \bar{\upsilon}_{i,2}\left(\frac{\bar{\upsilon}_{i,2}}{2\bar{m}^2_{1i}}\bar{\theta}_{\Delta,i}\delta^T_{\Delta,i}\delta_{\Delta,i} + \frac{\bar{\upsilon}_{i,2}}{2\bar{m}^2_{2i}}\vartheta_{i,1}\delta^T_i\delta_i\right) + \bar{\upsilon}_{i,2}(k_{i,2}\xi_{i,2} - \dot{\bar{w}}_{2,i}) + \bar{M}^*_i \tag{2.49}$$

Using $\bar{\alpha}_{2,i}$ in (2.33), one has

$$\dot{\bar{V}}_{i,2} \leq -k_{i,1}\bar{\upsilon}^2_{i,1} - k_{i,2}\bar{\upsilon}^2_{i,2} - \bar{\upsilon}_{i,2}b_i u_i(t) - \bar{\upsilon}_{i,2}\bar{\alpha}_{2,i} + \bar{M}^*_i + \frac{\bar{\upsilon}^2_{i,2}}{2\bar{m}^2_{2i}}\tilde{\vartheta}_{i,1}\delta^T_i\delta_i + \frac{\bar{\upsilon}^2_{i,2}}{2\bar{m}^2_{1i}}\tilde{\bar{\theta}}_{\Delta,i}\delta^T_{\Delta,i}\delta_{\Delta,i} \tag{2.50}$$

where $\tilde{\vartheta}_{i,1} = \vartheta_{i,1} - \hat{\vartheta}_{i,1}$, $\tilde{\bar{\theta}}_{\Delta,i} = \bar{\theta}_{\Delta,i} - \hat{\bar{\theta}}_{\Delta,i}$.

Then, the Lyapunov function $\bar{V}_i = \bar{V}_{i,2} + \frac{1}{2\varsigma_{\vartheta i}}\tilde{\vartheta}^2_{i,1} + \frac{1}{2\varsigma_{\theta i}}\tilde{\bar{\theta}}^2_{\Delta,i}$ is defined. Combining with (2.50), (2.35)–(2.36), the time derivative of $\bar{V}_i$ is

$$\begin{aligned}\dot{\bar{V}}_i &\leq -k_{i,1}\bar{\upsilon}_{i,1}^2 - k_{i,2}\bar{\upsilon}_{i,2}^2 - \bar{\upsilon}_{i,2}\alpha_{2,i} - \bar{\upsilon}_{i,2}b_i u_i(t) \\ &+ \bar{M}_i^* + \frac{\bar{\varsigma}_{\theta i}}{\varsigma_{\theta i}}\tilde{\bar{\theta}}_{\Delta,i}\hat{\bar{\theta}}_{\Delta,i} + \frac{\bar{\varsigma}_{\vartheta i}}{\varsigma_{\vartheta i}}\tilde{\vartheta}_{i,1}\hat{\vartheta}_{i,1}\end{aligned} \tag{2.51}$$

According to (2.34), then, (2.51) can be derived as

$$\dot{\bar{V}}_i \leq -k_{i,1}\bar{\upsilon}_{i,1}^2 - k_{i,2}\bar{\upsilon}_{i,2}^2 + \bar{M}_i^* + \frac{\bar{\varsigma}_{\theta i}}{\varsigma_{\theta i}}\tilde{\bar{\theta}}_{\Delta,i}\hat{\bar{\theta}}_{\Delta,i} + \frac{\bar{\varsigma}_{\vartheta i}}{\varsigma_{\vartheta i}}\tilde{\vartheta}_{i,1}\hat{\vartheta}_{i,1} \tag{2.52}$$

By the inequalities $\frac{\bar{\varsigma}_{\vartheta i}}{\varsigma_{\vartheta i}}\tilde{\vartheta}_{i,1}\hat{\vartheta}_{i,1} \leq \frac{\bar{\varsigma}_{\vartheta i}}{2\varsigma_{\vartheta i}}(\vartheta_{i,1}^2 - \tilde{\vartheta}_{i,1}^2)$, $\frac{\bar{\varsigma}_{\theta i}}{\varsigma_{\theta i}}\tilde{\bar{\theta}}_{\Delta,i}\hat{\bar{\theta}}_{\Delta,i} \leq \frac{\bar{\varsigma}_{\theta i}}{2\varsigma_{\theta i}}(\bar{\theta}_{\Delta,i}^2 - \tilde{\bar{\theta}}_{\Delta,i}^2)$, (2.52) can be

$$\dot{\bar{V}}_i \leq -k_{i,1}\bar{\upsilon}_{i,1}^2 - k_{i,2}\bar{\upsilon}_{i,2}^2 + \bar{N}_i^* - \frac{\bar{\varsigma}_{\vartheta i}}{2\varsigma_{\vartheta i}}\tilde{\vartheta}_{i,1}^2 - \frac{\bar{\varsigma}_{\theta i}}{2\varsigma_{\theta i}}\tilde{\bar{\theta}}_{\Delta,i}^2 \tag{2.53}$$

where $\bar{N}_i^* = \bar{M}_i^* + \frac{\bar{\varsigma}_{\vartheta i}\vartheta_{i,1}^2}{2\varsigma_{\vartheta i}} + \frac{\bar{\varsigma}_{\theta i}\bar{\theta}_{\Delta,i}^2}{2\varsigma_{\theta i}}$.

### *2.2.4 Vehicle Platoon Stability Performances Analysis and Proof*

Next, we will prove the effectiveness of the distributed resilient control design through the stability analysis of the connected vehicle platoon system.

***Proof*** Firstly, we will analyze the convergence performance of the compensated errors $E_{i,1}$ and $E_{i,2}$ for proving the closed-loop stability of the connected vehicle platoon system.

We consider the Lyapunov function candidate as $\bar{V}(t) = \sum_{i=1}^{n}(\bar{V}_i(t))$ for the whole vehicle platoon system. Based on (2.53), the time derivative of $\bar{V}(t)$ can be

$$\begin{aligned}\dot{\bar{V}}(t) &\leq 2\sum_{i=1}^{n}\left\{-k_{i,1}\left(\frac{|\bar{\upsilon}_{i,1}|^2}{2}\right) - k_{i,2}\left(\frac{|\bar{\upsilon}_{i,2}|^2}{2}\right)\right\} \\ &- \bar{\varsigma}_{\vartheta i}\sum_{i=1}^{n}\left\{\frac{\tilde{\vartheta}_{i,1}^2}{2\varsigma_{\vartheta i}}\right\} - \bar{\varsigma}_{\theta i}\sum_{i=1}^{n}\left\{\frac{\tilde{\bar{\theta}}_{\Delta,i}^2}{2\varsigma_{\theta i}}\right\} + \bar{N}^*\end{aligned} \tag{2.54}$$

where $\bar{N}^* = \sum_{i=1}^{n}\{\bar{N}_i^*\}$.

Then, (2.54) can be further derived as

$$\dot{V}(t) \leq -\bar{\mathrm{v}}^* \sum_{i=1}^{n} \left\{ \left( \frac{1}{2} \left| \bar{\upsilon}_{i,1} \right|^2 \right) + \left( \frac{1}{2} \left| \bar{\upsilon}_{i,2} \right|^2 \right) \right\} - \bar{\varsigma}_a \left\{ \sum_{i=1}^{n} \left( \frac{\tilde{\vartheta}_i^2}{2\varsigma_{\vartheta i}} + \frac{\tilde{\bar{\theta}}_{\Delta,i}^2}{2\varsigma_{\theta i}} \right) \right\} + \bar{N}^* \tag{2.55}$$

where $\bar{\varsigma}_a = \min\{\bar{\varsigma}_{\vartheta i}, \bar{\varsigma}_{\theta i}\}$, and $\bar{\mathrm{v}}^* = 2\min\{\min_{1\leq i\leq n} k_{i,1}, \min_{1\leq i\leq n} k_{i,2}\}$.

We can gets

$$\dot{\bar{V}}(t) \leq -\bar{\aleph}_a^* \bar{V}(t) + \bar{N}^* \tag{2.56}$$

where $\bar{\aleph}_a^* = \min\{\bar{\mathrm{v}}^*, \bar{\varsigma}_a\}$. According to the uniformly bounded stability theory, the vehicle platoon trajectories of $\bar{\upsilon}_{i,1}$, $\bar{\upsilon}_{i,2}$, $\tilde{\vartheta}_{i,1}$, and $\tilde{\bar{\theta}}_{\Delta,i}$ are uniformly bounded and can converge to zero. Thus, by the facts that $\vartheta_{i,1}$, $\bar{\theta}_{\Delta,i}$ are bounded constants, the boundedness of $\hat{\vartheta}_{i,1}$, and $\hat{\bar{\theta}}_{\Delta,i}$ can also be ensured.

Moreover, we will prove the boundedness of the compensation signals $\bar{\xi}_{i,1}$ and $\bar{\xi}_{i,2}$ in the whole closed-loop system. Selecting the Lyapunov function $\bar{V}_\xi = \sum_{i=1}^n \bar{V}_{\xi i}$, in which $V_{\xi i} = 0.5 \left\|\bar{\xi}_i\right\|^2$, $\bar{\xi}_i = [\bar{\xi}_{i,1}, \bar{\xi}_{i,2}]^T$. Taking the time derivative of $\bar{V}_\xi$,

$$\begin{aligned} \dot{\bar{V}}_\xi &= \sum_{i=1}^{n} (-k_{i,1}\bar{\xi}_{i,1}^2 - (k_{i,2}+1.5)\bar{\xi}_{i,2}^2 + \bar{\xi}_{i,1}(w_{2,i} - \alpha_{1,i})) \\ &\leq -k_{\bar{\xi}} V_{\xi i} \end{aligned} \tag{2.57}$$

where $k_\xi = \min_{1\leq i\leq n} \{2k_{i,1}, 2k_{i,2} + 3\}$. With $\bar{\xi}_{i,1}(0) = \bar{\xi}_{i,2}(0) = 0$, we can conclude that $\bar{\xi}_{i,1}$, $\bar{\xi}_{i,2}$ are uniformly bounded.

Then, we will prove the string stability of the connected vehicle platoon system. Through the fact that the uniformly boundedness of $\bar{\upsilon}_{i,1}$, $\bar{\upsilon}_{i,2}$, $\bar{\xi}_{i,1}$, and $\bar{\xi}_{i,2}$ in the whole closed-loop system, it follows that the signals $E_{i,1} = \bar{e}_i(t)$ and $E_{i,2}$ are also uniformly bounded and can converge into a small neighborhood of zero. Then, from the facts of $\dot{E}_{i,1} = E_{i,2} + \bar{w}_{2,i}$ in (2.28), the boundedness of $\bar{w}_{2,i}$ in (2.29) and $E_{i,2}$, we can obtain that $\dot{E}_{i,1} = \dot{\bar{e}}_i$ is also bounded. Combine with the conditions that $\bar{e}_i(0) = 0$, $\dot{\bar{e}}_i(0) = 0$, and by the definition of non-strict string stability in Chap. 1, there exists the constants $\bar{\phi} > 0$ and $\bar{\varphi} > 0$, with the condition $\sup_{i\in(1,2,\ldots,n)} \{|\bar{e}_i(0)|, |\dot{\bar{e}}_i(0)|\} < \bar{\phi}$, such that $\sup_{i\in(1,2,\ldots,n)} \{|\bar{e}_i(t)|, |\dot{\bar{e}}_i(t)|\} < \bar{\varphi}$ can be satisfied. Therefore, the non-strict string stability of the connected vehicle platoon system is proved. □

The feasibility demonstration of the proposed scheme by the simulations is provided. The connected vehicle platoon system consisting of five follower vehicles and one leader is introduced. The actual inter-vehicle spacing is $\bar{D}_i(t) = \bar{p}_{i-1}(t) -$

$\bar{p}_i(t) - L_{i-1}$, the desired inter-vehicle spacing $\bar{d}_{i-1,i} = 5$m, $L_{i-1} = [2.5, 3.5, 4, 5, 4]^T$ for $i = 1, 2, 3, 4, 5$, and $\bar{h}_{i1} = 1$. For the fuzzy membership functions, $\kappa_{F_i^l}(\bar{X}_i) = \exp[-(\bar{X}_i + 2l_s - 6)^2/2]$, $\bar{X}_i = [v_i, a_i]^T$, $l_s = 1, 2, \ldots, 5$, $l = 1, 2, \ldots, 9$. For the leader vehicle 0, the initial states are given as $\bar{p}_0(0) = 145$ m, $\bar{v}_0 = 1.2$ m/s, and $\bar{a}_0(0) = 0$ m/s$^2$. In addition, we supposes that the false data injection attack occurs on the controller-actuator channel of vehicle 3 with the attack function $\Phi_3(X_3) = 0.16\sqrt{v_3^2 + a_3^2}\cos(t) + 0.05$ when $t = [20, 23)$, the attack duration is 3s. The simulation results are shown in Figs. 2.8, 2.9, 2.10, 2.11, 2.12, 2.13 and 2.14.

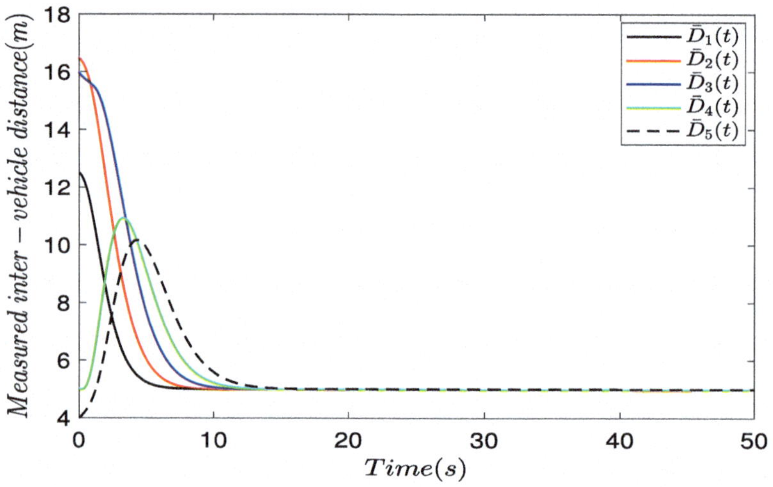


**Fig. 2.8** Actual measured inter-vehicle distance $\bar{D}_i(t)$ in the connected vehicle platoon system

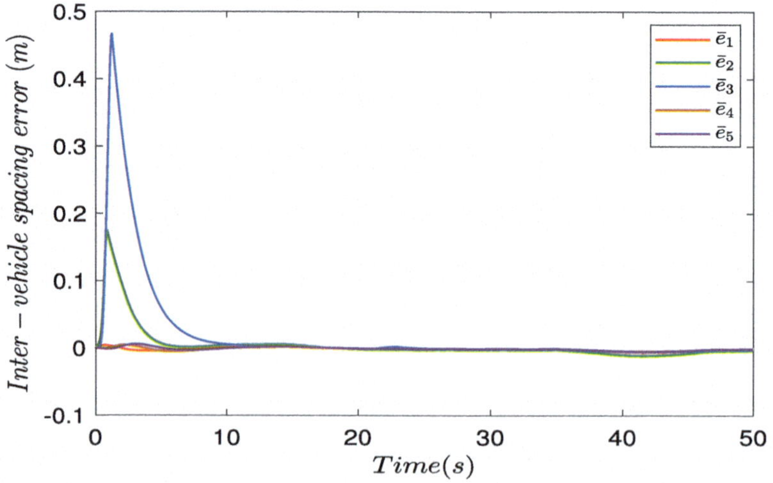


**Fig. 2.9** Inter-vehicle spacing error of the connected vehicle platoon system

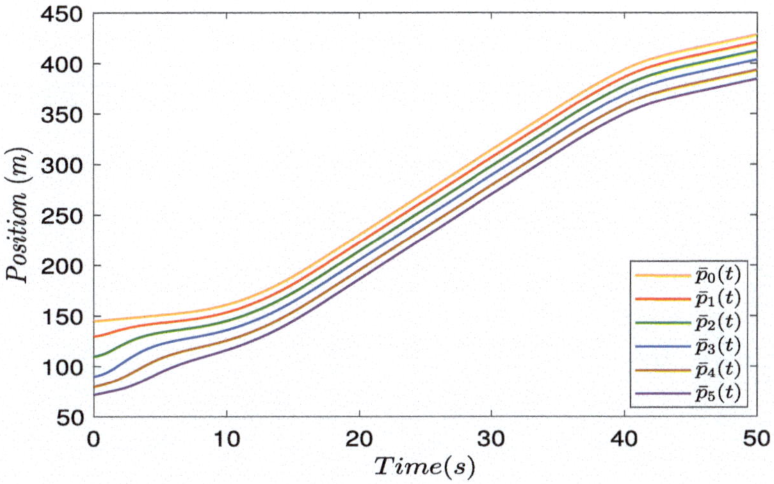


**Fig. 2.10** Longitudinal position of the connected vehicle platoon system

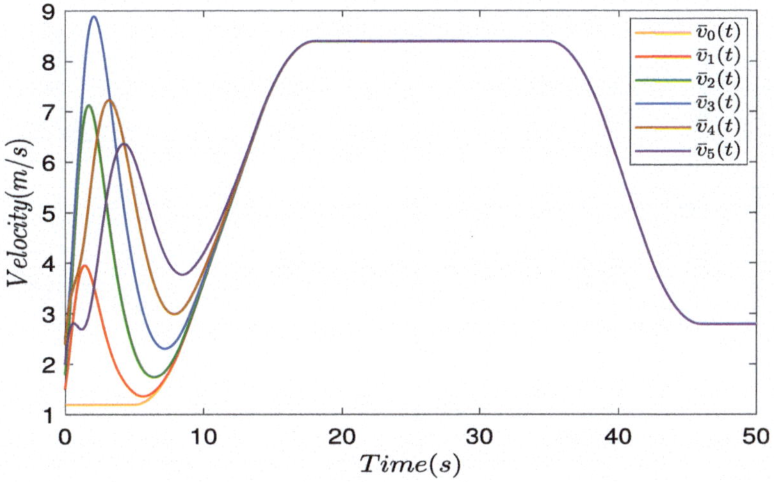


**Fig. 2.11** Longitudinal velocity of the connected vehicle platoon system

The actual measured inter-vehicle distance and the inter-vehicle spacing error of the connected vehicle platoon system are shown in Figs. 2.8 and 2.9, respectively. The Longitudinal position, velocity, acceleration, and the control input are shown in Figs. 2.10 and 2.13, respectively. Figure 2.14 show the adaptive updated and approximation for the unknown nonlinear functions and cyber-attack function by using FLSs. The results show the effectiveness of the distributed adaptive fuzzy resilient control scheme design.

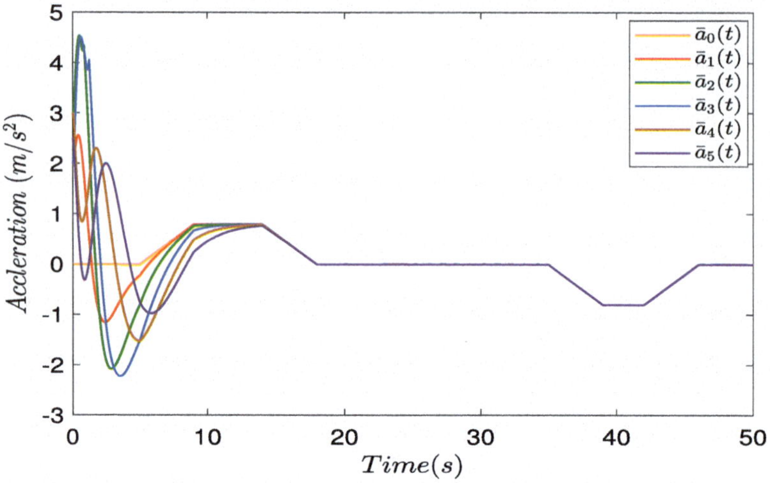


**Fig. 2.12** Longitudinal acceleration of the connected vehicle platoon system

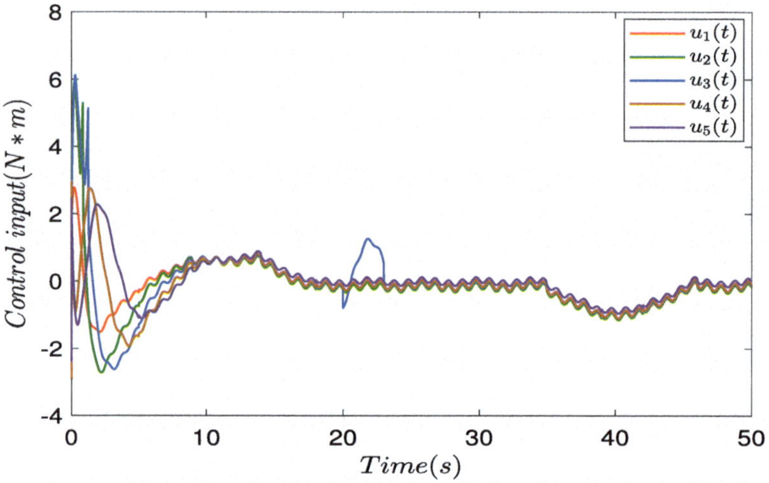


**Fig. 2.13** Control input of the connected vehicle platoon system

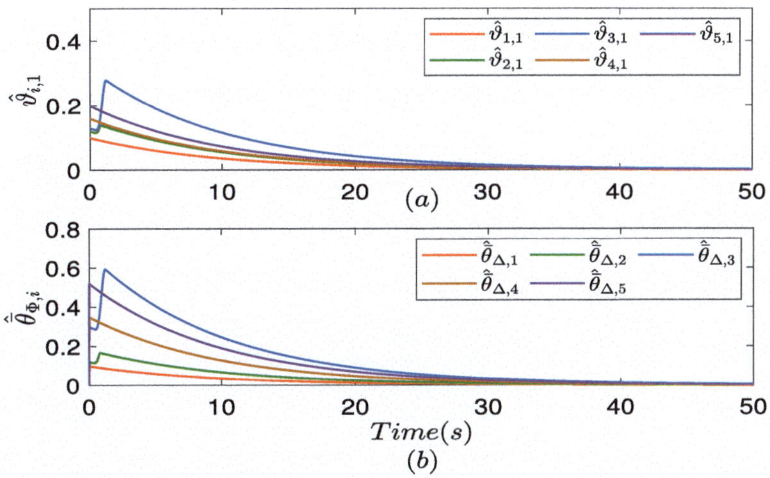


**Fig. 2.14** Adaptive approximation by using FLSs

# References


1. Kwon J, Chwa D(2014) Adaptive bidirectional platoon control using a coupled sliding mode control method. *IEEE Transactions on Intelligent Transportation Systems*, 15(5): 2040-2048.
2. Guo X, Wang J, Liao F, et al(2016) Distributed adaptive integrated-sliding-mode controller synthesis for string stability of vehicle platoons. *IEEE Transactions on Intelligent Transportation Systems*, 17(9): 2419-2429.
3. Liu Y, Xu B, and Ding Y(2017) Convergence analysis of cooperative braking control for interconnected vehicle systems. *IEEE Transactions on Intelligent Transportation Systems*, 18(7): 1894-1906.
4. Zhu Y, Wu J, Su H(2022) V2V-based cooperative control of uncertain, disturbed and constrained nonlinear CAVs platoon. *IEEE Transactions on Intelligent Transportation Systems*, 23(3): 1796-1806.
5. Zhu Y, Zhu F(2018) Distributed adaptive longitudinal control for uncertain third-order vehicle platoon in a networked environment. *IEEE Transactions on Vehicular Technology*, 67(10): 9183-9197.
6. Swaroop D, Hedrick J, Yip P, and Gerdes J(2000) Dynamic surface control for a class of nonlinear systems. *IEEE Transactions on Automatic Control*, 45(10): 1893-1899.
7. Farrell J. A, Polycarpou M, Sharma M, and Dong W(2009) Command filtered backstepping. *IEEE Transactions on Automatic Control*, 54(6), 1391-1395.
8. Dong W. J, Farrell J. A, Polycarpou M, et al(2012) Command filtered adaptive backstepping. *IEEE Transactions Control Systems and Technology*, 20(3): 566-580.
9. Stankovic S, Stanojevic M, and Siljak D(20000) Decentralized overlapping control of a platoon of vehicles. *IEEE Transactions Control Systems and Technology*, 8(5): 816-832.
10. Huang S, Tan K, and Lee T(2005) Autonomous cruise control using neural networks in platooning. *Advanced Robotics*, 19: 169-189.
11. Guo B, Dian S, Zhao T(2022) Event-driven-observer-based fuzzy fault tolerant control for nonlinear system with actuator fault. *Nonlinear Dynamics*, 107(4): 3505-3519.
12. Yao X, Sun Z, and Su X(2024) Disturbance-observer-based hierarchical control for vehicle front steering: Interval type-2 fuzzy system approach. *IEEE Transactions on Fuzzy Systems*, 32(1): 137-151.

# Chapter 3
# Distributed Adaptive Resilient Control for Heterogeneous Intelligent Connected Vehicle Platoon Systems with Fault and Saturation

Random and unpredictable actuator faults will seriously affect the closed-loop control stability and the string stability of the ICVPS, resulting in collision risk. As a large-scale multi-source complex connected cyber physical fusion system composed of multiple heterogeneous intelligent vehicles, it will not only suffer from physical constraints such as actuator saturation of the vehicle's dynamic system, but also suffer from complex environmental disturbances such as road slope, wind gust and strong sand, etc. In the physical layer of vehicle subsystem, the loss of effectiveness fault, stuck fault, abrupt fault, periodicity fault and other additive faults caused by actuator equipment such as wheel hub motor and hydraulic cylinder also occur frequently [1–3]. Therefore, it is urgent to carry out research on distributed adaptive platoon resilient control for heterogeneous connected vehicles. While increasing the system's resistance to actuator saturation and complex environment disturbances, improving the system's response and resilience recovery ability to faults, and maintaining the strict string stability is the basic problems that needs to be solved.

Based on the above analysis, in this chapter, a distributed adaptive platoon resilient control strategy [4] is proposed for the heterogeneous connected vehicles with actuator fault of single vehicle, input saturation and complex environment disturbances. The main contributions of this chapter are as follows. First, for reducing the conservatism of traditional spacing policies and avoiding large transient problems caused by faults, this chapter designs an improved quadratic spacing policy by combining the upper bounds of fault factors to ensure the strict string stability of heterogeneous ICVPS under actuator fault conditions. Then, the adaptive compensation laws are designed to deal with the unknown upper bound of the actuator saturation approximation function and the disturbance of complex environment, eliminating the assumption that the upper bound of the actuator saturation approximation function is known in existing literatures, such as in [5, 6]. Third, different from the uniformly bounded



C. Pan and Y. Chen, *Resilient Control of Intelligent Connected Vehicle Platoon Systems*,
https://doi.org/10.1007/978-981-97-8850-7_3

stability in the sense of asymptotic stability, a distributed adaptive resilient control strategy based on coupled sliding mode and the finite time stability is proposed to achieve resilience and fast stability under fault conditions.

## 3.1 Introduction

Since serious traffic congestion and air pollution, intelligent transportation system has become a research hotspot. As the key part of intelligent transportation system, heterogeneous intelligent connected vehicle platoon system (HICVPS) takes significant advantages in reducing fuel or energy consumption, alleviating urban congestion, speeding up traffic flow, etc. [7, 8]. However, in practical engineering applications, the single vehicle is prone to physical fault problems, which will seriously damage the stability and safety performance of the closed loop system of the entire vehicle platoon system. The problem of physical fault brings great challenge to the HICVPS control. Therefore, it is of great significance to study the rapid response and resilience ability of HICVPS control after suffering from the physical faults.

Due to the physical structure limitations of the motor and brake system of the intelligent connected vehicle, the problem of actuator input saturation is inevitable for the HICVPS, and complex environmental disturbances such as road slope, wind gust and strong sand will also seriously affect the control performance of the system. At the same time, the unpredictable actuator fault of the single vehicle such as the wheel motor will also worsen the stability of the HICVPS and cause vehicle collisions. The adaptive sliding mode fault tolerant coordination control is proposed to address the coordinate operation of the four driving motors with the actuator faults in the four wheel independently driven electric vehicle system in [9]. In [10], by designing the output feedback $H_\infty$ control method, the actuator fault and input time delay issues of the in-wheel motor-driven electric vehicle are solved. In [11], a finite-time fault-tolerant trajectory tracking control scheme was proposed to deal with the unknown actuator faults and disturbances of the autonomous vehicle. In [12], the autonomous vehicles suffering from the faults in the braking actuator are studied by designing the path following fault-tolerant control method. However, the single vehicle dynamic's physical fault was considered by the existing research works. The resilient control issue for the HICVPS with actuator fault of a single vehicle, actuator input saturation, and complex environment disturbances remains a challenge.

Sliding mode control method is widely studied in the vehicle dynamic system control because of the strong robustness, resilience performance and fast response to the physical faults and disturbances. Based on the sliding mode control method, Guo et al. [13] studied the actuator fault and input saturation issue of the high-speed train. In [14], the uniformly ultimately bounded stability in the sense of asymptotic stability was solved by developing the PID-type sliding mode control method for heterogeneous vehicular platoons with actuator fault and input saturation. Xu et al. [15] developed the fault-tolerant sliding-mode control scheme for the 2-D plane vehicular platoon systems against actuator faults with unknown time-varying fault direction and asymmetric nonlinear actuator saturation. Gao et al. [16] studied the prescribed performance control problem for a vehicular platoon with actuator faults and saturation combining with the sliding mode control. In [17], the prescribed time performance recovery sliding mode fault tolerant control method was developed to ensure the platoon stability performance under multiple faults, including actuator faults with deferred backup actuator switching and leader-follower link faults. In this chapter, the issue of distributed adaptive platoon sliding mode resilient control for the HICVPS with actuator fault, input saturation, and complex environment disturbances is studied.

Since the string stability property of the vehicle platoon can avoid the spacing errors between the two adjacent vehicles amplify when the spacing errors propagate upstream along the platoon. String stability is one of key transient properties of the HICVPS. In [18, 19], the constant spacing policy was designed to guarantee the closed-loop stability and the strict string stability. However, the constant spacing policy required that the initial spacing error and velocity errors of the vehicle platoon are zero and the acceleration of the preceding vehicle in the platoon should be known. Except the constant spacing policy, another policy named constant time headway policy has also attracted the attention of researchers, which only requires the velocity information of the vehicle in [20–22]. But for the constant spacing policy and constant time headway policy abovementioned, the actuator fault issue was not fully involved. The existing spacing policy may not work if any single vehicle of the platoons suffering from the actuator faults abruptly. Recently, for guaranteeing string stability and traffic flow stability performances under the actuator fault conditions, a new quadratic spacing policy is designed for the vehicular platoon system in [5, 6]. Based on the above analysis, by employing the quadratic spacing policy, a robust adaptive resilient control scheme is developed to achieve the string stability, traffic flow stability and resilient ability for the HICVPS with actuator fault of a single vehicle, actuator saturation, and complex environment disturbances in this chapter. The closed-loop finite-time stability of the platoon system can also be ensured.

## 3.2 Problem Statement

Since the on-board vehicle communication unit is highly integrated with GPS, sensors and embedded systems, which takes advantages in mobile VANETs communication. In this chapter, it is assumed that every intelligent connected vehicle is installed with OBU and realizes V2V and V2I communication interaction through cellular communication and dedicated short-range communication technology (DSRC) [23]. This chapter implements V2V communication interaction based on bidirectional information flow topology. As shown in Fig. 3.1, the HICVPS with actuator fault of a single vehicle, actuator saturation, and complex environment disturbances is studied in this chapter. The coupled sliding mode based distributed adaptive resilient controller is developed by combining with the quadratic spacing policy and Bidirectional

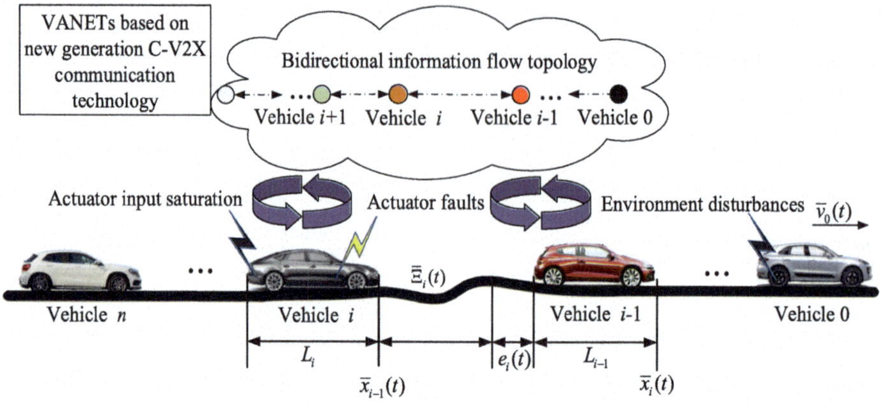


**Fig. 3.1** The composition and communication structure of HICVPS in this chapter

information flow topology. The robustness and resilience stability of the closed-loop system can be improved and the strict string stability of the system is ensured.

First, the longitudinal dynamics model of HICVPS is established based on the dynamics characteristics of the $i$th intelligent connected vehicle, $i = 1, 2, \ldots, n$. In this model, the longitudinal dynamics of the vehicle under the disturbances of complex environment are considered, and yaw and lateral dynamics are not considered. The longitudinal dynamics model of the $i$th intelligent connected vehicle in HICVPS considered is:

$$\begin{aligned}
\dot{\bar{x}}_i(t) &= \bar{v}_i(t) \\
\dot{\bar{v}}_i(t) &= \bar{a}_i(t) \\
\dot{\bar{a}}_i(t) &= \frac{\bar{\eta}_i}{\bar{M}_i R_i \tau_i} sat(\bar{u}_i(t)) - \frac{1}{\bar{M}_i \tau_i}\left[\rho_a A_i C_{ai}\left(\frac{1}{2}\bar{v}_i^2 + \tau_i \bar{v}_i \bar{a}_i(t)\right) + \bar{\Xi}_i\right] \\
&\quad - \frac{1}{\tau_i}\bar{a}_i(t) + \bar{D}_i
\end{aligned} \tag{3.1}$$

where $\bar{x}_i(t)$, $\bar{v}_i(t)$ and $\bar{a}_i(t)$ represent the longitudinal position, speed, and acceleration of the $i$th intelligent connected vehicle, respectively. $\bar{\eta}_i$ stands for the mechanical efficiency of the vehicle's drivetrain. $\bar{M}_i$ denotes the mass of vehicle, $\bar{R}_i$ is the tire radius of the vehicle, $\tau_i$ is the delay of vehicle motor. Moreover, $g$, $\rho_a$, and $C_{ai}$ denote the acceleration of gravity, the density of air, and the aerodynamic drag coefficient, respectively. $A_i$ is the cross-sectional area of the vehicle; $f_i$ is the rolling resistance coefficient. $\bar{\Xi}_i = \bar{M}_i g f_i \cos(\delta_i) + \bar{M}_i g \sin(\delta_i)$ is the road slope function, in which $\delta_i$ is the slope angle of the road. $\bar{x}_0(t)$, $\bar{v}_0(t)$ and $\bar{a}_0(t)$ are the longitudinal position, speed, and acceleration of the leader vehicle $i = 0$, respectively. It satisfied that $\dot{\bar{x}}_0(t) = \bar{v}_0(t)$ $\dot{\bar{v}}_0(t) = \bar{a}_0(t)$.

In Eq. (3.1), $sat(\bar{u}_i(t)) = \begin{cases} sign(\bar{u}_i(t))\bar{u}_{bi}, & |\bar{u}_i(t)| \geq \bar{u}_{bi} \\ \bar{u}_i(t), & |\bar{u}_i(t)| < \bar{u}_{bi} \end{cases}$, where $sign(\bullet)$ means sign function, $sat(\bar{u}_i(t))$ represents the control torque of the motor in the input saturated state, $\bar{u}_i(t)$ is the control input of vehicle, and $\bar{u}_{bi}$ is the upper bound of parameter. $\bar{D}_i = (1/\tau_i)d_i + \dot{d}_i$, we suppose that $\bar{D}_i$ is bounded, and satisfied that $\left|(\bar{M}_i R_i \tau_i \bar{D}_i)/\eta_i\right| \leq \tilde{\bar{D}}_i$, $\tilde{\bar{D}}_i$ is the unknown positive constant, and $d_i$ means the uncertainties and disturbances function, such as wind gusts, unknown acceleration disturbances, unmodeled dynamics, parametric perturbations and so on.

Considering the actuator fault caused by the control torque of the $i$th vehicle in the input saturated state, the actuator fault model is

$$\bar{u}_{fi}(t) = (1 - \bar{\lambda}_i)sat(\bar{u}_i(t)) + \bar{\Delta}_i(t), t \geq t_{i,F} \tag{3.2}$$

where $\bar{\lambda}_i$ is the partial loss of effectiveness fault factor, which can describes the fault degree of efficiency loss of actuator equipment such as hub motor, hydraulic cylinder, rectifier and inverter of the physical layer of HICVPS, and $0 < \bar{\lambda}_{i1} \leq \bar{\lambda}_i \leq \bar{\lambda}_{i2} < 1$. $\bar{\Delta}_i(t)$ represents the time-varying actuator additive fault function and is satisfied $\bar{\Delta}_i \leq \tilde{\bar{\Delta}}_i$. In the above description, $\bar{\lambda}_{i1}, \bar{\lambda}_{i2}$, and $\tilde{\bar{\Delta}}_i$ are the unknown positive constant. When no actuator fault occurs, $\bar{\lambda}_{i1} = \bar{\lambda}_{i2} = 0$, $\tilde{\bar{\Delta}}_i = 0$. $t_{i,F}$ means the unknown time instant when the fault occurs. Figure 3.2 shows the fault status of different types of actuators. The additive fault model [9, 24] described above can represent the following situations:

(i) Abrupt fault. Due to the leakage hole in the hydraulic cylinder or the constant torque caused by mechanical transmission locking fault, abrupt fault will occur. When the abrupt fault occurs, the control input changes from a nominal value ($\bar{\Delta}_i(t) = 0$, $t < t_{i,F}$) to an unknown fault constant value $\bar{\Delta}_i(t) = \bar{\alpha}_i$, $t \geq t_{i,F}$ after the time instant $t_{i,F}$.
(ii) Periodic fault. The fault of an insulated-gate bipolar transistor will result in the periodic fault of the drivetrain at a known frequency. The mathematical expression of the periodic fault can be $\bar{\Delta}_i(t) = \breve{\lambda}_i \sin(\omega t)$, $t \geq t_{i,F}$. When no fault occurs, $\bar{\Delta}_i(t) = 0$, $t < t_{i,F}$.$\breve{\lambda}_i$ is the non-zero unknown constant and $\omega$ is the known frequency.

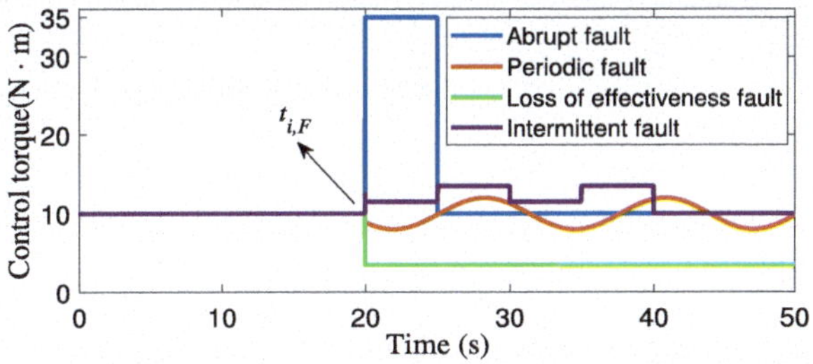


**Fig. 3.2** Different types of actuator fault

(iii) Intermittent fault. Intermittent faults usually occur intermittently in an unforeseeable form, and the same fault form may occur repeatedly in different time periods. The unpredictability of intermittent faults makes it extremely difficult to deal with. The mathematical expression is $\bar{\Delta}_i(t) = -0.5\, \widehat{\lambda}_{i,1}$ $sign(\sin(2\pi(t - t_{i,F}){T_w}^{-1})) + 0.5\, \widehat{\lambda}_{i,2}$, where $\widehat{\lambda}_{i,1}$ and $\widehat{\lambda}_{i,2}$ are the unknown fault constant, and $T_w$ indicates the fault period.

Based on the Eqs. (3.1) and (3.2), the longitudinal dynamics model of the $i$th vehicle in HICVPS can be rewritten as:

$$\begin{aligned}
\dot{\bar{x}}_i(t) &= \bar{v}_i(t)\\
\dot{\bar{v}}_i(t) &= \bar{a}_i(t)\\
\dot{\bar{a}}_i(t) &= \frac{\bar{\eta}_i(1-\bar{\lambda}_i)}{\bar{M}_i R_i \tau_i} sat(\bar{u}_i(t)) - \frac{1}{\bar{M}_i \tau_i}\left[\rho_a A_i C_{ai}\left(\frac{\bar{v}_i^2}{2} + \tau_i \bar{v}_i \bar{a}_i(t)\right) + \bar{\Xi}_i\right]\\
&\quad - \frac{\bar{a}_i(t)}{\tau_i} + \frac{\bar{\eta}_i \bar{\Delta}_i}{\bar{M}_i R_i \tau_i} + \bar{D}_i
\end{aligned} \tag{3.3}$$

In Eq. (3.1), the actuator input saturation nonlinear function $sat(\bar{u}_i(t))$ is a discontinuous function. When $|\bar{u}_i(t)| = \bar{u}_{bi}$, there exists a sharp corner between the saturation nonlinear function $sat(\bar{u}_i(t))$ and the control input signal $\bar{u}_i(t)$. Hence, a smooth function $h(\bar{u}_i(t))$ is introduced for approximating the saturation nonlinear problems. Then, the function $sat(\bar{u}_i(t))$ can be transformed into:

$$sat(\bar{u}_i(t)) = h(\bar{u}_i(t)) + \ell(\bar{u}_i(t)) \tag{3.4}$$

where $h(\bar{u}_i(t)) = \bar{u}_{bi}\tanh(\bar{u}_i(t)/\bar{u}_{bi})$, $\ell(\bar{u}_i(t))$ is the approximation errors and meet the conditions $|\ell(\bar{u}_i(t))| = |sat(\bar{u}_i(t)) - h(\bar{u}_i(t))| \le \bar{u}_{bi}(1 - \tanh(1)) = o_i$.

The smooth function $h(\bar{u}_i(t))$ can be rewritten as follow by the intermediate value theorem.

$$h(\bar{u}_i(t)) = h_{u_i}(t)(\bar{u}_i(t) - \bar{u}_{i0}) + h(\bar{u}_{i0}) \tag{3.5}$$

where $h_{u_i}(t) = \frac{\partial h(\bar{u}_i(t))}{\partial \bar{u}_i(t)})|_{\bar{u}_i = u_{\mu i}}$, and deriving it as $h_{u_i}(t) = \frac{4}{(e^{\bar{u}_i/\bar{u}_{bi}} + e^{-\bar{u}_i/\bar{u}_{bi}})^2}|_{\bar{u}_i = u_{\mu i}} \le 1$, and $u_{\mu i} = \mu\bar{u}_i(t) + (1-\mu)\bar{u}_{i0} (0 < \mu < 1)$.

When $\bar{u}_{i0} = 0$, and meet the condition that $h(\bar{u}_{i0}) = 0$, (3.4) can be

$$sat(\bar{u}_i(t)) = h_{u_i}(t)\bar{u}_i(t) + \ell(\bar{u}_i(t)) \tag{3.6}$$

where the condition $0 < h_{u_i}(t) \le 1$ is satisfied for any $\bar{u}_i(t)$. It is therefore reasonable to assume that there is an unknown positive definite constant $\bar{h}_{mi}$ that satisfies the condition $0 < \bar{h}_{mi} \le h_{u_i} \le 1$.

Combined with formula (3.6), the longitudinal dynamics model (3.3) of the $i$th heterogeneous intelligent connected vehicle can be transformed into:

$$
\begin{aligned}
\dot{\bar{x}}_i(t) &= \bar{v}_i(t) \\
\dot{\bar{v}}_i(t) &= \bar{a}_i(t) \\
\dot{\bar{a}}_i(t) &= \frac{\bar{\eta}_i}{\bar{M}_i R_i \tau_i}\left(1-\bar{\lambda}_i\right) h_{u_i}(t)\bar{u}_i(t) - \frac{1}{\bar{M}_i \tau_i}\left[\rho_a A_i C_{ai}\left(\frac{1}{2}\bar{v}_i^2\right.\right. \\
&\quad \left.\left. + \tau_i \bar{v}_i \bar{a}_i(t)\right) + \bar{\Xi}_i\right] - \frac{1}{\tau_i}\bar{a}_i(t) + \frac{\bar{\eta}_i}{\bar{M}_i R_i \tau_i}\bar{\aleph}_i
\end{aligned} \tag{3.7}
$$

where $\bar{\aleph}_i = (1-\bar{\lambda}_i)\ell(\bar{u}_i(t)) + \bar{\Delta}_i + (\bar{M}_i R_i \tau_i \bar{D}_i)/\bar{\eta}_i$, which be used to describe the approximation redundancy errors, actuator bias faults, uncertainties or external disturbances.

For avoiding the large transient errors, overcoming the influence of non-zero initial spacing errors on the closed-loop system stability, and solving the problem of acceleration sudden change caused by actuator fault. Based on the bidirectional information flow topology (as shown in Fig. 3.1), this chapter introduces the quadratic spacing policy, whose expression is

$$
\begin{aligned}
\bar{e}_i(t) &= \bar{e}_{xi}(t) - \bar{\Upsilon}_i(t) \\
\bar{e}_{xi}(t) &= \Delta\bar{x}_i(t) - L_{i-1} - \bar{e}_{yi}(t) \\
\bar{e}_{yi}(t) &= \varphi_d + \varepsilon\bar{v}_i^2/(2\bar{\phi}_{\max}(1-\bar{\lambda}_{i2})) + h\bar{v}_i(t)
\end{aligned} \tag{3.8}
$$

where the exponential decay function is $\bar{\Upsilon}_i(t) = (\bar{e}_{xi}(0) + [\varsigma_i \bar{e}_{xi}(0) + \dot{\bar{e}}_{xi}(0)]t + (1/2)[\varsigma_i^2 \bar{e}_{xi}(0) + 2\varsigma_i \dot{\bar{e}}_{xi}(0) + \ddot{\bar{e}}_{xi}(0)]t^2)\exp(-\varsigma_i t)$, $\Delta\bar{x}_i(t) = \bar{x}_{i-1}(t) - \bar{x}_i(t)$ is the actual inter-vehicle distance between the $i-1$th vehicle and the $i$th vehicle. $\bar{e}_i(t)$ indicates the inter-vehicle spacing errors, $L_{i-1}$ is the length of the $i$th vehicle. $\varphi_d$ represents the desired inter-vehicle distance when the platoon system is in steady state. $\varepsilon$ indicates the safety factor of driving vehicle. $\bar{\phi}_{\max} > 0$ is the maximum deceleration velocity of the vehicles. $h$ denotes the time headway constant, and $h\bar{v}_i(t)$ is the inter-vehicle safety distance. Furthermore, $\bar{e}_{xi}(0) = \bar{e}_{xi}(t)|_{t=0}$, $\dot{\bar{e}}_{xi}(0) = \dot{\bar{e}}_{xi}(t)|_{t=0}$, $\ddot{\bar{e}}_{xi}(0) = \ddot{\bar{e}}_{xi}(t)|_{t=0}$, $\varsigma_i$ is strict positive constant. It can derives that $\bar{e}_i(t)|_{t=0} = 0$, $\dot{\bar{e}}_i(t)|_{t=0} = 0$, and $\ddot{\bar{e}}_i(t)|_{t=0} = 0$ for any initial spacing error. Therefore, by Eq. (3.2), the string stability of the system is ensured and the assumption limitation of zero initial spacing error is eliminated.

**Assumption 3.1** According to the above conditions that the actuator loss of effectiveness fault factor satisfy $0 < \bar{\lambda}_{i1} \le \bar{\lambda}_i \le \bar{\lambda}_{i2} < 1$, the input saturation nonlinear function error meet that $|\ell(\bar{u}_i(t))| \le o_i$, additive fault satisfy that $\bar{\Delta}_i \le \bar{\bar{\Delta}}_i$, and $\left|(\bar{M}_i R_i \tau_i \bar{D}_i)/\bar{\eta}_i\right| \le \bar{\bar{D}}_i$, it can be deduced that the terms of the redundant function $\bar{\aleph}_i$ is bounded. Therefore, it is reasonable to assume that there is an unknown positive definite constant $\bar{\bar{\aleph}}_i$ satisfies the condition $\left|\dot{\bar{\aleph}}_i\right| \le \bar{\bar{\aleph}}_i$.

**Lemma 3.1** ([5, 18]) *For all $q > 0$, and $r \in \mathrm{R}$, there exists that $0 \le |r| - r\tanh(r/\varepsilon) \le q\varepsilon$ is true, where $v = 0.2785$, $\varepsilon$ is the positive constant.*

**Lemma 3.2** ([13]) *For any* $\phi$, $\varphi \in \mathrm{R}$, *the following inequality exist that* $\phi\varphi \leq \frac{g^{q_1}}{q_1}|\phi|^{q_1} + \frac{1}{q_2 g^{q_2}}|\varphi|^{q_2}$, *whereg* $> 0$, $q_1 > 1$, $q_2 > 1$, *and* $(q_1 - 1)(q_2 - 1) = 1$.

**Lemma 3.3** ([6]) *For* $k \in \mathrm{R}$ *and any real number* $h > 0$, *there exists an inequality* $0 \leq |k| - \frac{k^2}{\sqrt{k^2+h^2}} < h$.

**Lemma 3.4** (Globally finite time stability [25]) *Considering the nonlinear system* $\dot{x} = F(x, t)$, $x(0) = 0$, *where* $F\colon \Omega \times \mathrm{R}^+$ *is the nonlinear function, and* $\Omega$ *is the neighborhood of the origin. If there is a positive definite Lyapunov function* $V(x)$ *that makes the inequality* $\dot{V}(x) + \chi V^r(x) \leq 0$ *true, where* $\chi > 0$ *and* $0 < r < 1$, *then, the nonlinear system* $\dot{x} = F(x, t)$ *is globally finite time stability, and the convergence time meet that* $T \leq V^{1-r}(x(0))(\chi(1-r))^{-1}$.

**Definition 3.1** (*Strict string stability* [5, 6]) If the condition $|\bar{e}_n(t)| \leq |\bar{e}_{n-1}(t)| \leq \cdots \leq |\bar{e}_1(t)|$ is true, that is, the inter-vehicle distance error propagation transfer function $H_i(\mathrm{s}) := \bar{e}_{i+1}(\mathrm{s})\bar{e}_i^{-1}(\mathrm{s})$ is satisfied $|H_i(\mathrm{s})| \leq 1$ for all $i \in N$, then the intelligent connected vehicle platoon system (3.7) is string stable in the sense of strict. $\bar{e}_i(\mathrm{s})$ denotes the Laplace transform of $\bar{e}_i(t)$, $\bar{e}_i(t)$ is the Longitudinal inter-vehicle spacing error.

**Control objective** For the HICVPS, a distributed adaptive platoon resilient control method is proposed to maintain the desired inter-vehicle distance under the single vehicle actuator fault, actuator input saturation and complex environment disturbances. The main control objectives are as follows:

(1) The acceleration and velocity of the follower vehicle can be consistent with that of the leader vehicle, and the whole HICVPS can achieve a relatively ideal safety spacing error during the movement.
(2) The closed-loop finite-time stability and strict string stability of the HICVPS are guaranteed.

## 3.3 Resilient Control Design

As shown in Fig. 3.3, in this section, a distributed adaptive platoon resilient controller will be designed based on the quadratic spacing policy and bidirectional information flow topology. Based on Lyapunov stability theory, the closed-loop control stability and the string stability of HICVPS under actuator fault, input saturation and environmental disturbances will be analyzed and proved. The satisfying condition also will be discussed.

### *3.3.1 Design of Distributed Adaptive Resilient Controller*

Based on the quadratic spacing policy (3.2), a finite-time robust adaptive coupled sliding mode resilient controller is proposed in this section to achieve finite-time

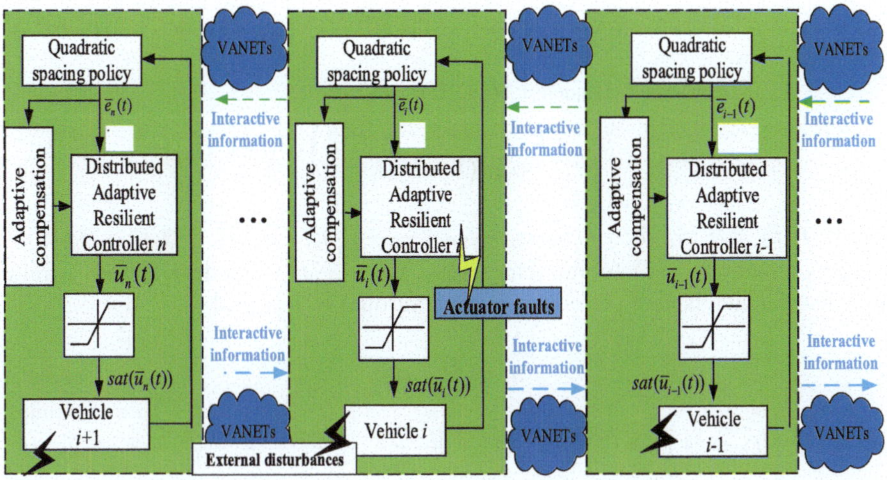


**Fig. 3.3** Resilient control structure of HICVPS under actuator fault and input saturation

convergence of spacing errors of the $i$th vehicle, while ensuring the strict string stability under fault, saturation and environmental disturbances. The sliding surface of the $i$th follower vehicle is designed as:

$$p_i(t) = \alpha_{i0}\dot{\bar{e}}_i(t) + \alpha_i \bar{e}_i(t) + \int_0^t \alpha_{i1}\bar{e}_i(\tau)d\tau \tag{3.9}$$

where $\alpha_{i0}$, $\alpha_i$, and $\alpha_{i1}$ are designed positive constant.

For ensuring the string stability of HICVPS, the coupling sliding mode surfaces are defined as follows:

$$P_i(t) = \breve{\hbar}_1 p_i - \breve{\hbar}_2 p_{i+1} \tag{3.10}$$

where $\breve{\hbar}_1$ and $\breve{\hbar}_2$ are positive definite weight parameter, and meet that $0 < \breve{\hbar}_1/\breve{\hbar}_2 \leq 1$. For the $n$th follower vehicle of the HICVPS, we have that $P_{n+1} = 0$, $p_{n+1} = \dot{p}_{n+1} = 0$.

At the same time, combined with the quadratic spacing policy (3.2), for $i = 1, 2, \ldots, n-1$, the derivative of the coupled sliding mode surface,

$$\begin{aligned} \dot{P}_i(t) &= \breve{\hbar}_1(\alpha_{i0}\ddot{\bar{e}}_i + \alpha_i\dot{\bar{e}}_i + \alpha_{i1}\bar{e}_i) - \breve{\hbar}_2\dot{p}_{i+1} \\ &= -\breve{\hbar}_1\alpha_{i0}\left(\frac{\varepsilon\bar{v}_i}{\bar{\phi}_{\max}(1-\bar{\lambda}_{i2})} + h\right)\dot{\bar{a}}_i + \bar{O}_i \end{aligned} \tag{3.11}$$

where $\bar{O}_i(t) = \breve{\hbar}_1\alpha_{i0}(\bar{a}_{i-1} - \bar{a}_i - \varepsilon(\bar{\phi}_{\max}(1-\bar{\lambda}_{i2}))^{-1}\bar{a}_i^2 - \ddot{\Upsilon}_i) + \breve{\hbar}_1(\alpha_i\dot{\bar{e}}_i + \alpha_{i1}\bar{e}_i) - \breve{\hbar}_2\dot{p}_{i+1}$.

For the $n$th follower vehicle at the tail of the HICVPS, the time derivative of $P_n$ is

$$\begin{aligned}\dot{P}_n(t) &= \breve{\hbar}_1(\alpha_{n0}\ddot{\bar{e}}_n + \alpha_n\dot{\bar{e}}_n + \alpha_{n1}\bar{e}_n)\\ &= -\breve{\hbar}_1\alpha_{n0}\left(\frac{\varepsilon\bar{v}_n}{\phi_{\max}(1-\bar{\lambda}_{n2})} + h\right)\dot{\bar{a}}_n(t) + \bar{O}_n(t)\end{aligned} \tag{3.12}$$

where $\bar{O}_n(t) = \breve{\hbar}_1(\alpha_{n0}\bar{a}_{n-1} - \alpha_{n0}\bar{a}_n - \alpha_{n0}\frac{\varepsilon}{\phi_{\max}(1-\bar{\lambda}_{n2})}\bar{a}_n^2 - \alpha_{n0}\ddot{\Upsilon}_n + \alpha_n\dot{\bar{e}}_n + \alpha_{n1}\bar{e}_n)$.

In combination (3.11), for the vehicles $i = 1, 2, \ldots, n-1$ in the HICVPS, a finite time distributed adaptive platoon resilient controller is designed as,

$$\bar{u}_i = \hat{\bar{\Omega}}_i\bar{u}_{i1}(t) \tag{3.13}$$

where $\bar{u}_{i1}(t) = \frac{\hat{\bar{k}}_{1i}}{\bar{Z}_i}\tanh\left(\frac{P_i}{\mu_i}\right) + \frac{\bar{k}_{2i}}{\bar{Z}_i}P_i + \frac{(1-\bar{\lambda}_{i2})\Theta_i}{\bar{Z}_i}\frac{\bar{O}_i^2 P_i}{|\bar{O}_i P_i| + b_{oi}} + \frac{R_i}{\bar{\eta}_i(1-\bar{\lambda}_{i2})}\hat{\bar{\Xi}}_i + \bar{u}_{i1y}(t)$, $\bar{u}_{i1y}(t) = \frac{1}{(1-\bar{\lambda}_{i2})}\hat{\bar{\aleph}}_i - \frac{\bar{M}_i R_i \tau_i}{\bar{\eta}_i(1-\bar{\lambda}_{i2})}\left[-\frac{1}{\tau_i}\bar{a}_i(t) - \frac{1}{\bar{M}_i\tau_i}\left(\rho_a A_i C_{ai}\left(\frac{1}{2}\bar{v}_i^2 + \tau_i\bar{v}_i\bar{a}_i\right)\right)\right]$, and $\bar{Z}_i = \breve{\hbar}_1\alpha_{i0} \times \left(\frac{\varepsilon\bar{v}_i}{\phi_{\max}(1-\bar{\lambda}_{i2})} + h\right)\frac{\bar{\eta}_i}{\bar{M}_i R_i \tau_i}(1-\bar{\lambda}_{i2})$, $1 \le 1/(1-\bar{\lambda}_{i2}) \le \bar{\Theta}_i$, $\bar{k}_{2i}$, $\bar{\Theta}_i$, $b_{oi} > 0$ and $\mu_i$ are all very small positive definite constants.

The design adaptive compensation laws are

$$\dot{\hat{\bar{k}}}_{1i} = b_{ki}P_i(t)\tanh\left(\frac{P_i}{\mu_i}\right) - \varsigma_{ki}\hat{\bar{k}}_{1i} \tag{3.14}$$

$$\dot{\hat{\bar{\Omega}}}_i = b_{\Omega i}\bar{Z}_i\bar{u}_{i1}(t)P_i(t) - \varsigma_{\Omega i}\hat{\bar{\Omega}}_i \tag{3.15}$$

At the same time, considering that it is difficult to measure and obtain parameters of road slope function $\bar{\Xi}_i$ and redundant error term $\bar{\aleph}_i$ in engineering practice, adaptive estimation law is designed for them,

$$\dot{\hat{\bar{\Xi}}}_i = b_{\Xi i}\bar{Z}_i\frac{R_i}{\bar{\eta}_i(1-\bar{\lambda}_{i2})}P_i(t) - \varsigma_{\Xi i}\hat{\bar{\Xi}}_i \tag{3.16}$$

$$\dot{\hat{\bar{\aleph}}}_i = -b_{\aleph i}\frac{\bar{Z}_i}{(1-\bar{\lambda}_{i2})}P_i(t) - \varsigma_{\aleph i}\hat{\bar{\aleph}}_i \tag{3.17}$$

where $\hat{\bar{k}}_{1i}$, $\hat{\bar{\Omega}}_i$, $\hat{\bar{\Xi}}_i$, and $\hat{\bar{\aleph}}_i$ are the estimates of the parameters $\bar{k}_{1i}$, $\bar{\Omega}_i$, $\bar{\Xi}_i$, and $\bar{\aleph}_i$, respectively. $\hat{\bar{k}}_{1i}(0) \ge 0$, $\hat{\bar{\Omega}}_i(0) \ge 0$, $\hat{\bar{\Xi}}_i(0) \ge 0$, $\hat{\bar{\aleph}}_i(0) \ge 0$. $\bar{\Omega}_i = \bar{h}_{mi}^{-1}$ is the inverse of positive definite constant $\bar{h}_{mi}$. Meanwhile, $\tilde{\bar{k}}_{1i} = \bar{k}_{1i} - \hat{\bar{k}}_{1i}$, $\tilde{\bar{\Omega}}_i = \bar{\Omega}_i - \hat{\bar{\Omega}}_i$, $\tilde{\bar{\Xi}}_i = \bar{\Xi}_i - \hat{\bar{\Xi}}_i$, and $\tilde{\bar{\aleph}}_i = \bar{\aleph}_i - \hat{\bar{\aleph}}_i$ are the estimation errors. $\bar{k}_{1i}, b_{ki}, b_{\Omega i}, b_{\Xi i}, b_{\aleph i}$ are positive constant. $\varsigma_{ki} > 0$, $\varsigma_{\Omega i} > 0$, $\varsigma_{\Xi i} > 0$, $\varsigma_{\Omega i} > 0$ are all very small positive definite constants.

For the $n$th follower vehicle, by (3.12), the finite time distributed adaptive platoon resilient controller is

$$\bar{u}_n = \hat{\bar{\Omega}}_n \bar{u}_{n1}(t) \tag{3.18}$$

where $\bar{u}_{n1}(t) = \frac{\hat{\bar{k}}_{1n}}{\bar{Z}_n} \tanh(\frac{P_n}{\mu_n}) + \frac{\bar{k}_{2n}}{\bar{Z}_n} P_n + \frac{(1-\bar{\lambda}_{n2})\Theta_n}{\bar{Z}_n} \frac{\bar{O}_n^2 P_n}{|\bar{O}_n P_n| + b_{on}} + \frac{R_i \hat{\bar{\Xi}}_n}{\bar{\eta}_n(1-\bar{\lambda}_{n2})} + \bar{u}_{n1y}(t)$,
$\bar{u}_{n1y}(t) = -\frac{\hat{\bar{\aleph}}_n}{(1-\bar{\lambda}_{n2})} - \frac{\bar{M}_n R_n \tau_n}{\bar{\eta}_n(1-\bar{\lambda}_{n2})}[-\frac{1}{\tau_n}\bar{a}_n(t) - \frac{1}{\bar{M}_n \tau_n}(\rho_a A_n C_{an}(\frac{1}{2}\bar{v}_n^2 + \tau_n \bar{v}_n \bar{a}_n))]$,
and $\bar{Z}_n = \breve{\hbar}_1 \alpha_{n0}(\frac{\varepsilon \bar{v}_n}{\phi_{\max}(1-\bar{\lambda}_{n2})} + h)\frac{\bar{\eta}_n}{\bar{M}_n R_n \tau_n}(1 - \bar{\lambda}_{n2})$.

The adaptive compensation laws are

$$\dot{\hat{\bar{k}}}_{1n} = b_{kn} P_n(t) \tanh(P_n \mu_n^{-1}) - \varsigma_{kn} \hat{\bar{k}}_{1n} \tag{3.19}$$

$$\dot{\hat{\bar{\Omega}}}_n = b_{\Omega n} \bar{Z}_n \bar{u}_{n1}(t) P_n(t) - \varsigma_{\Omega n} \hat{\bar{\Omega}}_n \tag{3.20}$$

$$\dot{\hat{\bar{\Xi}}}_n = b_{\Xi n} \bar{Z}_n \frac{R_n}{\bar{\eta}_n(1-\bar{\lambda}_{n2})} P_n(t) - \varsigma_{\Xi n} \hat{\bar{\Xi}}_n \tag{3.21}$$

$$\dot{\hat{\bar{\aleph}}}_n = -b_{\aleph n} \frac{\bar{Z}_n}{(1-\bar{\lambda}_{n2})} P_n(t) - \varsigma_{\aleph n} \hat{\bar{\aleph}}_n . \tag{3.22}$$

### 3.3.2 Analysis of System Stability

**Theorem 3.1** *Considering the HICVPS (3.3), suffering from the actuator fault and actuator input saturation problems (3.2), as well as complex environmental disturbance problems. The distributed finite-time adaptive platoon resilient controller (3.13), (3.18) and the adaptive compensation laws (3.14)–(3.17), (3.19)–(3.22) are designed to ensure that the spacing error* $\bar{e}_i(t)$ *(3.2) can converge to the minimal neighborhood of the origin in finite time,* $T \leq \wp^{-1}\sqrt{2}\bar{\Gamma}^{0.5}(0)$*. Moreover, the strict string stability can be guaranteed when the positive definite weight parameters* $\breve{\hbar}_1$ *and* $\breve{\hbar}_2$ *meet the conditions* $0 < \breve{\hbar}_1/\breve{\hbar}_2 \leq 1$.

***Proof*** For proving the closed-loop control stability of HICVPS, the candidate Lyapunov function is defined as

$$\Gamma(t) = \left(\sum_{i=1}^{n-1} \Gamma_i(t)\right) + \Gamma_n(t) \tag{3.23}$$

where $\Gamma_i = \frac{1}{2}P_i^2 + \sum_{\widehat{m}=1}^{4} \Gamma_{i\widehat{m}}, \Gamma_{i1} = \frac{1}{2}b_{ki}\tilde{\bar{k}}_{1i}^2, \Gamma_{i2} = \frac{1}{2}b_{\Xi i}\tilde{\bar{\Xi}}_i^2, \Gamma_{i3} = \frac{1}{2}b_{\aleph i}\tilde{\bar{\aleph}}_i^2, \Gamma_{i4} = \frac{1}{2}b_{\Omega i}\bar{h}_{mi}\tilde{\bar{\Omega}}_i^2$.

For the heterogeneous follower vehicles $i = 1, 2, \ldots, n-1$ in the HICVPS, the time derivation of $\Gamma_i$ is

$$\dot{\Gamma}_i = P_i \dot{P}_i + \dot{\Gamma}_{i1} + \dot{\Gamma}_{i2} + \dot{\Gamma}_{i3} + \dot{\Gamma}_{i4} \tag{3.24}$$

Then, using Eqs. (3.7), (3.2) and (3.11), $P_i \dot{P}_i$ can be further deduced as

$$\begin{aligned}\dot{P}_i(t)P_i(t) \leq & -\breve{\hbar}_1\alpha_{i0}\left(\frac{\varepsilon\bar{v}_i}{\bar{\phi}_{\max}(1-\bar{\lambda}_{i2})}+h\right)\left[\frac{\bar{\eta}_i(1-\bar{\lambda}_{i2})\bar{h}_{mi}(t)\bar{u}_i(t)}{\bar{M}_i R_i \tau_i} - \frac{\bar{a}_i(t)}{\tau_i}\right.\\ & \left. - \frac{1}{\bar{M}_i\tau_i}\left[\bar{\Xi}_i + \left(\rho_a A_i C_{ai}\left(\frac{1}{2}\bar{v}_i^2 + \tau_i\bar{v}_i\bar{a}_i\right)\right)\right] + \frac{\bar{\eta}_i \aleph_i}{\bar{M}_i R_i \tau_i}\right]P_i(t)\\ & + (1-\bar{\lambda}_{i2})\bar{\Theta}_i\left|\bar{\mathrm{O}}_i P_i\right| \end{aligned} \tag{3.25}$$

By using the distributed adaptive platoon resilient controller (3.13), for the vehicles $i = 1, 2, \ldots, n-1$, we have

$$\begin{aligned} & -\breve{\hbar}_1\alpha_{i0}\left(\frac{\varepsilon\bar{v}_i}{\bar{\phi}_{\max}(1-\bar{\lambda}_{i2})}+h\right)\frac{\bar{\eta}_i}{\bar{M}_i R_i \tau_i}(1-\bar{\lambda}_i)\bar{h}_{mi}(t)\bar{u}_i(t)P_i\\ & = -\bar{\mathrm{Z}}_i\bar{h}_{mi}(t)\hat{\bar{\Omega}}_i\bar{u}_{i1}P_i\\ & = -\bar{\mathrm{Z}}_i\bar{h}_{mi}(t)(\bar{\Omega}_i - \tilde{\bar{\Omega}}_i)u_{i1}P_i\\ & = -\bar{\mathrm{Z}}_i\bar{u}_{i1}P_i + \bar{\mathrm{Z}}_i\bar{h}_{mi}(t)\tilde{\bar{\Omega}}_i\bar{u}_{i1}P_i \end{aligned} \tag{3.26}$$

And then through the Eqs. (3.13), (3.26) and (3.25), we can further derive that,

$$\begin{aligned}\dot{P}_i(t)P_i(t) \leq & -\hat{\bar{k}}_{1i}P_i(t)\tanh\left(\frac{P_i}{\mu_i}\right) + \breve{\hbar}_1\alpha_{i0}\left(\frac{\varepsilon\bar{v}_i}{\bar{\phi}_{\max}(1-\bar{\lambda}_{i2})}+h\right)\frac{P_i(t)}{\bar{M}_i\tau_i}(\bar{\Xi}_i - \hat{\bar{\Xi}}_i)\\ & - \bar{k}_{2i}P_i^2 + (1-\bar{\lambda}_{i2})\bar{\Theta}_i b_{oi} - \breve{\hbar}_1\alpha_{i0}\left(\frac{\varepsilon\bar{v}_i}{\bar{\phi}_{\max}(1-\bar{\lambda}_{i2})}+h\right)\times\\ & \frac{\bar{\eta}_i P_i(t)(\bar{\aleph}_i - \hat{\bar{\aleph}}_i)}{\bar{M}_i R_i \tau_i} + \bar{\mathrm{Z}}_i\bar{h}_{mi}\tilde{\bar{\Omega}}_i\bar{u}_{i1}P_i \end{aligned} \tag{3.27}$$

Based on the Lemma 3.3, we have that $-\bar{\mathrm{O}}_i^2 P_i^2(\left|\bar{\mathrm{O}}_i P_i\right| + b_{oi})^{-1} \leq -\left|\bar{\mathrm{O}}_i P_i\right| + b_{oi}$. Then, by Eqs. (3.14)–(3.17), we can obtain that

$$\begin{aligned}
&\dot{\Gamma}_{i1}+\dot{\Gamma}_{i2}+\dot{\Gamma}_{i3}+\dot{\Gamma}_{i4}\\
&=-\frac{\tilde{\bar{k}}_{1i}\dot{\hat{\bar{k}}}_{1i}}{b_{ki}}-\frac{\tilde{\bar{\Xi}}_i\dot{\hat{\bar{\Xi}}}_i}{b_{\Xi i}}-\frac{\tilde{\bar{\aleph}}_i\dot{\hat{\bar{\aleph}}}_i}{b_{\aleph i}}-\frac{\bar{h}_{mi}\tilde{\bar{\Omega}}_i\dot{\hat{\bar{\Omega}}}_i}{b_{\Omega i}}\\
&=-\tilde{\bar{k}}_{1i}P_i(t)\tanh\left(\frac{P_i}{\mu_i}\right)-\tilde{\bar{\Xi}}_i\breve{h}_1\alpha_{i0}\left(\frac{\varepsilon\bar{v}_i}{\bar{\phi}_{\max}(1-\bar{\lambda}_{i2})}+h\right)\frac{P_i(t)}{\bar{M}_i\tau_i}\\
&+\breve{h}_1\alpha_{i0}\left(\frac{\varepsilon\bar{v}_i}{\bar{\phi}_{\max}(1-\bar{\lambda}_{i2})}+h\right)\frac{\bar{\eta}_i P_i(t)\tilde{\bar{\aleph}}_i}{\bar{M}_i R_i\tau_i}-\bar{h}_{mi}\tilde{\bar{\Omega}}_i\bar{Z}_i\bar{u}_{i1}(t)P_i(t)\\
&+\frac{\varsigma_{ki}}{b_{ki}}\tilde{\bar{k}}_{1i}\hat{\bar{k}}_{1i}+\frac{\varsigma_{\Xi i}}{b_{\Xi i}}\tilde{\bar{\Xi}}_i\hat{\bar{\Xi}}_i+\frac{\varsigma_{\aleph i}}{b_{\aleph i}}\tilde{\bar{\aleph}}_i\hat{\bar{\aleph}}_i+\frac{\varsigma_{\Omega i}}{b_{\Omega i}}\bar{h}_{mi}\tilde{\bar{\Omega}}_i\hat{\bar{\Omega}}_i
\end{aligned}\tag{3.28}$$

By combining the above Eqs. (3.27) and (3.28), (3.24) can be rewritten as follows

$$\begin{aligned}
\dot{\Gamma}_i\le&-\bar{k}_{1i}P_i(t)\tanh\left(\frac{P_i}{\mu_i}\right)-\bar{k}_{2i}P_i^2+(1-\bar{\lambda}_{i2})\bar{\Theta}_i b_{oi}\\
&+\frac{\varsigma_{ki}}{b_{ki}}\tilde{\bar{k}}_{1i}\hat{\bar{k}}_{1i}+\frac{\varsigma_{\Xi i}}{b_{\Xi i}}\tilde{\bar{\Xi}}_i\hat{\bar{\Xi}}_i+\frac{\varsigma_{\aleph i}}{b_{\aleph i}}\tilde{\bar{\aleph}}_i\hat{\bar{\aleph}}_i+\frac{\varsigma_{\Omega i}}{b_{\Omega i}}\bar{h}_{mi}\tilde{\bar{\Omega}}_i\hat{\bar{\Omega}}_i
\end{aligned}\tag{3.29}$$

According to Lemma 3.2 Young's inequality, Eq. (3.29) can be further expressed as

$$\begin{aligned}
\dot{\Gamma}_i\le&-\bar{k}_{1i}P_i(t)\tanh\left(\frac{P_i}{\mu_i}\right)-\bar{k}_{2i}P_i^2+(1-\bar{\lambda}_{i2})\bar{\Theta}_i b_{oi}\\
&-\frac{\varsigma_{ki}}{2b_{ki}}\tilde{\bar{k}}_{1i}^2+\frac{\varsigma_{ki}}{2b_{ki}}\bar{k}_{1i}^2-\frac{\varsigma_{\Xi i}}{2b_{\Xi i}}\tilde{\bar{\Xi}}_i^2+\frac{\varsigma_{\Xi i}}{2b_{\Xi i}}\bar{\Xi}_i^2-\frac{\varsigma_{\aleph i}}{2b_{\aleph i}}\tilde{\bar{\aleph}}_i^2\\
&+\frac{\varsigma_{\aleph i}}{2b_{\aleph i}}\bar{\aleph}_i^2-\frac{\varsigma_{\Omega i}}{2b_{\Omega i}}\bar{h}_{mi}\tilde{\bar{\Omega}}_i^2+\frac{\varsigma_{\Omega i}}{2b_{\Omega i}}\bar{h}_{mi}\bar{\Omega}_i^2
\end{aligned}\tag{3.30}$$

Using Lemma 3.1, the following conclusion can be obtained,

$$-\bar{k}_{1i}P_i(t)\tanh\left(\frac{P_i}{\mu_i}\right)\le 0.2785\bar{k}_{1i}\mu_i-\bar{k}_{1i}\left|P_i(t)\right|\tag{3.31}$$

Therefore, when $i=1,2,\ldots,n-1$, $\dot{\Gamma}_i$ can be derived as

$$\begin{aligned}
\dot{\Gamma}_i\le&\ 0.2785\bar{k}_{1i}\mu_i-\bar{k}_{2i}P_i^2+(1-\bar{\lambda}_{i2})\bar{\Theta}_i b_{oi}\\
&-\frac{\varsigma_{ki}}{2b_{ki}}\tilde{\bar{k}}_{1i}^2+\frac{\varsigma_{ki}}{2b_{ki}}\bar{k}_{1i}^2-\frac{\varsigma_{\Xi i}}{2b_{\Xi i}}\tilde{\bar{\Xi}}_i^2+\frac{\varsigma_{\Xi i}}{2b_{\Xi i}}\bar{\Xi}_i^2-\frac{\varsigma_{\aleph i}}{2b_{\aleph i}}\tilde{\bar{\aleph}}_i^2\\
&+\frac{\varsigma_{\aleph i}}{2b_{\aleph i}}\bar{\aleph}_i^2-\frac{\varsigma_{\Omega i}}{2b_{\Omega i}}\bar{h}_{mi}\tilde{\bar{\Omega}}_i^2+\frac{\varsigma_{\Omega i}}{2b_{\Omega i}}\bar{h}_{mi}\bar{\Omega}_i^2
\end{aligned}\tag{3.32}$$

Similarly, for the $i=n$ follower vehicle, if the Lyapunov function $\Gamma_n=(1/2)P_n^2+\sum_{\widehat{m}=1}^{4}\Gamma_{n\widehat{m}}$ is selected, its time derivative of $\dot{\Gamma}_n$ can be

$$
\begin{aligned}
\dot{\Gamma}_n \leq\ & 0.2785\bar{k}_{1n}\mu_n - k_{2n}P_i^2 + (1-\bar{\lambda}_{n2})\bar{\Theta}_n b_{on} \\
& - \frac{\varsigma_{kn}}{2b_{kn}}\tilde{\bar{k}}_{1n}^2 + \frac{\varsigma_{kn}}{2b_{kn}}\bar{k}_{1n}^2 - \frac{\varsigma_{\Xi n}}{2b_{\Xi n}}\tilde{\bar{\Xi}}_n^2 + \frac{\varsigma_{\Xi n}}{2b_{\Xi n}}\bar{\Xi}_n^2 - \frac{\varsigma_{\aleph n}}{2b_{\aleph n}}\tilde{\bar{\aleph}}_n^2 \\
& + \frac{\varsigma_{\aleph n}}{2b_{\aleph n}}\bar{\aleph}_n^2 - \frac{\varsigma_{\Omega n}}{2b_{\Omega n}}\bar{h}_{mn}\tilde{\bar{\Omega}}_n^2 + \frac{\varsigma_{\Omega n}}{2b_{\Omega n}}\bar{h}_{mn}\bar{\Omega}_n^2
\end{aligned}
\tag{3.33}
$$

For the whole HICVPS $i = 1, 2, \ldots, n$, the time derivative of $\Gamma(t)$ is

$$
\dot{\Gamma} \leq -\chi_1\left(\sum_{i}^{n}\Gamma_i\right) + \chi_2 \tag{3.34}
$$

where $\chi_2 = \sum_{i=1}^{n}\{0.2785\bar{k}_{1i}\mu_i + (1-\bar{\lambda}_{i2})\bar{\Theta}_i b_{oi} + \frac{\varsigma_{ki}\bar{k}_{1i}^2}{2b_{ki}} + \frac{\bar{\Xi}_i^2\varsigma_{\Xi i}}{2b_{\Xi i}} + \frac{\bar{\aleph}_i^2\varsigma_{\aleph i}}{2b_{\aleph i}} + \frac{\bar{h}_{mi}\bar{\Omega}_i^2\varsigma_{\Omega i}}{2b_{\Omega i}}\}$, $\chi_1 = \min\{2\bar{k}_{2i}, \min_{1\leq i\leq n}\varsigma_{ki}, \min_{1\leq i\leq n}\varsigma_{\Xi i}, \min_{1\leq i\leq n}\varsigma_{\aleph i}, \min_{1\leq i\leq n}\varsigma_{\Omega i}\}$.

Multiply both sides of the Eq. (3.34) by $e^{-\chi_1 t}$ and integrate over$[0, t]$, we get

$$
\begin{aligned}
\Gamma &\leq \left(\Gamma(0) - \frac{\chi_2}{\chi_1}\right)e^{-\chi_1 t} + \frac{\chi_2}{\chi_1} \\
&\leq \Gamma(0) + \frac{\chi_2}{\chi_1} \\
&= \sum_{i=1}^{n}\left(\frac{1}{2}P_i^2(0) + \sum_{\widehat{m}=1}^{3}\Gamma_{i\widehat{m}}(0)\right) + \frac{\chi_2}{\chi_1}
\end{aligned}
\tag{3.35}
$$

Therefore, under the conditions $\Gamma(0) \geq 0$, we know that the whole HICVPS is uniformly ultimately boundedness, where $P_i$, $\tilde{\bar{k}}_{1i}$, $\tilde{\bar{\Xi}}_i$, $\tilde{\bar{\aleph}}_i$, and $\tilde{\bar{\Omega}}_i \in L_\infty$. When the initial conditions are bounded, $P_i$, $\hat{\bar{k}}_{1i}$, $\hat{\bar{\Xi}}_i$, $\hat{\bar{\aleph}}_i$, $\hat{\bar{\Omega}}_i$ will also be uniformly ultimately boundedness.

By selecting suitable design parameters, reducing $\chi_2$ and increasing $\chi_1$, the coupled sliding mode surface $P_i$ can converge to the minimum neighborhood of the origin. Thus, from Eqs. (3.10), and (3.11), we can see that the spacing error $\bar{e}_i(t)$ can also converge to the minimum neighborhood of the origin.

Further, for proving the closed-loop finite-time stability of the whole HICVPS control, the Lyapunov function is chosen as

$$
\bar{\Gamma} = \bar{\Gamma}_1 + \bar{\Gamma}_2 = \sum_{i=1}^{n-1}\left(\frac{1}{2}P_i^2\right) + \frac{1}{2}P_n^2 \tag{3.36}
$$

where $\bar{\Gamma}_1 = \sum_{i=1}^{n-1}(\frac{1}{2}P_i^2)$, $\bar{\Gamma}_2 = \frac{1}{2}P_n^2$.

By using the Eqs. (3.13) and (3.18), the time derivatives of $\bar{\Gamma}_1$ and $\bar{\Gamma}_2$ are,

$$\begin{aligned}\dot{\bar{\Gamma}}_1 \le& \sum_{i=1}^{n-1}\{-\hat{\bar{k}}_{1i}P_i(t)\tanh\left(\frac{P_i}{\mu_i}\right) - \bar{k}_{2i}P_i^2 + (1-\bar{\lambda}_{i2})\bar{\Theta}_i b_{oi} \\ &+ \frac{\bar{Z}_i\bar{h}_{mi}(t)\tilde{\bar{\Omega}}_i\bar{u}_i(t)P_i}{\hat{\bar{\Omega}}_i} + \breve{\hbar}_1\alpha_{i0}\left(\frac{\varepsilon\bar{v}_i}{\bar{\phi}_{\max}(1-\bar{\lambda}_{i2})}+h\right)\frac{P_i(t)\tilde{\bar{\Xi}}_i}{\bar{M}_i\tau_i} \\ &- \breve{\hbar}_1\alpha_{i0}\left(\frac{\varepsilon\bar{v}_i}{\bar{\phi}_{\max}(1-\bar{\lambda}_{i2})}+h\right)\frac{\eta_i}{\bar{M}_iR_i\tau_i}P_i(t)\tilde{\bar{\aleph}}_i\}\end{aligned} \tag{3.37}$$

$$\begin{aligned}\dot{\bar{\Gamma}}_2 \le& -\hat{\bar{k}}_{1n}P_n(t)\tanh\left(\frac{P_n}{\mu_n}\right) - \bar{k}_{2n}P_n^2 + (1-\bar{\lambda}_{n2})\bar{\Theta}_n b_{on} \\ &+ \frac{\bar{Z}_n\bar{h}_{mn}\tilde{\bar{\Omega}}_n\bar{u}_nP_n}{\hat{\bar{\Omega}}_n} + \breve{\hbar}_1\alpha_{n0}\left(\frac{\varepsilon\bar{v}_n}{\bar{\phi}_{\max}(1-\bar{\lambda}_{n2})}+h\right)\frac{P_n(t)\tilde{\bar{\Xi}}_n}{\bar{M}_n\tau_n} \\ &- \breve{\hbar}_1\alpha_{n0}\left(\frac{\varepsilon\bar{v}_n}{\bar{\phi}_{\max}(1-\bar{\lambda}_{n2})}+h\right)\frac{\eta_n}{\bar{M}_nR_n\tau_n}P_n(t)\tilde{\aleph}_n\end{aligned} \tag{3.38}$$

Based on Eq. (3.31), Lemma 3.1, and Eqs. (3.37) and (3.38), the time derivative of $\bar{\Gamma}$ can be derived as

$$\begin{aligned}\dot{\bar{\Gamma}} \le& \sum_{i=1}^{n}\{0.2785\hat{\bar{k}}_{1i}\mu_i - \hat{\bar{k}}_{1i}\,|P_i(t)| - \bar{k}_{2i}P_i^2 + (1-\bar{\lambda}_{i2})\bar{\Theta}_i b_{oi} \\ &+ \frac{\bar{Z}_i\bar{h}_{mi}(t)\tilde{\bar{\Omega}}_i\bar{u}_i(t)P_i}{\hat{\bar{\Omega}}_i} + \breve{\hbar}_1\alpha_{i0}\left(\frac{\varepsilon\bar{v}_i}{\bar{\phi}_{\max}(1-\bar{\lambda}_{i2})}+h\right)\frac{P_i(t)\tilde{\bar{\Xi}}_i}{\bar{M}_i\tau_i} \\ &- \breve{\hbar}_1\alpha_{i0}\left(\frac{\varepsilon\bar{v}_i}{\bar{\phi}_{\max}(1-\bar{\lambda}_{i2})}+h\right)\frac{\eta_i}{\bar{M}_iR_i\tau_i}P_i(t)\tilde{\bar{\aleph}}_i\}\end{aligned} \tag{3.39}$$

According to the above analysis of $\tilde{\bar{k}}_{1i}$, $\tilde{\bar{\Xi}}_i$, $\tilde{\bar{\aleph}}_i$, $\tilde{\bar{\Omega}}_i \in L_\infty$, we know that there exists four positive constant $\bar{\bar{k}}_{1i}$, $\bar{\bar{\Xi}}_i$, $\bar{\bar{\Omega}}_i$, $\breve{\bar{\aleph}}_i$ can meet the conditions that $\left|\tilde{\bar{k}}_{1i}\right| \le \bar{\bar{k}}_{1i}$, $\left|\tilde{\bar{\Xi}}_i\right| \le \bar{\bar{\Xi}}_i$, $\tilde{\bar{\Omega}}_i \le \bar{\bar{\Omega}}_i$, and $\tilde{\bar{\aleph}}_i \le \breve{\bar{\aleph}}_i$, respectively. Hence, we obtain that $\left|\frac{\tilde{\bar{\Omega}}_i}{\hat{\bar{\Omega}}_i}\right| \le \frac{\bar{\bar{\Omega}}_i}{|\bar{\Omega}_i|+\bar{\bar{\Omega}}_i}$, and $\hat{\bar{k}}_{1i}$ is bounded and meet that $\left|\hat{\bar{k}}_{1i}\right| \le \bar{k}_{1i} + \bar{\bar{k}}_{1i}$. Meanwhile, through the input saturation function approximation design of Eqs. (3.4) and (3.6), we can see that $\bar{h}_{m_i}(t)\bar{u}_i(t)$ is bounded, and $\left|\bar{h}_{mi}(t)\bar{u}_i(t)\right| \le \bar{U}$, $\bar{U}$ is positive constant. When the following conditions meet that

$$\bar{k}_{1i} + \bar{\bar{k}}_{1i} \geq 0.2785 \left(\bar{k}_{1i} + \bar{\bar{k}}_{1i}\right) \mu_i + (1 - \bar{\lambda}_{i2})\bar{\Theta}_i b_{oi} + \bar{\wp}$$
$$+ \frac{\bar{Z}_i \bar{U} \bar{\bar{\Omega}}_i}{|\bar{\Omega}_i| + \bar{\bar{\Omega}}_i} + \left| \breve{\hbar}_1 \alpha_{i0} \left( \frac{\varepsilon \bar{v}_i}{\bar{\phi}_{\max}(1 - \bar{\lambda}_{i2})} + h \right) \frac{1}{\bar{M}_i \tau_i} \right| \bar{\bar{\Xi}}_i$$
$$- \left| \breve{\hbar}_1 \alpha_{i0} \left( \frac{\varepsilon \bar{v}_i}{\bar{\phi}_{\max}(1 - \bar{\lambda}_{i2})} + h \right) \frac{\bar{\eta}_i}{\bar{M}_i R_i \tau_i} \right| \breve{\breve{\aleph}}_i \tag{3.40}$$

where $\bar{\wp} > 0$ is positive constant. Then, (3.39) can be

$$\dot{\bar{\Gamma}} \leq \sum_{i=1}^{n} \{-\bar{k}_{2i} P_i^2 - \wp |P_i|\}$$
$$\leq -\wp \sqrt{2\bar{\Gamma}} \tag{3.41}$$

Therefore, based on the global finite-time stability theory of Lemma 3.4, the coupled sliding mode surface $P_i(t)(i = 1, 2, \ldots, n)$ can converge to the neighborhood of the origin in finite time. So the spacing error (3.2) can be ensured to converge to the minimal neighborhood of the origin in finite time with $T \leq \wp^{-1}\sqrt{2}\bar{\Gamma}^{0.5}(0)$, and the closed-loop HICVPS also can be globally finite-time stable.

**Remark 3.1** The control precision of spacing error depends on the selection of control parameters in the formula (3.34). Based on the formula (3.35) and $\bar{\Gamma}(t) \leq \bar{\Gamma}(0) + \chi_2 \chi_1^{-1}$, we have that the control precision of HICVPS spacing error depends on the selection of constant $\chi_1$ and $\chi_2$. By increasing $\bar{k}_{2i}$, $b_{ki}$, $b_{\Xi i}$, $b_{\aleph i}$ and $b_{\Omega i}$, and simultaneously decreasing $\varsigma_{ki}$, $\varsigma_{\Xi i}$, $\varsigma_{\aleph i}$, $\varsigma_{\Omega i}$ and $\bar{\Theta}_i$, higher accurate spacing steady-state errors can be achieved. However, when $\chi_1$ is too large and $\chi_2$ is too small, the required control energy and chattering will increase accordingly. Therefore, the adjustment of control parameters needs to be combined with the actual application requirements, and the control precision, available control energy and precision should be balanced.

For eliminating the coupling effect of the system and reducing the transmission and amplification of the inter-vehicle distance error along the whole platoon system, it is important to ensure the string stability of HICVPS. According to the above proof and analysis, it can be seen that the coupled sliding mode surface $P_i(t)$ can converge to the neighborhood of the origin in a finite time, and the combined Eqs. (3.9) and (3.10),

$$\hbar_1(\alpha_{i0}\dot{\bar{e}}_i(t) + \alpha_i \bar{e}_i + \int_0^t \alpha_{i1}\bar{e}_i(\tau)d\tau)$$
$$\approx \hbar_2(\alpha_{(i+1)0}\dot{\bar{e}}_{i+1}(t) + \alpha_{i+1}\bar{e}_{i+1} + \int_0^t \alpha_{(i+1)1}\bar{e}_{i+1}(\tau)d\tau) \tag{3.42}$$

Because of $\bar{e}_i(0) = 0$, and $\int_{-\infty}^{0} \bar{e}_i(t)dt = 0$. By the Laplace transform, (3.42) can be

$$\begin{aligned}&\hbar_1(\alpha_{i0} s\bar{E}_i(s) + \alpha_i \bar{E}_i(s) + \alpha_{i1}\bar{E}_i(s)/s)\\&\approx \hbar_2(\alpha_{(i+1)0} s\bar{E}_{i+1}(s) + \alpha_{i+1}\bar{E}_{i+1}(s) + \alpha_{(i+1)1}\bar{E}_{i+1}(s)/s)\end{aligned} \tag{3.43}$$

Therefore, $G(s) = \frac{\bar{E}_{i+1}(s)}{\bar{E}_i(s)} = \frac{\alpha_{i0}s^2+\alpha_i s+\alpha_{i1}}{\alpha_{(i+1)0}s^2+\alpha_{i+1}s+\alpha_{(i+1)1}} \approx \frac{\hbar_1}{\hbar_2}$, when the condition $0 < \frac{\hbar_1}{\hbar_2} \leq 1$ is satisfied, $|G_i(s)| \leq 1$, the spacing error decreases layer by layer along the length of the platoon system under the influence of faults and disturbances can be ensured. According to Definition 3.1, the strict string stability of HICVPS is guaranteed. □

## 3.4 Method Verification and Analysis

To verify the feasibility of the proposed platoon resilient control scheme, in this section, the HICVPS composed of seven heterogeneous intelligent connected vehicles as a longitudinal control model is used for comparative experimental design. Considering the sensitivity of vehicles at different positions in the platoon system to actuator faults, the comparison experiments between the proposed scheme and the traditional control method are designed in Case 1 and Case 2, respectively. It should be pointed out that the HICVPS suffered from actuator input saturation and complex environmental disturbances in the comparative experimental cases were considered. The dynamics parameters of the vehicles involved in HICVPS are as follows: $\rho_a = 1\,\text{N/m}^3$, $C_{ai} = 0.35$, $A_i = 2.2\,\text{m}^2$, $f_i = 0.02$, $R_i = 0.32\,\text{m}$, $g = 9.8\,\text{m/s}^2$, $\varphi_d = 8\,\text{m}$, $i = 1, 2, \ldots, 6$. Table 3.1 lists the vehicle heterogeneous characteristics in HICVPS, and Table 3.2 lists the initial states of the vehicle. Other required parameters such as vehicle safety factor $\varepsilon = 0.2$, maximum deceleration $\bar{\phi}_{\max} = 7$, time headway constant $h = 0.1$. The complex external disturbance function is $d_i = 0.05\sin(2t + 1)$. The road slope function is selected as $\delta_i = \begin{cases} 2\%, 180 \leq \bar{x}_i < 320 \\ 1\%, 320 \leq \bar{x}_i < 1000 \\ 0\%, \text{otherwise} \end{cases}$.

**Table 3.1** The vehicle heterogeneous characteristics in HICVPS

| Vehicle $i = 1, 2, \ldots, 6$ | 1 | 2 | 3 | 4 | 5 | 6 |
|---|---|---|---|---|---|---|
| $\bar{M}_i(\times 10^3 kg)$ | 1.8 | 2.0 | 1.75 | 1.85 | 1.6 | 1.35 |
| $\tau_i(s)$ | 0.26 | 0.8 | 0.4 | 0.47 | 0.35 | 0.65 |
| $\bar{\eta}_i$ | 0.92 | 0.90 | 0.91 | 0.92 | 0.89 | 0.91 |
| Vehicle $i = 0, 1, \ldots, 5$ | 0 | 1 | 2 | 3 | 4 | 5 |
| $L_{i-1}(m)$ | 4 | 4.5 | 5 | 5 | 4.5 | 4.5 |

**Table 3.2** The initial states of the vehicle in HICVPS

| Vehicle $i = 0, 1, 2, \ldots, 6$ | 0 | 1 | 2 | 3 | 4 | 5 | 6 |
|---|---|---|---|---|---|---|---|
| $\bar{x}(0)(m)$ | 150 | 125.5 | 112.5 | 99.5 | 87 | 75.5 | 63.5 |
| $\bar{v}(0)(m/s)$ | 1 | 2 | 2.5 | 5 | 3.5 | 3 | 2 |
| $\bar{a}(0)(m/s^2)$ | 0 | 5 | 2 | 1 | 3 | 4 | 2 |

### 3.4.1 *Case 1*

In the longitudinal driving process of HICVPS, under the condition that no actuator fault occurs, the system suffers from actuator input saturation constraint and complex external environment disturbances. For the vehicle $i = 1, 2, \ldots, 6$ in HICVPS, the actuator saturation constraint parameters are assumed to be $\bar{u}_{bi} = 55$ . The simulation results are shown in Figs. 3.4, 3.5, and 3.6. The acceleration of the leader vehicle is $\bar{a}_0(t) = \begin{cases} 2.45 \text{ m/s}^2, 5 \text{ s} \le t < 10 \text{ s} \\ 0 \text{ m/s}^2, \text{otherwise} \end{cases}$.

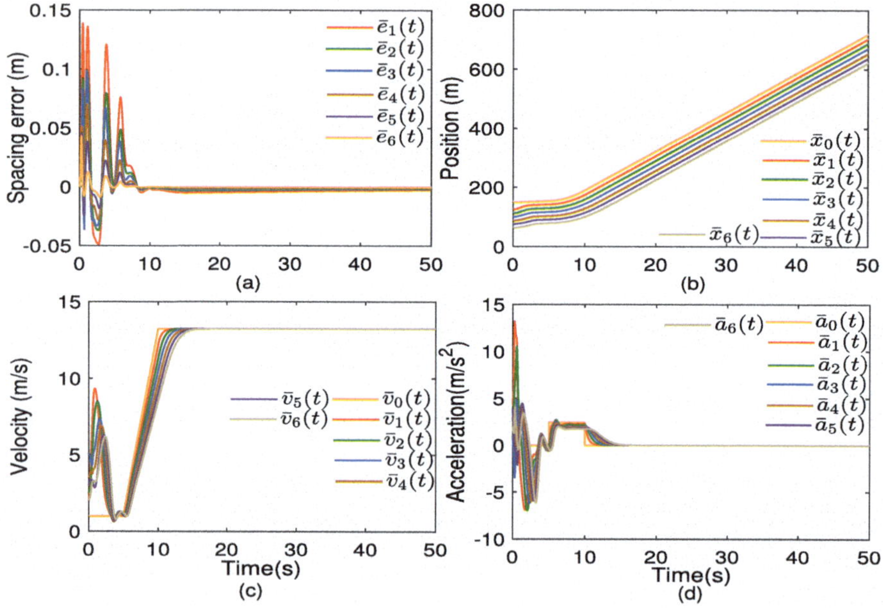


**Fig. 3.4** Experimental results of the proposed scheme under the condition of no faults. **a** Spacing error $\bar{e}_i(t)$, **b** Longitudinal position $\bar{x}_i(t)$, **c** Longitudinal velocity $\bar{v}_i(t)$, **d** Longitudinal acceleration $\bar{a}_i(t)$

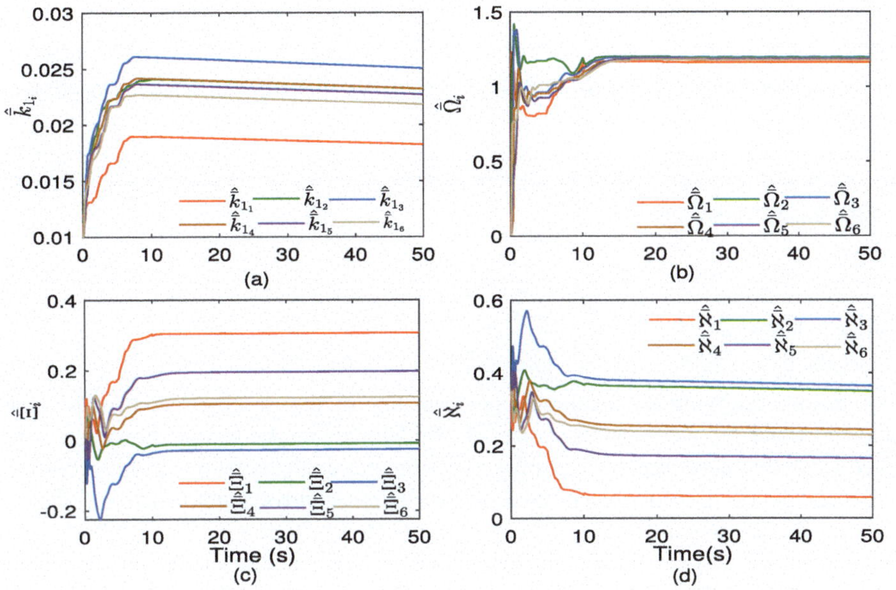


**Fig. 3.5** Adaptive updated laws of the proposed scheme under the condition of no faults. **a** Adaptive estimation of $\hat{\hat{k}}_{1i}(t)$, **b** Adaptive estimation of $\hat{\hat{\Omega}}_i(t)$, **c** Adaptive estimation of $\hat{\hat{\Xi}}_i(t)$, **d** Adaptive estimation of $\hat{\hat{\aleph}}_i(t)$

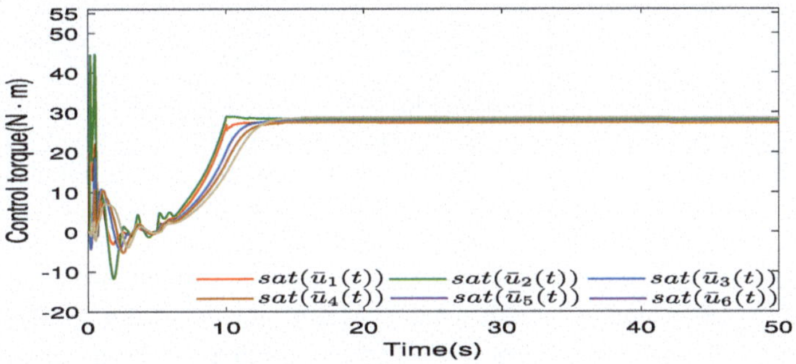


**Fig. 3.6** Control torque of the proposed scheme under the condition of no faults

As shown in Fig. 3.4a, based on the proposed control scheme, the inter-vehicle spacing error of HICVPS can converge to zero in a finite time under actuator saturation constraints and external disturbances, which can verify the finite time stability of the closed-loop system. The spacing error of each vehicle in the platoon system obviously keeps the characteristics of decreasing layer by layer from the leader vehicle to the end vehicle, that is $|\bar{e}_6(t)| \leq |\bar{e}_5(t)| \leq \cdots \leq |\bar{e}_1(t)|$. Under the action of the quadratic spacing policy, all spacing errors in the system can be kept at zero from the initial time to the last time, which indicates the strict string stability of

HICVPS. The position and trajectory in Fig. 3.4b indicates that all vehicles in the system always travel in the form of longitudinal platoon, during which no collision occurs. As shown in Fig. 3.4c and d, all vehicles can track the leader vehicle and keep the same velocity and acceleration as the leader vehicle within a limited time ($T \leq 5$ s). Figure 3.5 shows the adaptive estimation of unknown functions under the proposed method. Figure 3.6 describes the control input torques of each vehicle under the proposed method.

### 3.4.2 Case 2

For further verifying the resilience performance of the proposed control scheme and the rapid recovery performance of HICVPS under fault and saturation conditions, in this case, the comparative experimental analysis between the proposed scheme and the traditional coupled sliding mode control method [5, 6] is conducted. It is assumed that the intelligent connected vehicle $i = 3$ in HICVPS suffers from the actuator fault at different times. The actuator saturation constraint parameters of each vehicle in the platoon system are $u_{bi} = 60$. For the vehicle $i = 3$ with faults, the additive periodic fault function in the whole experiment process is: $\Delta_3 = 0.05\cos(5t)$. The partial loss of effectiveness actuator fault function at different times is $\lambda_3 = \begin{cases} 0.1, & 15 \leq t < 20;\ 0.25, \quad 35 \leq t < 40 \\ 0, & otherwise \end{cases}$. The acceleration of the leader vehicle is $\bar{a}_0(t) = \begin{cases} 2.5\ \text{m/s}^2, 5\ \text{s} \leq t < 10\ \text{s} \\ -1.5\ \text{m/s}^2, 20\ \text{s} \leq t < 25\ \text{s} \\ 0\ \text{m/s}^2, \text{otherwise} \end{cases}$.

The comparative experimental results are shown in Figs. 3.7, 3.8, 3.9 and 3.10. Considering that all vehicles in HICVPS suffer from actuator input saturation, environmental disturbances and actuator periodic fault, as well as that vehicles $i = 3$ suffer from actuator partial loss of effectiveness fault during $t = 15$ s~20s and during $t = 35$ s~40s respectively. The strict string stability of HICVPS is guaranteed, and the inter-vehicle spacing error can achieve rapid convergence (convergence time is less than 10 s), as shown in Fig. 3.7a. The longitudinal position tracking trajectory in Fig. 3.7b shows that even if a sudden fault occurs in the vehicle platoon, the collision can still be avoided, which proves the safety of the proposed method.

Figure 3.7c and d show that the considered actuator faults and disturbances problem do not have a significant impact on the velocity and acceleration tracking in the vehicle platoon system with the proposed control scheme. Figure 3.8 shows adaptive compensation and estimation laws. Compared with the experimental results of the proposed scheme in Figs. 3.7 and 3.9, under the traditional control method, the actuator fault has a greater impact on the closed-loop control stability and string stability of the system. Figure 3.10a, b reveal the control input torque of HICVPS under the proposed scheme and the traditional control method, respectively.

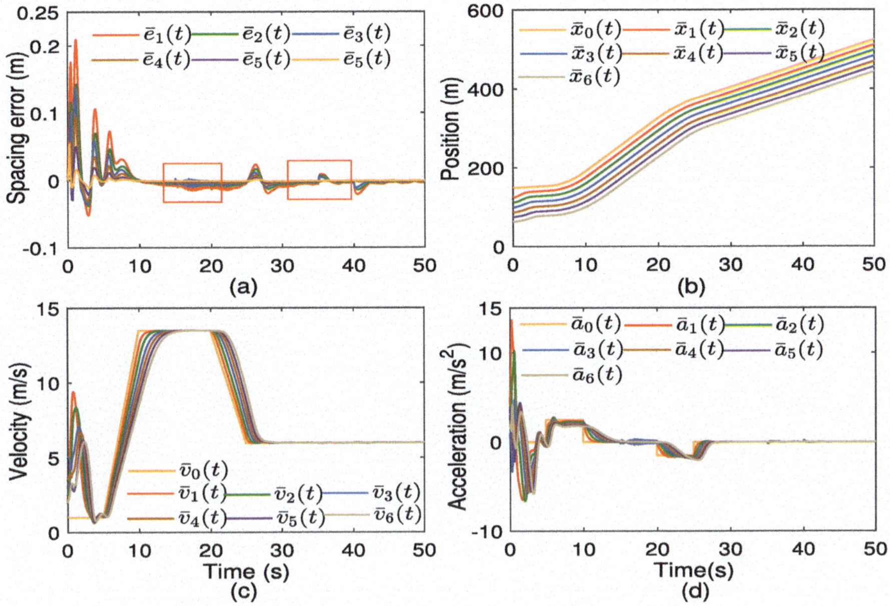


**Fig. 3.7** Experimental results of the proposed scheme under the condition of the vehicle $i = 3$ with faults. **a** Spacing error $\bar{e}_i(t)$, **b** Longitudinal position $\bar{x}_i(t)$, **c** Longitudinal velocity $\bar{v}_i(t)$, **d** Longitudinal acceleration $\bar{a}_i(t)$

The experimental results show that the finite time adaptive robust resilient control scheme proposed in this chapter is feasible and effective in improving the fault recovery response speed of the system, improving the accuracy and robustness of HICVPS resilient control, and ensuring the strict string stability under actuator fault and saturation conditions.

## 3.5 Summary

In this chapter, a distributed adaptive finite time robust platoon resilient control strategy is proposed for HICVPS with single vehicle actuator fault, actuator input saturation and complex road environment disturbances. First, in order to overcome the sudden acceleration of vehicles caused by faults, reduce the conservatism of traditional spacing policy, and avoid large transient problems, the quadratic spacing policy is used to ensure the strict string stability of HICVPS under actuator fault conditions by combining the upper bound of fault factors. Adaptive compensation laws

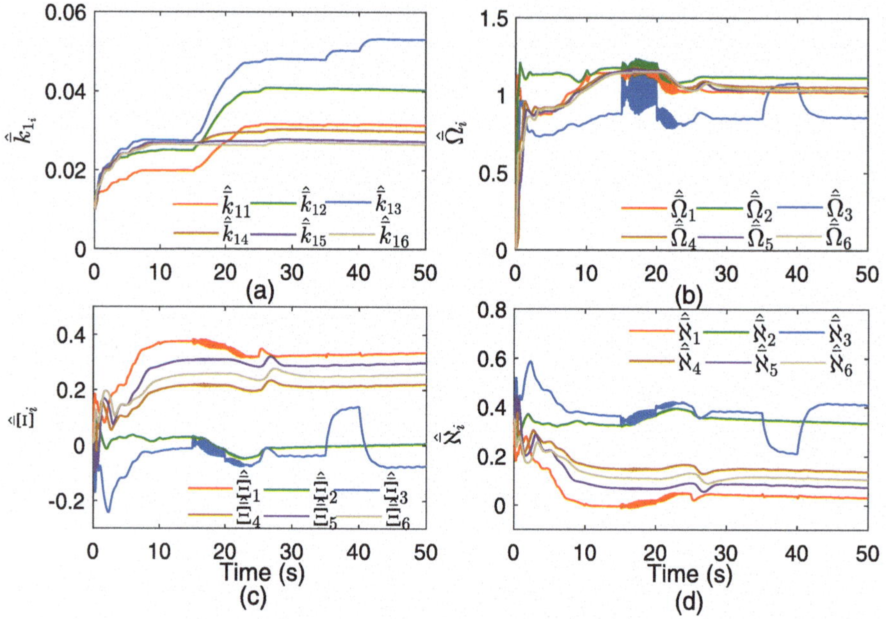


**Fig. 3.8** Adaptive updated laws of the proposed scheme under the condition of the vehicle $i = 3$ with faults. **a** Adaptive estimation of $\hat{\hat{k}}_{1i}(t)$, **b** Adaptive estimation of $\hat{\hat{\Omega}}_i(t)$, **c** Adaptive estimation of $\hat{\hat{\Xi}}_i(t)$, **d** Adaptive estimation of $\hat{\hat{\aleph}}_i(t)$

are designed to deal with the unknown upper bound of the actuator saturation approximation function and the dsiturbances of complex environment, and to eliminate the assumption that the upper bound of the actuator saturation approximation function is known in the traditional method. Then a finite time adaptive coupled sliding mode robust resilient controller is designed to ensure the resilience performance and fast stability of the system under fault conditions. The analysis proves that the stability of strict string can be guaranteed and the conditions are given. Based on Lyapunov stability theory, it is proved that the closed-loop system is globally finite time stable under fault, saturation and environmental disturbances.

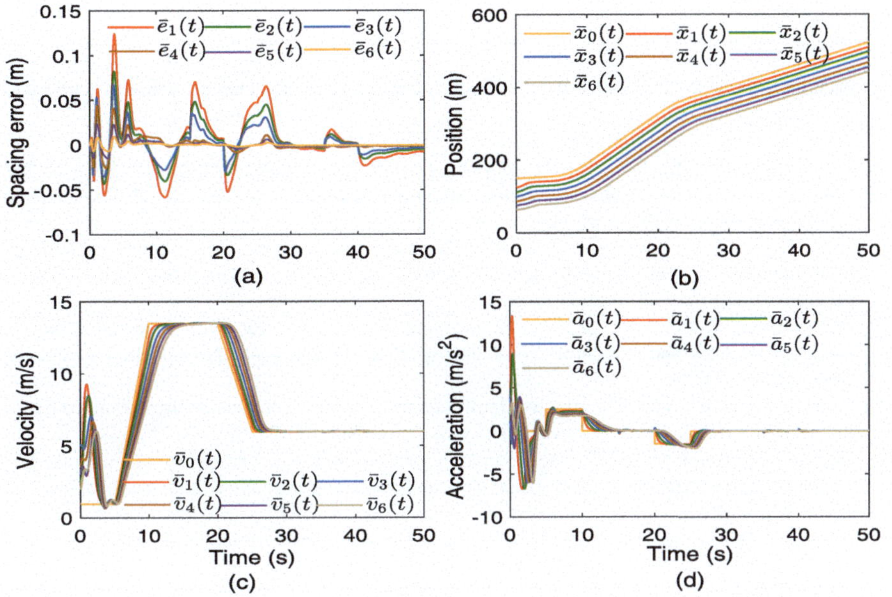


**Fig. 3.9** Experimental results of the traditional method under the condition of the vehicle $i = 3$ with faults. **a** Spacing error $\bar{e}_i(t)$, **b** Longitudinal position $\bar{x}_i(t)$, **c** Longitudinal velocity $\bar{v}_i(t)$, **d** Longitudinal acceleration $\bar{a}_i(t)$

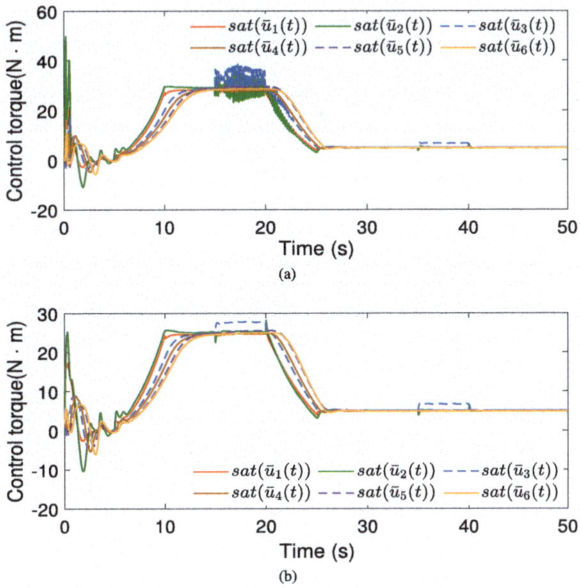


**Fig. 3.10** Control torques under the condition of the vehicle $i = 3$ with faults. **a** Proposed scheme, **b** Traditional method

## References


1. Zhang H, Zhao W and Wang J(2019) Fault-tolerant control for electric vehicles with independently driven in-wheel motors considering individual driver steering characteristics. *IEEE Transactions on Vehicular Technology*, 68(5): 4527-4536.
2. Zhao W, Wang A, Zou S, et al(2021) Individual auxiliary and fault-tolerant control of steer-by-wire system considering different drivers steering characteristics. *IEEE/ASME Transactions on Mechatronics*, 26(3): 1558-1569.
3. Deng H, Zhao Y, Nguyen A T, et al(2023) Fault-tolerant predictive control with deep-reinforcement-learning-based torque distribution for four in-wheel motor drive electric vehicles. *IEEE/ASME Transactions on Mechatronics*, 28(2): 668-680.
4. Pan C, Chen Y, Liu Y, et al(2022) Adaptive resilient control for interconnected vehicular platoon with fault and saturation. *IEEE Transactions on Intelligent Transportation Systems*, 23(8): 10210-10222.
5. Guo G, Li P, Hao L Y(2020) Adaptive fault-tolerant control of platoons with guaranteed traffic flow stability. *IEEE Transactions on Vehicular Technology*, 69(7): 6916-6927.
6. Guo G, Li P, Hao L Y(2022) A new quadratic spacing policy and adaptive fault-tolerant platooning with actuator saturation. *IEEE Transactions on Intelligent Transportation Systems*, 23(2): 1200-1212.
7. Bichiou Y and Rakha H A(2019) Developing an optimal intersection control system for automated connected vehicles. *IEEE Transactions on Intelligent Transportation Systems*, 20(5): 1908-1916.
8. Lv C, et al(2018) Simultaneous observation of hybrid states for cyber-physical systems: A case study of electric vehicle powertrain. *IEEE Transactions on Cybernetics*, 48(8): 2357-2367.
9. Zhang D, Liu G, Zhou H, et al(2018) Adaptive sliding mode fault-tolerant coordination control for four-wheel independently driven electric vehicles. *IEEE Transactions on Industrial Electronics*, 65(11): 9090-9100.
10. Shao X, Naghdy F, Du H, et al(2019) Output feedback $H_\infty$ control for active suspension of in-wheel motor-driven electric vehicle with control faults and input delay. *ISA Transactions*, 92: 94-108.
11. Wang N, Pan X, Su S(2020) Finite-time fault-tolerant trajectory tracking control of an autonomous surface vehicle. *Journal of the Franklin Institute*, 357(16): 11114-11135.
12. Cao X, Tian Y, Ji X, et al(2021) Fault-tolerant controller design for path following of the autonomous vehicle under the faults in braking actuators. *IEEE Transactions on Transportation Electrification*, 7(4): 2530-2540.
13. Guo X, Ahn C K(2021) Adaptive fault-tolerant pseudo-PID sliding-mode control for high-speed train with integral quadratic constraints and actuator Saturation. *IEEE Transactions on Intelligent Transportation Systems*, 22(12): 7421-7431.
14. Guo G, Li P, and Hao L Y(2022) A new quadratic spacing policy and adaptive fault-tolerant platooning with actuator saturation. *IEEE Transactions on Intelligent Transportation Systems*, 23(2): 1200-1212.
15. Xu W D, Guo X G, Wang J L, et al(2022) Nonlinear disturbance observer-based fault-tolerant sliding-mode control for 2-D plane vehicular platoon with UTVFD and ANAS. *IEEE Transactions on Cybernetics*, https://doi.org/10.1109/TCYB.2022.3222496.
16. Gao Z, Zhang Y and Guo G(2023) Finite-time fault-tolerant prescribed performance control of connected vehicles with actuator saturation. *IEEE Transactions on Vehicular Technology*, 72(2): 1438-1448.
17. Han J. et al(2023) Prescribed-time performance recovery fault tolerant control of platoon with nominal constraints guarantee. *IEEE Transactions on Intelligent Transportation Systems*, 24(3): 3501-3513.
18. Dunbar W, Caveney D(2012) Distributed receding horizon control of vehicle platoons: Stability and string stability. *IEEE Transactions on Automatic Control*, 57(3): 620-633.

19. Zheng Y, Li E S, Wang J, et al(2016) Stability and scalability of homogeneous vehicular platoon: Study on the influence of information flow topologies. *IEEE Transactions on Intelligent Transportation Systems*, 17(1): 14-26.
20. Ali A, Garcia G, and Martinet P(2015) The flatbed platoon towing model for safe and dense platooning on highways. *IEEE Intelligent Transportation Systems Magazine*, 7(1): 58-68.
21. Li S, et al(2017) Dynamical modeling and distributed control of connected and automated vehicles: Challenges and opportunities. *IEEE Intelligent Transportation Systems Magazine*, 9(3): 46-58.
22. Guo X, Wang J, Liao F, et al(2017) Distributed adaptive sliding mode control strategy for vehicle-following systems with nonlinear acceleration uncertainties. *IEEE Transactions on Vehicular Technology*, 66(2): 981-991.
23. Lesch V, Breitbach M, Segata M, et al(2022) An overview on approaches for coordination of platoons. *IEEE Transactions on Intelligent Transportation Systems*, 23(8): 10049-10065.
24. Yu M, Wang D(2014) Model-based health monitoring for a vehicle steering system with multiple faults of unknown type. *IEEE Transactions on Industrial Electronics*, 61(7): 3574-3586.
25. Haddad W M, L'Afflitto A(2016) Finite-time stabilization and optimal feedback control. *IEEE Transactions on Automatic Control*, 61(4): 1069-1074.

# Chapter 4
# Distributed Finite Time Resilient Control for Intelligent Connected Vehicle Platoon Systems with Exponential Policy

String stability and the traffic flow stability are the two important characteristics of ICVPS, as mentioned in Chap. 1. External disturbances have a good influence on the spacing error and vehicle error between vehicles. But string stability can ensure the external disturbances of upstream vehicles will not be amplified and propagated among downstream vehicles of the platoon system. So that the risk of collision can be avoided, and the safety of longitudinal driving of the platoon can be achieved. For the traffic flow stability, which can avoid the situation that the when traffic flow density is increasing and the traffic flow velocity will decrease under the influence of disturbances. Thus, the traffic flow stability can ensure that the threat of traffic congestion be eliminated. From the perspective of control, the string stability depends heavily on the inter-vehicle spacing policy and the controller design. The characteristics of traffic flow stability are closely related to the steady state process of the vehicle platoon system. And the inter-vehicle spacing policy will also directly affect the stability of the traffic flow of the platoon system. In order to ensure the string stability and traffic flow stability of the vehicle platoon system, scholars focus on the inter-vehicle spacing error policy, such as constant spacing policy, constant time headway spacing policy, and exponential spacing policy, etc (as mentioned in Chap. 1).

On the one hand, only the constant spacing policy under the condition of the desired distance constant can improve the traffic flow capacity, but it can not effectively ensure the string stability of the vehicle platoon system in complex environments. The constant time headway spacing policy can ensure the string stability, but the inter-vehicle distance will continue to increase at high velocity. In that way, the instability of traffic flow will occur. The traditional exponential spacing policy does not consider the problem of string instability caused by non-zero initial spacing error. At the same time, under the influence of sudden acceleration caused by the external disturbances, the exponential spacing policy can not effectively ensure the string

 
C. Pan and Y. Chen, *Resilient Control of Intelligent Connected Vehicle Platoon Systems*,
https://doi.org/10.1007/978-981-97-8850-7_4

stability and traffic flow stability. On the other hand, the study of the inter-vehicle spacing policy does not consider the effects of random and unpredictable physical faults. The existing spacing policy may fail, and the string stability of the vehicle platoon system and traffic flow stability will be degraded. The existing exponential spacing policy in [1] can ensure the string stability and traffic flow stability under the condition of actuator fault by introducing fault upper bound factor. However, the exponential spacing policy is designed based on the constant headway constant of the vehicle control system. When the system is in the steady state, if the velocity is too low and the constant headway constant is too small, the string may be unstable. If the constant headway constant is too large, the traffic flow velocity will also be reduced. The traffic flow instability will occur. Therefore, by considering the influence of physical faults and disturbances, how to improve the inter-vehicle spacing policy and reduce the conservative of the existing inter-vehicle spacing policy has important scientific significance and theoretical value for improving the traffic flow stability while ensuring the string stability of the vehicle platoon system. In addition, on the basis of balancing the relationship between the string stability and traffic flow stability, how to design the platoon resilient controller to improve the fault recovery response speed and fault tolerance performance of closed-loop system is still a challenging problem.

Based on the above analysis, this chapter focuses on the distributed adaptive finite time platoon resilient control of heterogeneous under the improved exponential spacing policy [2]. The main contributions of this chapter are listed as follows. Firstly, based on Chap. 3, an improved exponential spacing strategy is proposed by taking fault factors into account to improve the string stability and traffic flow stability of the vehicle platoon system. At the same time, the design limitations of the time headway interval constant [1, 3] of the quadratic spacing policy are also removed. In addition, the designed improved spacing policy also eliminates the assumption of zero initial spacing error. Secondly, a nonlinear observer is designed to estimate the redundant errors caused by saturation approximation, bias fault and disturbances. Different from the traditional disturbance observer in [4], the designed observer can further ensure the stability of the closed-loop system through the error compensator. Finally, different from the resilient control methods in the sense of uniformly bounded stability, this chapter designs a distributed adaptive finite-time platoon resilient control strategy. The influence of actuator fault, saturation and disturbances are compensated based on the adaptive update law of unknown parameters. Thus, the closed-loop global finite-time stability of the vehicle platoon system is realized. In addition, the effectiveness of the improved exponential spacing policy is analyzed. It is proved that it can effectively improve the of the traffic flow stability and ensure the string stability of the platoon system.

## 4.1 Introduction

Especially, as one part of intelligent transportation systems and connected vehicle highway systems, the vehicular platoon system takes a great advantage in reducing fuel or energy consumption, increasing the transportation throughput, mitigating the urban traffic congestion, and so on [5]. The vehicular platoon system not only requires that the intelligent connected vehicles from the same lane should be organized automatically into the platoons. But also a predefined small inter-vehicle spacing should be kept [6]. The basic control components of the vehicular platoon system [7] have been extensively studied such as homogeneous and heterogeneous control, the spacing policy, information flow topology, string stability, traffic flow stability, and so on.

Due to the inter-vehicle spacing error of the platoon under high speed condition is more sensitive to fault and disturbances. Collision accidents are easy to occur, resulting in safety losses. As the important characteristic of the vehicle platoon control, string stability guarantees that the bounded disturbances or faults in the downstream of the platoon are not amplified in the upstream. Under the condition of traffic flow stability, the traffic flow reduction and traffic jam will not occur with the bounded disturbances in traffic flow density [8]. When the traffic density is below the critical characteristic density, the gradient of traffic flow rate along the traffic density is positive. It indicates that the traffic flow is stable. Most of existing research works focus on spacing error policy design of vehicular platoon systems control for achieving the string stability and maintaining the smaller inter-vehicle spacing while ensuring higher traffic density. The constant spacing policy can achieve the higher traffic capacity in [9]. However, there exists the restrictive condition that the preceding vehicles acceleration is known in the constant spacing policy. To eliminate restrictions and improve the applicability of the constant spacing policy, the constant time headway policy is studied in [8]. Then, the modified constant time headway policy was proposed to remove the assumption of initial error is zero in [10]. Although the above policies have advantages in ensuring the string stability, these policies based on the constant time headway may lead to instability and lower traffic flow and even cause jams. In [11], the exponential spacing policy was firstly studied, and the traffic flow instability was proved under the constant time headway spacing policy since the traffic flow gradient is always negative. Based on the above policy, the maximum acceleration of intelligent connected vehicle will be changed due to the actuator fault issue was not involved. For this issue, the improved quadratic spacing policy with the lower bound of fault factor was developed in [3] based on the time headway constant of the control system. However, when the time headway constant is set low enough, the higher risk will increase of the string unstability at low velocities. When the time headway constant is set high enough, the traffic flow rate of the vehicle platoon system will decrease [12, 13].

The studied vehicular platoon system usually is highly vulnerable to the complex environment disturbances, which could cause by wind gusts, strong sand, road slopes, and so on. In [14], the disturbance issue of bidirectional vehicle platoons with different asymmetry in position and velocity coupling was solved. In [15], the external disturbances problem of the vehicle platoons was dealt with by the delay-based spacing policy. For the issue of the heterogeneous vehicle platoons on curved roads subject to disturbances including the up-down road slopes, the integrated robust control scheme and modeling method were proposed in [16]. In addition, physical device safety and the architectural constraints, such as actuator input saturation, are common in the vehicular platoon system. A nonlinear controller was designed for the input saturation problem of the platoon of vehicles in [17]. In [18], a distributed sliding mode control method was presented for the actuator saturation issue of the vehicular platoons. For dealing with the problems of the actuator input saturation or disturbances in the vehicular platoons, the existing research works have developed a lot of control schemes. However, based on the issues of both input saturation and disturbances, the vehicle platoon system is also extremely sensitive and vulnerable to the actuator fault. In the real engineering practice environment, the actuator of the intelligent connected vehicles is prone to physical fault because of the problems of motor aging, flux-weakening, wear out, overcurrent, overheat and overvoltage, and so on [19]. Based on the aforementioned analysis, it can be concluded that the control problem of the vehicular platoons against the actuator fault, input saturation, and disturbances is still not completely solved and deserves further study.

Through the finite time control and sliding mode control method, the string stability and resilience performance of the vehicle platoon can be heavily enhanced. Finite time control scheme takes great advantages in transient performance of closed-loop system such as fast stability, and has been studied extensively (see [20, 21]). Sliding mode control method have strong robustness and resilience performance and insensitivity against actuator fault and disturbances. In [22], for the vehicle platoon with the nonlinearities and disturbances, asymptotic stability based on the sliding mode control was ensured in [22]. In [23], an adaptive sliding mode control method was proposed to realize the string stability in the platoon with the uncertain interaction topology. A sliding mode control approach based on the fuzzy logic system was developed to maintain the string stability of the vehicle platoon under the parameter uncertainties and disturbances in [24]. However, it is still an open and hot issue to design the sliding mode control method and ensure the string stability performance in the sense of finite-time for the vehicle platoon system against the actuator fault, input saturation, and external disturbances.

## 4.2 Problem Statement

Compared with the study of vehicle platoon system composed of homogeneous intelligent vehicles, the study of heterogeneous vehicle platoon system control is closer to the engineering practice by considering different dynamic parameters of intelligent connected vehicles. In this chapter, the heterogeneous vehicle platoon system control scenario is shown in Fig. 4.1. This chapter is based on the fact that the heterogeneous vehicle platoon system is composed of $n + 1$ intelligent connected vehicles driving

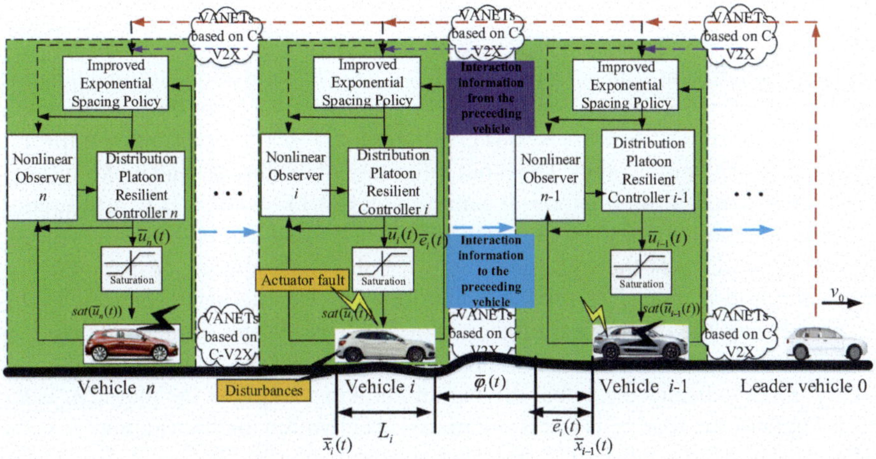


**Fig. 4.1** The control structure of heterogeneous vehicle platoon system scenario

on the rugged road. Based on the actuator fault, input saturation, and external disturbances problem of the heterogeneous vehicle platoon system, a distributed adaptive platoon finite time resilient control scheme including improved exponential spacing policy, nonlinear observer and distributed controller is designed in the control layer of the system. Furthermore, information communication interaction is carried out in the communication layer of the platoon system. Based on VANETs under C-V2X communication technology, the bi-directional information flow topology with the leader is adopted to carry out inter-vehicle information exchange. That is, intelligent connected vehicle $i$ can communicate with the leader vehicle 0, the follower vehicle $i-1$ and the follower vehicle $i+1$.

As shown in Eq. (4.1), considering the longitudinal platoon dynamics model of heterogeneous vehicles against external disturbances caused by environmental factors such as wind gusts and bumpy road surfaces. All intelligent connected vehicles have different dynamic parameters, such as vehicle mass, inertial delay, mechanical efficiency and so on. In engineering practice, the problem of actuator input saturation constraints is inevitable and needs to be considered due to the safety constraints and physical structure limitations of the throttle/brake system of the vehicles. The yaw angle and lateral motion of the vehicle are not considered in this chapter.

$$\begin{aligned}
\dot{\bar{x}}_i(t) &= \bar{v}_i(t) \\
\dot{\bar{v}}_i(t) &= \frac{\eta_i \bar{F}_i(t)}{R_i M_i} - \frac{\rho_a A_i C_{ai} \bar{v}_i^2}{2M_i} - g f_i \cos(\bar{\delta}_i) - g \sin(\bar{\delta}_i) + \bar{d}_i(t) \\
sat(\bar{u}_i(t)) &= \tau_i \dot{\bar{F}}_i(t) + \bar{F}_i(t)
\end{aligned} \tag{4.1}$$

where $sat(\bar{u}_i(t)) = \begin{cases} sign(\bar{u}_i(t))\bar{u}_{bi}, & |\bar{u}_i(t)| \geq \bar{u}_{bi} \\ \bar{u}_i(t), & |\bar{u}_i(t)| < u_{bi} \end{cases}$, and $i = 1, 2, \ldots, n$ represent the $i$th intelligent connected vehicle. $\bar{x}_i(t)\bar{v}_i(t)$ denote the longitudinal position and velocity of $i$th intelligent connected vehicle, respectively. For the leader vehicle 0, the dynamic model is $\dot{\bar{x}}_0(t) = \bar{v}_0(t)$, $\dot{\bar{v}}_0(t) = \bar{a}_0(t)$. $\bar{u}_i(t)$ represents the throttle and brake control torque input. The input saturation of the throttle and brake systems is also considered in the platoon dynamic model (4.1). $sat(\bar{u}_i(t))$ is a saturation nonlinear function affected by the input signal of throttle and brake control torque $\bar{u}_i(t)$, $sign(\bullet)$ denotes the sign function, $u_{bi}$ is the constraint upper parameter for input saturation. $\bar{a}_i(t) = \dot{\bar{v}}_i(t)$is the acceleration of the $i$th intelligent connected vehicle. $\eta_i$ indicates the mechanical efficiency of the drive system of the vehicles, $\tau_i$ is the inertia delay of the motor. $M_i$ represents the mass of the vehicles, $\rho_a$is the air density. $\bar{F}_i(t)$ indicates the actual driving and braking torque of the motor. $\bar{\delta}_i$ is the slope Angle of the road. $g$ is the gravitational acceleration coefficient, and $f_i$ is the rolling resistance coefficient. $C_{ai}$ is the aerodynamic drag coefficient, $A_i$ is the front cross-sectional area of the vehicle. $R_i$ represents the tire radius of the vehicle. $\bar{d}_i$ is the external disturbance function, such as parameter disturbance, gust effect, etc.

In engineering practice, the actuator of intelligent connected vehicles may lose part of its efficiency and occur faults due to wear out, aging, weakening of magnetic flux, overvoltage, overcurrent or overheating and other problems [3, 19]. These problems may seriously affect the control performance of the closed-loop system and even destroy the stability of the vehicle platoon system. When the actuator fault occurs on the $i$th vehicle, the mathematical form of its control input torque is

$$\bar{u}_{fi}(t) = (1 - \lambda_i)sat(\bar{u}_i(t)) + \bar{\sigma}_i\bar{\Delta}_i(t) \tag{4.2}$$

where $0 < \lambda_{i1} \leq \lambda_i \leq \lambda_{i2} < 1$ denote the actuator fault factor of loss of effectiveness, $\lambda_{i1}$ and $\lambda_{i2}$ is the lower bound and upper bound of $\lambda_i$, respectively. $\bar{\sigma}_i \in \{0, 1\}$, when $\bar{\sigma}_i = 1$, the $i$th vehicle has an actuator bias fault, and there is no faults occur when $\bar{\sigma}_i = 0$. $\bar{\Delta}_i(t)$denotes the bias fault function, and meet the condition that $\bar{\Delta}_i \leq \bar{\bar{\Delta}}_i$,where $\bar{\bar{\Delta}}_i$ is the unknown positive constant.

Through Eqs. (4.1) and (4.2), the longitudinal dynamics model of vehicle platoon system can be further expressed as

$$\begin{aligned} \dot{\bar{a}}_i(t) = & \frac{\eta_i}{M_i R_i \tau_i}(1 - \lambda_i)sat(\bar{u}_i(t)) - \frac{\bar{a}_i(t)}{\tau_i} + \bar{D}_i \\ & - \frac{1}{M_i \tau_i}\left[\rho_a A_i C_{ai}\left(\frac{1}{2}\bar{v}_i^2 + \tau_i \bar{v}_i \bar{a}_i(t)\right) + \bar{\Xi}_i\right] + \frac{\eta_i \bar{\sigma}_i \bar{\Delta}_i}{M_i R_i \tau_i} \end{aligned} \tag{4.3}$$

where $\bar{D}_i = (\bar{d}_i/\tau_i) + \dot{\bar{d}}_i$, and $|(M_i R_i \tau_i D_i)/\eta_i| \leq \bar{D}_i$, $\bar{\Xi}_i = M_i g f_i \cos(\bar{\delta}_i) + M_i g \sin(\bar{\delta}_i)$, $\bar{D}_i$ is the unknown positive constant, and $\bar{\Xi}_i$ is the road slope function.

If $|\bar{u}_i(t)| = \bar{u}_{bi}$ is true, there exists the unsmooth phenomenon between the function $sat(\bar{u}_i(t))$ and the control input torque signal $\bar{u}_i(t)$. The saturation function $sat(\bar{u}_i(t))$ can be approximated by the smooth hyperbolic tangent function $h(\bar{u}_i(t)) = \bar{u}_{bi}\tanh(\bar{u}_i(t)/\bar{u}_{bi})$. Thus, it can be obtained that $sat(\bar{u}_i(t)) = h(\bar{u}_i(t)) + \bar{\ell}$, $\bar{\ell}$ is the approximation error, and satisfies the condition that $\left|\bar{\ell}\right| = |sat(\bar{u}_i(t)) - h(\bar{u}_i(t))| \leq \bar{u}_{bi}(1 - \tanh(1)) = \bar{o}_i$, $\bar{o}_i$ is a positive definite constant.

By using the mean value theorem, $h(\bar{u}_i(t))$ can be expressed as

$$h(\bar{u}_i) = h_{\bar{u}_i}(t)(\bar{u}_i - \bar{u}_{i0}) + h(\bar{u}_{i0}) \tag{4.4}$$

where $h_{\bar{u}_i}(t) = \partial h(\bar{u}_i)/\partial\bar{u}_i|_{\bar{u}_i=\bar{u}_{\mu i}} = 4/(e^{\bar{u}_i/\bar{u}_{bi}} + e^{-\bar{u}_i/\bar{u}_{bi}})^2|_{\bar{u}_i=\bar{u}_{\mu i}}$, $h_{\bar{u}_i}(t) \leq 1$. If $\bar{u}_{i0} = 0$ is true, then $h(0) = 0$, and $\bar{u}_{\mu i} = \mu\bar{u}_i(t) + (1-\mu)\bar{u}_{i0}(0 < \mu < 1)$. Based on formula (4.4), the following relationship exists

$$sat(\bar{u}_i(t)) = h_{\bar{u}_i}(t)\bar{u}_i(t) + \bar{\ell} \tag{4.5}$$

Based on formula (4.5), the dynamics model of the \$i\$th vehicle in the platoon system (4.3) is

$$\begin{aligned}\dot{\bar{a}}_i(t) = &\frac{\eta_i}{M_iR_i\tau_i}(1-\lambda_i)h_{\bar{u}_i}(t)\bar{u}_i(t) - \frac{1}{\tau_i}\bar{a}_i(t) + \frac{\eta_i}{M_iR_i\tau_i}\bar{\Lambda}_i \\ &- \frac{1}{M_i\tau_i}\left[\rho_aA_iC_{ai}\left(\frac{1}{2}\bar{v}_i^2 + \tau_i\bar{v}_i\bar{a}_i(t)\right) + \Xi_i\right]\end{aligned} \tag{4.6}$$

where $\bar{\Lambda}_i = (1-\lambda_i)\bar{\ell} + \bar{\sigma}_i\bar{\Delta}_i(t) + M_iR_i\tau_i\bar{D}_i/\eta_i$.

**Remark 4.1** In this section, firstly, a nonlinear longitudinal dynamics model of heterogeneous intelligent vehicle with actuator saturation constraint is presented in Eq. (4.1). The actuator physical fault model (4.2) of the $i$th vehicle is considered in the longitudinal dynamics model (4.3). However, since there is an unsmooth phenomenon between the function $sat(\bar{u}_i(t))$ and the control input torque signal $\bar{u}_i(t)$, the smooth hyperbolic tangent function $h(\bar{u}_i(t))$ is used to approximate the saturation function $sat(\bar{u}_i(t))$ in Eqs. (4.4) and (4.5). Finally, the longitudinal dynamics model of the $i$th vehicle in the platoon system with actuator physical faults, input saturation and external disturbances was established in Formula (4.6), which will be used for the subsequent design of distributed adaptive finite-time platoon resilient control schemes.

**Assumption 3.1** Based on the above analysis, it can be seen that $\left|\bar{\ell}\right| \leq \bar{o}_i$, $\bar{\Delta}_i \leq \bar{\bar{\Delta}}_i$ and $\left|M_iR_i\tau_i\bar{D}_i/\eta_i\right| \leq \bar{\bar{D}}_i$, $\bar{\Lambda}_i(t)$ can be derived as a bounded function. Then, it is reasonable to assume that there is an unknown positive definite constant $\bar{\bar{\Lambda}}_i$ that makes $\left|\dot{\bar{\Lambda}}_i\right| \leq \bar{\bar{\Lambda}}_i$.

The following lemmas and definitions are given for control design and stability analysis.

**Definition 4.1** (*String stability*) [9, 15] If for any given $w > 0$, there is a constant $g > 0$ that makes $\sup_i \max\left\{|\bar{e}_i(0)|, \left|\dot{\bar{e}}_i(0)\right|\right\} < g \Rightarrow \sup_i \sup_{t \geq 0} \max\left\{|\bar{e}_i(t)|, \left|\dot{\bar{e}}_i(t)\right|\right\} < w$ true, then the vehicle platoon system as shown in Eq. (4.1) is string stable, where $\bar{e}_i(t)$ is the inter-vehicle spacing error, $i = 1, 2, \ldots, n$ represents the $i$th vehicle.

**Definition 4.2** ([11]) If the gradient $\partial F/\partial g$ is positive definite and the condition $\partial F/\partial g > 0$ can be met, then the traffic flow is stable, where the traffic flow rate is $F$, and $g$ represents the traffic flow density

**Definition 4.3** ([3, 9]) If for a given arbitrary constant $w > 0$, and there is positive constant $g > 0$ makes $\sup_i \max\left\{|\bar{e}_i(0)|, \left|\dot{\bar{e}}_i(0)\right|\right\} < g \Rightarrow \sup_i \sup_{t \geq 0} \max\{|\bar{e}_i(t)|, \left|\dot{\bar{e}}_i(t)\right|\} < w$ true, $i = 1, 2, \ldots, n$ represents the $i$th vehicle, then the inter-vehicle spacing error of the platoon system is uniformly bounded.

**Control objective** The control objectives of heterogeneous vehicle platoon system affected by the actuator physical faults, input saturation constraints and external disturbances are as follows:

(1) The stability of each follower vehicle in the vehicle platoon system is guaranteed. The velocity and acceleration of each follower vehicle can quickly be consistent with the leader vehicle 0. At the same time, in the process of platoon system control, the ideal inter-vehicle spacing distance can be achieved, then $\lim_{t \to \infty} |\bar{e}_i(t)| = 0$, $\lim_{t \to \infty} |\bar{v}_{i-1}(t) - \bar{v}_i(t)| = 0$, and $\lim_{t \to \infty} |\bar{a}_{i-1}(t) - \bar{a}_i(t)| = 0$.
(2) The traffic flow stability of the heterogeneous vehicle platoon system can be ensured, that is $\partial F/\partial g > 0$.
(3) The string stability of the entire closed-loop vehicle platoon system can be satisfied, that is the inter-vehicle spacing distance error of the follower vehicle $\bar{e}_i(t)$ and $\dot{\bar{e}}_i(t)$ are uniformly bounded and meets the conditions $|\bar{e}_i(t)| < \varepsilon$, $\left|\dot{\bar{e}}_i(t)\right| < \varepsilon$, where $\varepsilon > 0$. The closed-loop finite-time stability of heterogeneous vehicle platoon system can be realized.

## 4.3 Resilient Control Design

Figure 4.1 shows the contents of this section. When heterogeneous vehicle platoon system suffers from the external disturbances, actuator physical fault and input saturation constraint, based on the improved exponential spacing policy, a nonlinear observer is designed to estimate the saturation approximation error, bias fault and external disturbances. On this basis, the distributed adaptive platoon finite-time resilient control strategy is designed. Finally, the closed-loop system bounded stability, finite-time stability, string stability and traffic flow stability of closed-loop vehicle platoon system are analyzed and proved.

### 4.3.1 Designe of Improved Exponential Policy

For removing the limitations of the time headway constant design of the control system in the existing quadratic spacing policy [1, 3] and overcoming the assumption of zero initial spacing error in the traditional exponential spacing policy [11], in this section, the upper bound of the fault factor and the velocity information of the leader vehicle are introduced. An improved exponential spacing policy is designed to ensure the string stability and high traffic flow density of heterogeneous vehicle platoon system (4.6). The improved exponential spacing policy is

$$\begin{aligned} \bar{e}_i(t) &= \overset{\frown}{\bar{e}}_i(t) - \bar{\Pi}_i(t) \\ \overset{\frown}{\bar{e}}_i(t) &= \bar{x}_{i-1}(t) - \bar{x}_i(t) - L_i - \bar{\varphi}_d - \frac{\varepsilon(\bar{v}_i^2 - \bar{v}_0^2)}{2\phi(1-\lambda_{i2})} - \bar{\kappa}\left(1 - \exp\left(-\frac{\overset{\frown}{\bar{v}}_i}{\bar{\zeta}}\right)\right) \end{aligned} \tag{4.7}$$

where $\bar{\Pi}_i(t) = (\overset{\frown}{\bar{e}}_i(0) + [\varsigma_i \overset{\frown}{\bar{e}}_i(0) + \dot{\overset{\frown}{\bar{e}}}_i(0)]t + 0.5[\varsigma_i^2 \overset{\frown}{\bar{e}}_i(0) + 2\varsigma_i \dot{\overset{\frown}{\bar{e}}}_i(0) + \ddot{\overset{\frown}{\bar{e}}}_i(0)]t^2) \times \exp(-\varsigma_i t)$, $\bar{e}_i(t)$ is the inter-vehicle spacing error, $\overset{\frown}{\bar{v}}_i(t) = \bar{v}_i(t) - \bar{v}_0(t)$, $\bar{\varphi}_i = \bar{x}_{i-1}(t) - \bar{x}_i(t) - L_i$. $L_i$ the length of the vehicle, $\bar{\varphi}_d$ represents the ideal inter-vehicle distance between two adjacent vehicles in the platoon system, $\varepsilon$ is the safety factor based on road environment, $\phi$ is the maximum deceleration of the vehicle. $\bar{\kappa}$ and $\bar{\zeta}$ both are positive definite constants. $\varsigma_i$ is strictly positive definite constant. Further, according to Eq. (4.7), $\bar{e}_i(0) = 0$, $\dot{\bar{e}}_i(0) = 0$, and $\ddot{\bar{e}}_i(0) = 0$, which show that by the introduced function $\bar{\Pi}_i(t)$, for any non-zero initial inter-vehicle spacing error $\overset{\frown}{\bar{e}}_i(t)$, $\bar{e}_i(t)$ can always be attenuated to zero.

**Remark 4.2** In [11], the traditional exponential spacing policy is designed to achieve the string stability and traffic flow stability. However, the hypothesis of zero initial spacing error and the threat of string instability caused by physical fault and disturbances are not considered. In order to achieve string stability and traffic flow stability under fault conditions, in [1], an improved quadratic spacing policy is proposed. However, there are some limitations in the design of time headway constant of the control system. When the time headway parameter is too small, the stability of the system can not be guaranteed at low velocity. When the time headway parameter is too large, the traffic flow capacity will decrease. Therefore, for improving the above shortcomings, in this chapter, an improved exponential spacing policy is proposed. By introducing the function $\bar{\Pi}_i(t)$ into the Eq. (4.7), any non-zero initial spacing error $\overset{\frown}{\bar{e}}_i(t)$, $\bar{e}_i(t)$ can always be attenuated to zero. Different from the policy in [1], the upper bound of the fault factor and the velocity information of the leader's vehicle are introduced in (4.7). No matter the heterogeneous vehicle platoon system drives at high velocity or low velocity, the exponential function terms can always ensure the closed-loop system robustness, string stability and high traffic flow rate under the influence of physical faults, input saturation and disturbances.

### 4.3.2 Design of Nonlinear Observer

In this section, a nonlinear observer is designed to estimate the function term $\bar{\Lambda}_i$ in Eq. (4.6). The form of the designed nonlinear observer is as follows:

$$
\begin{aligned}
\dot{W}_i &= l_i \left\{ \bar{a}_{i-1}(t) - \bar{a}_i(t) - \frac{\varepsilon(\bar{a}_i^2 - \bar{a}_0^2)}{\phi(1-\lambda_{i2})} + \bar{\Phi}_i^1 \frac{\bar{a}_i(t)}{\tau_i} - \bar{\Phi}_i^1 \frac{\eta_i \hat{\bar{\Lambda}}_i}{M_i R_i \tau_i} \right\} \\
&\quad + W_i^1 - l_i \ddot{\bar{\Pi}}_i - \bar{\wp}_i \\
\hat{\bar{\Lambda}}_i &= W_i - l_i \dot{\bar{e}}_i(t)
\end{aligned}
\tag{4.8}
$$

where $\hat{\bar{\Lambda}}_i$ is the estimation of the function $\bar{\Lambda}_i$, $W_i^1 = l_i \bar{\kappa}\, \dot{\widehat{\bar{v}}}_i^2 \bar{\xi}^{-2} \exp(-\,\widehat{\bar{v}}_i\, \bar{\xi}^{-1}) - l_i \bar{\Phi}_i^1 \frac{\eta_i(1-\lambda_i)h_{\bar{u}_i}(t)\bar{u}_i(t)}{M_i R_i \tau_i} + l_i \bar{\Phi}_i^1 \frac{[\rho_a A_i C_{ai}(0.5\bar{v}_i^2 + \tau_i \bar{v}_i \bar{a}_i) + \bar{\Xi}_i]}{M_i \tau_i}$, $W_i$ is the internal state variable, $l_i > 0$ is the observer gain, and $\bar{\Phi}_i^1 = \frac{\varepsilon \bar{v}_i}{\phi(1-\lambda_{i2})} + \bar{\kappa} \bar{\zeta}^{-1} \exp(-\,\widehat{\bar{v}}_i\, \bar{\zeta}^{-1})$.

The error compensator for the constructed the observer (4.8) is

$$
\bar{\wp}_i = l_i \bar{\Phi}_i^1 \frac{\eta_i}{M_i R_i \tau_i} \tilde{\bar{\Lambda}}_i - \tilde{\bar{\Lambda}}_i \tag{4.9}
$$

where $\bar{\wp}_i$ is the compensation of the observer estimation error, which can improve the observation accuracy and ensure the stability of the observer. $\tilde{\bar{\Lambda}}_i = \bar{\Lambda}_i - \hat{\bar{\Lambda}}_i$ is an observation error.

**Theorem 4.1** *By means of the designed nonlinear observer (4.8) and the error compensator (4.9), the observation errors can be guaranteed the convergence performance and kept within the domain* $Q_1 = \{\tilde{\bar{\Lambda}}_i : \tilde{\bar{\Lambda}}_i \leqslant \sqrt{\tilde{\bar{\Lambda}}_i(0) + 2\bar{b}_{\wp}}\}$, $\bar{b}_{\wp} = (1/2)\bar{\bar{\Lambda}}_i^{\,2}$.

**Remark 4.3** The function term $\bar{\Lambda}_i$ is composed of saturation approximation error, bias fault and external disturbances. and the upper bound of $\bar{\Lambda}_i$ is usually not easy to obtain, so the traditional adaptive compensation law method will not be applicable. Compared with the traditional disturbance observer such as in [4], the nonlinear observer designed in this section can estimate the redundant function term and construct the observation error compensation term to ensure the stability of the observer.

***Proof*** By taking the time derivative of $\hat{\bar{\Lambda}}_i$, and based on Eqs. (4.6), (4.8), and (4.9), it can be derived,

$$
\dot{\hat{\bar{\Lambda}}}_i = l_i \bar{\Phi}_i^1 \frac{\eta_i \tilde{\bar{\Lambda}}_i}{M_i R_i \tau_{i\,i}} - \bar{\wp}_i \tag{4.10}
$$

For the $i$th vehicle, by selecting the Lyapunov candidate function $V_{i\bar{\Lambda}} = (1/2)\tilde{\bar{\Lambda}}_i^2$, we can get

$$\begin{aligned}
\dot{V}_{i\tilde{\Lambda}} &= -l_i\bar{\Phi}_i^1\frac{\eta_i}{M_iR_i\tau_i}\tilde{\bar{\Lambda}}_i^2 + \tilde{\bar{\Lambda}}_i\bar{\wp}_i + \tilde{\bar{\Lambda}}_i\dot{\bar{\Lambda}}_i \\
&= -l_i\bar{\Phi}_i^1\frac{\eta_i\tilde{\bar{\Lambda}}_i^2}{M_iR_i\tau_i} + \tilde{\bar{\Lambda}}_i\left(l_i\bar{\Phi}_i^1\frac{\eta_i\tilde{\bar{\Lambda}}_i}{M_iR_i\tau_i} - \tilde{\bar{\Lambda}}_i\right) + \tilde{\bar{\Lambda}}_i\dot{\bar{\Lambda}}_i \\
&= -\tilde{\bar{\Lambda}}_i\tilde{\bar{\Lambda}}_i + \tilde{\bar{\Lambda}}_i\dot{\bar{\Lambda}}_i
\end{aligned} \tag{4.11}$$

By using Young's inequality, it can be deduced

$$\begin{aligned}
\dot{V}_{i\tilde{\Lambda}} &\leqslant -\frac{1}{2}\tilde{\bar{\Lambda}}_i\tilde{\bar{\Lambda}}_i + \frac{1}{2}\dot{\bar{\Lambda}}_i^{\,2} \\
&\leqslant -\frac{1}{2}\left|\tilde{\bar{\Lambda}}_i\right|^2 + \bar{b}_{\wp}
\end{aligned} \tag{4.12}$$

where $\dot{V}_{i\tilde{\Lambda}} \leqslant -V_{i\tilde{\Lambda}} + \bar{b}_{\wp}$, after a long enough stable time, we can be obtained that $V_{i\tilde{\Lambda}} \leqslant (V_{i\tilde{\Lambda}}(0) - \bar{b}_{\wp})e^{-t} + \bar{b}_{\wp} \leqslant V_{i\tilde{\Lambda}}(0) + \bar{b}_{\wp}$. So we know that $V_{i\tilde{\Lambda}}$ is ultimately uniformly bounded. The observation error $\tilde{\bar{\Lambda}}_i$ will also converge and remain within the domain $Q_1 = \{\tilde{\bar{\Lambda}}_i : \tilde{\bar{\Lambda}}_i \leqslant \sqrt{\tilde{\bar{\Lambda}}_i(0) + 2\bar{b}_{\wp}}\}$. □

### 4.3.3 Design of Distributed Finite-Time Resilient Controller

For the $i$th vehicle in the heterogeneous vehicle platoon system, the non-singular fast terminal sliding mode surface is designed to reduce the transient errors and steady-state errors caused by physical faults, input saturation and external disturbances. The designed sliding mode surface is

$$r_i(t) = \dot{\bar{e}}_i(t) + \bar{\alpha}_i\bar{e}_i + \int_0^t [\bar{\alpha}_{i1}sign(\bar{e}_i(\tau))|\bar{e}_i(\tau)|^{1+\beta} + \bar{\alpha}_{i2}sign(\bar{e}_i(\tau))|\bar{e}_i(\tau)|^{1-\beta}]d\tau \tag{4.13}$$

where $\bar{\alpha}_i, \bar{\alpha}_{i1}, \bar{\alpha}_{i2}$ is the design positive definite constant, and there is constant $\beta$ such that $0 < \beta < 1, \forall i = 1, 2, \ldots, n$.

However, the sliding surface $r_i(t)$ can only ensure the convergence of the sapcing error of the $i$th vehicle. In order to realize the string stability of heterogeneous vehicle platoon system, this chapter further introduces the coupled sliding mode surface, whose form is

$$J_i(t) = \breve{\lambda}_1 r_i(t) - \breve{\lambda}_2 r_{i+1}(t) \tag{4.14}$$

where $\breve{\lambda}_1, \breve{\lambda}_2$ both are designed positive definite weights, $0 < \breve{\lambda}_1/\breve{\lambda}_2 \leq 1$. For the $n$th vehicle, let $J_{n+1} = 0$, and $r_{n+1} = \dot{r}_{n+1} = 0$. For the $i = 1, 2, \ldots, n$ vehicle, $J(t) = \mathrm{M}r(t)$, the matrix M is invertible, $J(t) = [J_1(t), \ldots, J_n(t)]^T$, $r(t) = [r_1(t), \ldots, r_n(t)]^T$. Based on the above, it can be seen that the coupled sliding mode surface $J_i(t)$ has the same convergence with $r_i(t)$.

Based on the improved exponential spacing policy (4.7), the time derivative of the coupled sliding mode surface $J_i(t)$ (4.14) is taken, $i = 1, 2, \ldots, n-1$, then

$$\dot{J}_i(t) = -\bar{\Phi}_i^1 \dot{\bar{a}}_i(t) + \bar{\Gamma}_i(t) \tag{4.15}$$

where $\bar{\Gamma}_i(t) = \breve{\lambda}_1(\bar{a}_{i-1} - \bar{a}_i - \frac{\varepsilon(\bar{a}_i^2 - \bar{a}_0^2)}{\phi(1-\lambda_{i2})} + \frac{\kappa \dot{\bar{v}}_i^2}{\zeta^2}\exp(-\frac{\widehat{\bar{v}_i}}{\zeta}) - \ddot{\Pi}_i) + \bar{\Gamma}_{iA}(t)$, $\bar{\Gamma}_{iA}(t) = \breve{\lambda}_1(\bar{\alpha}_i \dot{\bar{e}}_i + \bar{\alpha}_{i1} sign(\bar{e}_i)\left|\bar{e}_i\right|^{1+\beta} + \bar{\alpha}_{i2} sign(\bar{e}_i)\left|\bar{e}_i\right|^{1-\beta}) - \breve{\lambda}_2 \dot{r}_{i+1}$.

Secondly, by using Eqs. (4.6) and (4.15), we can get

$$\begin{aligned}
\dot{J}_i(t)J_i(t) = &-\bar{\Phi}_i^1\left[\frac{\eta_i(1-\lambda_i)h_{\bar{u}_i}(t)\bar{u}_i(t)}{M_i R_i \tau_i} - \frac{1}{M_i\tau_i}[\rho_a A_i C_{ai}(\frac{1}{2}\bar{v}_i^2 + \tau_i \bar{v}_i \bar{a}_i) + \bar{\Xi}_i\right. \\
&\left. - \frac{\bar{a}_i(t)}{\tau_i} + \frac{\eta_i \Lambda_i}{M_i R_i \tau_i}]J_i(t) + \bar{\Gamma}_i J_i\right] \\
\le &-\bar{\Phi}_i[\frac{\eta_i(1-\lambda_{i2})h_{\bar{u}_i}\bar{u}_i(t)}{M_i R_i \tau_i} - \frac{[(\rho_a A_i C_{ai}(\frac{1}{2}\bar{v}_i^2 + \tau_i \bar{v}_i \bar{a}_i)) + \bar{\Xi}_i]}{M_i \tau_i} \\
&- \frac{\bar{a}_i(t)}{\tau_i} + \frac{\eta_i \bar{\Lambda}_i}{M_i R_i \tau_i}]J_i(t) + (1-\lambda_{i2})\bar{\Theta}_i\left|\bar{\Gamma}_i J_i\right|
\end{aligned} \tag{4.16}$$

where $1 \le 1/(1-\lambda_{i2}) \le \bar{\Theta}_i$, $\bar{\Phi}_i = \breve{\lambda}_1 \frac{\varepsilon \bar{v}_i}{\phi(1-\lambda_{i2})}$.

For the rear vehicle $i = n$ in the heterogeneous vehicle platoon system, the time derivative $J_n$ is obtained,

$$\dot{J}_n(t) = -\bar{\Phi}_n \dot{\bar{a}}_n(t) + \dot{\Gamma}_n(t) \tag{4.17}$$

where $\bar{\Gamma}_n(t) = \breve{\lambda}_1(\bar{a}_{n-1} - \bar{a}_n - \frac{\varepsilon(\bar{a}_n^2 - \bar{a}_0^2)}{\phi(1-\lambda_{n2})} + \frac{\kappa \dot{\widehat{\bar{v}_n^2}}}{\zeta^2}\exp(-\frac{\widehat{\bar{v}_n}}{\zeta}) - \ddot{\Pi}_n + \bar{\Gamma}_{nA}(t)$, $\bar{\Gamma}_{nA}(t) = \bar{\alpha}_n \dot{\bar{e}}_n + \bar{\alpha}_{n1} sign(\bar{e}_n)\left|\bar{e}_n\right|^{1+\beta} + \bar{\alpha}_{n2} sign(\bar{e}_n)\left|\bar{e}_n\right|^{1-\beta})$.

Therefore, based on the bi-directional information flow topology with leader, a distributed adaptive platoon finite time resilient controller is constructed for the heterogeneous vehicle platoon system $i = 1, 2, \ldots, n-1$,

$$\begin{aligned}
\bar{u}_i(t) = &\hat{\bar{\mathrm{H}}}_i\left\{\frac{\hat{\bar{k}}_{1i}}{\bar{\Phi}_i \bar{\mathrm{B}}_i}\tanh\left(\frac{J_i}{\mu_i}\right) + \frac{\bar{k}_{2i}J_i}{\bar{\Phi}_i \bar{\mathrm{B}}_i} + \frac{(1-\lambda_{i2})\bar{\Theta}_i}{\bar{\Phi}_i \bar{\mathrm{B}}_i}\frac{\Gamma_i^2 J_i}{|\Gamma_i J_i| + \bar{b}_{oi}}\right\} \\
&+ \hat{\bar{\mathrm{H}}}_i\left\{-\frac{\eta_i \hat{\bar{\Lambda}}_i}{M_i R_i \tau_i \bar{\mathrm{B}}_i} + \frac{1}{\bar{\mathrm{B}}_i M_i \tau_i}\left[(\rho_a A_i C_{ai}(\frac{\bar{v}_i^2}{2} + \tau_i \bar{v}_i \bar{a}_i)) + \hat{\bar{\Xi}}_i\right] + \frac{\bar{a}_i}{\bar{\mathrm{B}}_i \tau_i}\right\}
\end{aligned} \tag{4.18}$$

where $\hat{\bar{\mathrm{H}}}_i$, $\hat{\bar{\Xi}}_i$ and $\hat{\bar{k}}_{1i}$ are the estimations of $\bar{\mathrm{H}}_i$, $\bar{\Xi}_i$ and $\bar{k}_{1i}$, respectively. $\bar{k}_{1i}$ is an unknown positive definite control parameter, $\bar{\mathrm{H}}_i = h_{\bar{u}_i}^{-1}$, $\bar{\mathrm{B}}_i = \frac{\eta_i(1-\lambda_{i2})}{M_i R_i \tau_i}$, $\bar{b}_{oi} > 0$ and $\mu_i$ are all very small positive definite constants.

In the joint formula (4.17), a distributed adaptive platoon finite time resilient controller is constructed for the heterogeneous vehicle platoon system $i = n$,

$$\bar{u}_n(t) = \hat{\bar{\mathrm{H}}}_n \left\{ \frac{\hat{\bar{k}}_{1n}}{\bar{\Phi}_n \bar{\mathrm{B}}_n} \tanh\left(\frac{J_n}{\mu_n}\right) + \frac{(1-\lambda_{n2})\bar{\Theta}_n}{\bar{\Phi}_n \bar{\mathrm{B}}_n} \frac{\bar{\Gamma}_n^2 J_n}{|\bar{\Gamma}_n J_n| + \bar{b}_{on}} + \frac{\bar{a}_n}{\bar{\mathrm{B}}_n \tau_n} \right\}$$
$$+ \hat{\bar{\mathrm{H}}}_n \left\{ -\frac{\eta_n \hat{\bar{\Lambda}}_n}{M_n R_n \tau_n \bar{\mathrm{B}}_n} + \frac{\bar{k}_{2n} J_n}{\bar{\Phi}_n \bar{\mathrm{B}}_n} + \frac{[(\rho_a A_n C_{an}(\frac{\bar{v}_n^2}{2} + \tau_n \bar{v}_n \bar{a}_n)) + \hat{\bar{\Xi}}_n]}{\bar{\mathrm{B}}_n M_n \tau_n} \right\} \tag{4.19}$$

For the autonomous vehicle $i = 1, 2, \ldots, n$ of heterogeneous vehicle platoon system, the adaptive law is constructed to realize the update and estimation of unknown design constant $\bar{k}_{1i}$, and its expression is

$$\dot{\hat{\bar{k}}}_{1i} = b_{ki} \tanh\left(\frac{J_i}{\mu_i}\right) J_i - \varsigma_{ki} \hat{\bar{k}}_{1i} \tag{4.20}$$

where $\hat{\bar{k}}_{1i}(0) \geq 0$, $\tilde{\bar{k}}_{1i} = \bar{k}_{1i} - \hat{\bar{k}}_{1i}$ represents the estimation error, $b_{ki}$ is a positive definite design constant, and $\varsigma_{ki} > 0$ is a constant.

Because the road slope function is difficult to obtain, and the saturation approximation parameter $h_{\bar{u}_i}$ is difficult to know precisely, it has a great influence on the control performance. Therefore, the adaptive compensation laws of $\hat{\bar{\Xi}}_i$ and $\hat{\bar{\mathrm{H}}}_i$ are constructed,

$$\dot{\hat{\bar{\Xi}}}_i = b_{\Xi i} \frac{\bar{\Phi}_i}{M_i \tau_i} J_i - \varsigma_{\Xi i} \hat{\bar{\Xi}}_i \tag{4.21}$$

$$\dot{\hat{\bar{\mathrm{H}}}}_i = b_{\mathrm{H}i} \bar{\Phi}_i \bar{\mathrm{B}}_i \frac{\bar{u}_i(t)}{\hat{\bar{\mathrm{H}}}_i} J_i - \varsigma_{\mathrm{H}i} \hat{\bar{\mathrm{H}}}_i \tag{4.22}$$

where $\hat{\bar{\mathrm{H}}}_i(0) \geq 0$, $\hat{\bar{\Xi}}_i(0) \geq 0$, $\tilde{\bar{\Xi}}_i = \bar{\Xi}_i - \hat{\bar{\Xi}}_i$, $\tilde{\bar{\mathrm{H}}}_i = \bar{\mathrm{H}}_i - \hat{\bar{\mathrm{H}}}_i$ both are estimation errors. $b_{\mathrm{H}i}$, $b_{\Xi i}$ are positive definite constants. $\varsigma_{\mathrm{H}i} > 0$, $\varsigma_{\Xi i} > 0$ are constants.

**Remark 4.4** Different from the ultimately uniformly bounded stability of the closed-loop system realized in existing literatures such as [1, 3], the proposed Theorem 4.2 can ensure the finite time stability of the closed-loop system. The string stability and traffic flow stability under the improved exponential spacing policy can also be ensured for the heterogeneous vehicle platoon system with physical faults, input saturation and external disturbances.

### 4.3.4 Analysis of System Stability

**Theorem 4.2** *Considering a heterogeneous vehicle platoon system nonlinear longitudinal dynamics model (4.1) with actuator input saturation and external disturbances, which suffers from physical actuator fault (4.2). Based on the improved exponential spacing policy (4.7), the nonlinear observer (4.8), the distributed adaptive platoon finite-time resilient controller (4.18) and (4.19), and the adaptive compensation laws (4.20)–(4.22) are designed. Based on the above design scheme, for all follower vehicles* $i = 1, 2, \ldots, n$ *in the heterogeneous vehicle platoon system, the convergence performance of spacing error* $\bar{e}_i(t)$ *can be ensured by selecting appropriate design parameters. Meanwhile, the closed-loop finite-time stability of the whole heterogeneous vehicle platoon system is guaranteed. When* $\bar{e}_i(t)$ *and* $\dot{\bar{e}}_i(t)$ *are uniformly bounded, the string stability of heterogeneous vehicle platoon system can also be guaranteed. Further, the traffic flow stability is guaranteed.*

***Proof*** For verify the effectiveness of Theorem 4.2, first, the ultimately uniformly bounded stability of the entire heterogeneous vehicle platoon system is proved. For the heterogeneous vehicle platoon system $i = 1, 2, \ldots, n$, the Lyapunov candidate function is constructed as

$$\bar{V} = \left(\sum_{i=1}^{n-1} \bar{V}_i\right) + \bar{V}_n \tag{4.23}$$

where $\bar{V}_i = \frac{1}{2}J_i^2 + \sum_{\widehat{m}=1}^{3} \bar{V}_{i\widehat{m}}$, $\bar{V}_{i1} = \frac{1}{2b_{ki}}\tilde{\bar{k}}_{1i}^2$, $\bar{V}_{i2} = \frac{1}{2b_{\Xi i}}\tilde{\bar{\Xi}}_i^2$, and $\bar{V}_{i3} = \frac{1}{2b_{\mathrm{H}i}}h_{\bar{u}_i}\tilde{\bar{H}}_i^2$.

By taking the time derivative with respect to $\bar{V}$, we get

$$\dot{\bar{V}} = \{\sum_{i=1}^{n-1}(J_i\dot{J}_i + \dot{\bar{V}}_{i1} + \dot{\bar{V}}_{i2} + \dot{\bar{V}}_{i3})\} + \dot{\bar{V}}_n \tag{4.24}$$

Secondly, by formula (4.6), (4.7) and (4.16), it can be obtained that

$$\begin{aligned}\dot{\bar{V}} \le &\left\{\sum_{i=1}^{n-1}\left(-\bar{\Phi}_i\left[\frac{\eta_i(1-\lambda_{i2})h_{\bar{u}_i}\bar{u}_i(t)}{M_iR_i\tau_i} - \frac{\bar{a}_i(t)}{\tau_i} + \frac{\eta_i\bar{\Lambda}_i}{M_iR_i\tau_i}\right]J_i(t)\right)\right\} \\ &+\left\{\sum_{i=1}^{n-1}\left(-\bar{\Phi}_i\left[-\frac{1}{M_i\tau_i}\left[\left(\rho_aA_iC_{ai}\left(\frac{\bar{v}_i^2}{2}+\tau_i\bar{v}_i\bar{a}_i\right)\right)+\bar{\Xi}_i\right]\right]J_i(t)\right)\right\} \\ &+\left\{\sum_{i=1}^{n-1}((1-\lambda_{i2})\bar{\Theta}_i\left|\bar{\Gamma}_iJ_i\right| - \tilde{\bar{k}}_{1i}\dot{\hat{\bar{k}}}_{1i}b_{ki}^{-1} - \tilde{\bar{\Xi}}_i\dot{\hat{\bar{\Xi}}}_ib_{\Xi i}^{-1} - h_{\bar{u}_i}\tilde{\bar{\mathrm{H}}}_i\dot{\hat{\bar{\mathrm{H}}}}_ib_{\mathrm{H}i}^{-1})\right\} + \dot{\bar{V}}_n\end{aligned} \tag{4.25}$$

By estimating the error $\tilde{\bar{\mathrm{H}}}_{\mathrm{i}} = \bar{\mathrm{H}}_i - \hat{\bar{\mathrm{H}}}_{\mathrm{i}}$, Eq. (4.25) can be further expressed as

$$\begin{aligned}\dot{V} \leq &\left\{\sum_{i=1}^{n-1}\left(-\bar{\Phi}_i \frac{\eta_i(1-\lambda_{i2})h_{\bar{u}_i}\bar{\mathrm{H}}_i}{M_i R_i \tau_i}\frac{\bar{u}_i(t)}{\hat{\bar{\mathrm{H}}}_i}J_i + \frac{\bar{\Phi}_i\bar{a}_i(t)J_i}{\tau_i}\right)\right\} \\ &+\left\{\sum_{i=1}^{n-1}\left(\frac{\bar{\Phi}_i}{M_i\tau_i}\left[\left(\rho_a A_i C_{ai}\left(\frac{1}{2}\bar{v}_i^2 + \tau_i\bar{v}_i\bar{a}_i\right)\right) + \bar{\Xi}_i\right]J_i - \frac{\bar{\Phi}_i\eta_i}{M_i R_i \tau_i}\bar{\Lambda}_i J_i\right)\right\} \\ &+\left\{\sum_{i=1}^{n-1}\left((1-\lambda_{i2})\bar{\Theta}_i\left|\bar{\Gamma}_i J_i\right| - \tilde{\bar{k}}_{1i}\dot{\hat{\bar{k}}}_{1i}b_{ki}^{-1} - \tilde{\bar{\Xi}}_i\dot{\hat{\bar{\Xi}}}_i b_{\Xi i}^{-1} - h_{\bar{u}_i}\tilde{\bar{\mathrm{H}}}_i\dot{\hat{\bar{\mathrm{H}}}}_i b_{\mathrm{H}i}^{-1}\right)\right\} \\ &+\left\{\sum_{i=1}^{n-1}\left(\frac{\bar{\Phi}_i\eta_i(1-\lambda_{i2})h_{\bar{u}_i}\tilde{\bar{\mathrm{H}}}_i}{M_i R_i\tau_i}\frac{\bar{u}_i(t)}{\hat{\bar{\mathrm{H}}}_i}J_i\right)\right\} + \dot{V}_n \end{aligned} \tag{4.26}$$

By substituting the adaptive compensation laws (4.18) and (4.19) into the above formula (4.26), we get

$$\begin{aligned}\dot{V} \leq &\left\{\sum_{i=1}^{n}\left(-\bar{\Phi}_i\bar{\mathrm{B}}_i h_{\bar{u}_i}\bar{\mathrm{H}}_i\left[\frac{\hat{\bar{k}}_{1i}}{\bar{\Phi}_i\bar{\mathrm{B}}_i}\tanh\left(\frac{J_i}{\mu_i}\right) + \frac{(1-\lambda_{i2})\bar{\Theta}_i}{\bar{\Phi}_i\bar{\mathrm{B}}_i}\frac{\bar{\Gamma}_i^2 J_i}{\left|\bar{\Gamma}_i J_i\right| + \bar{b}_{oi}} + \frac{\bar{a}_i(t)}{\bar{\mathrm{B}}_i\tau_i}\right]J_i\right)\right\} \\ &-\left\{\sum_{i=1}^{n}\left(\bar{\Phi}_i\bar{\mathrm{B}}_i h_{\bar{u}_i}\bar{\mathrm{H}}_i\left[\frac{\left(\rho_a A_i C_{ai}\left(\frac{1}{2}\bar{v}_i^2 + \tau_i\bar{v}_i\bar{a}_i\right)\right)}{\bar{\mathrm{B}}_i M_i\tau_i} + \frac{\hat{\bar{\Xi}}_i}{\bar{\mathrm{B}}_i M_i \tau_i} - \frac{\eta_i\hat{\bar{\Lambda}}_i}{M_i R_i\tau_i\bar{\mathrm{B}}_i}\right]J_i\right)\right\} \\ &+\left\{\sum_{i=1}^{n}\left(\frac{\bar{\Phi}_i\bar{a}_i(t)J_i}{\tau_i} + \frac{\bar{\Phi}_i}{M_i\tau_i}\left(\left(\rho_a A_i C_{ai}\left(\frac{1}{2}\bar{v}_i^2 + \tau_i\bar{v}_i\bar{a}_i\right)\right) + \bar{\Xi}_i\right)J_i - \bar{k}_{2i}J_i^2\right)\right\} \\ &+\left\{\sum_{i=1}^{n}((1-\lambda_{i2})\bar{\Theta}_i\left|\bar{\Gamma}_i J_i\right| - \tilde{\bar{k}}_{1i}\dot{\hat{\bar{k}}}_{1i}b_{ki}^{-1} - \tilde{\bar{\Xi}}_i\dot{\hat{\bar{\Xi}}}_i b_{\Xi i}^{-1} - h_{\bar{u}_i}\tilde{\bar{\mathrm{H}}}_i\dot{\hat{\bar{\mathrm{H}}}}_i b_{\mathrm{H}i}^{-1})\right\} \\ &+\left\{\sum_{i=1}^{n}\left(-\frac{\bar{\Phi}_i\eta_i\bar{\Lambda}_i J_i}{M_i R_i\tau_i} + \bar{\Phi}_i\bar{\mathrm{B}}_i h_{\bar{u}_i}\tilde{\bar{\mathrm{H}}}_i\frac{\bar{u}_i(t)}{\hat{\bar{\mathrm{H}}}_i}J_i\right)\right\}\end{aligned} \tag{4.27}$$

Then, after deduction and simplification, Eq. (4.27) can be further expressed as

$$\begin{aligned}\dot{V} \leq &\left\{\sum_{i=1}^{n}\left(-\bar{\Phi}_i\bar{\mathrm{B}}_i\left[\frac{\hat{\bar{k}}_{1i}}{\bar{\Phi}_i\bar{\mathrm{B}}_i}\tanh\left(\frac{J_i}{\mu_i}\right)\right]J_i - \bar{k}_{2i}J_i^2 + (1-\lambda_{i2})\bar{\Theta}_i\bar{b}_{oi}\right)\right\} \\ &+\left\{\sum_{i=1}^{n}\left(\bar{\Phi}_i\frac{1}{M_i\tau_i}\tilde{\bar{\Xi}}_i J_i - \bar{\Phi}_i\frac{\eta_i}{M_i R_i\tau_i}\tilde{\bar{\Lambda}}_i J_i + \bar{\Phi}_i\bar{\mathrm{B}}_i h_{\bar{u}_i}\tilde{\bar{\mathrm{H}}}_i\frac{\bar{u}_i(t)}{\hat{\bar{\mathrm{H}}}_i}J_i\right)\right\} \\ &+\left\{\sum_{i=1}^{n}(-\tilde{\bar{k}}_{1i}\dot{\hat{\bar{k}}}_{1i}b_{ki}^{-1} - \tilde{\bar{\Xi}}_i\dot{\hat{\bar{\Xi}}}_i b_{\Xi i}^{-1} - h_{\bar{u}_i}\tilde{\bar{\mathrm{H}}}_i\dot{\hat{\bar{\mathrm{H}}}}_i b_{\mathrm{H}i}^{-1})\right\}\end{aligned} \tag{4.28}$$

However, by substituting the adaptive laws (4.20)–(4.22) into the above Eq. (4.28), it can be derived as

$$\dot{V} \le \left\{ \sum_{i=1}^{n} \left( -\bar{k}_{1i} \tanh\left(\frac{J_i}{\mu_i}\right) J_i - \bar{k}_{2i} J_i^2 + (1-\lambda_{i2})\bar{\Theta}_i \bar{b}_{oi} \right) \right\}$$
$$+ \left\{ \sum_{i=1}^{n} \left( -\frac{\bar{\Phi}_i \eta_i \tilde{\bar{\Lambda}}_i J_i}{M_i R_i \tau_i} + \frac{\varsigma_{ki} \tilde{\bar{k}}_{1i} \hat{\bar{k}}_{1i}}{b_{ki}} + \frac{\varsigma_{\Xi i} \tilde{\bar{\Xi}}_i \hat{\bar{\Xi}}_i}{b_{\Xi i}} + \frac{\varsigma_{Hi} h_{\bar{u}_i} \tilde{\bar{H}}_i \hat{\bar{H}}_i}{b_{Hi}} \right) \right\} \tag{4.29}$$

Further, through inequality reduction and transformation, Eq. (4.29) is rewritten as

$$\dot{V} \le \left\{ \sum_{i=1}^{n} \left( -\bar{k}_{1i} \tanh\left(\frac{J_i}{\mu_i}\right) J_i - \Im_i J_i^2 + (1-\lambda_{i2})\bar{\Theta}_i \bar{b}_{oi} \right) \right\}$$
$$+ \left\{ \sum_{i=1}^{n} \left( \frac{\bar{\Phi}_i \eta_i}{2 M_i R_i \tau_i} \tilde{\bar{\Lambda}}_i^2 + \frac{\varsigma_{ki}}{b_{ki}} \tilde{\bar{k}}_{1i} \hat{\bar{k}}_{1i} + \frac{\varsigma_{\Xi i}}{b_{\Xi i}} \tilde{\bar{\Xi}}_i \hat{\bar{\Xi}}_i + \frac{\varsigma_{Hi}}{b_{Hi}} h_{\bar{u}_i} \tilde{\bar{H}}_i \hat{\bar{H}}_i \right) \right\} \tag{4.30}$$

where $\Im_i = \bar{k}_{2i} - \bar{\Phi}_i \eta_i / 2 M_i R_i \tau_i > 0$.

Meanwhile, based on formula (4.12), (4.30) can be transformed into

$$\dot{V} \le \left\{ \sum_{i=1}^{n} (-\Im_i J_i^2 - \bar{k}_{1i} \tanh\left(\frac{J_i}{\mu_i}\right) J_i + (1-\lambda_{i2})\bar{\Theta}_i \bar{b}_{oi}) \right\}$$
$$+ \left\{ \sum_{i=1}^{n} \left( \frac{(\tilde{\bar{\Lambda}}_i(0) + 2\bar{b}_{\wp})\eta_i \bar{\Phi}_i}{2 M_i R_i \tau_i} + \frac{\tilde{\bar{k}}_{1i} \hat{\bar{k}}_{1i} \varsigma_{ki}}{b_{ki}} + \frac{\tilde{\bar{\Xi}}_i \hat{\bar{\Xi}}_i \varsigma_{\Xi i}}{b_{\Xi i}} + \frac{h_{\bar{u}_i} \tilde{\bar{H}}_i \hat{\bar{H}}_i \varsigma_{Hi}}{b_{Hi}} \right) \right\} \tag{4.31}$$

By using Youngs inequality, and Eqs. (4.20) to (4.22), we can get

$$\dot{V} \le \left\{ \sum_{i=1}^{n} (-\Im_i J_i^2 - \bar{k}_{1i} \tanh\left(\frac{J_i}{\mu_i}\right) J_i + (1-\lambda_{i2})\bar{\Theta}_i \bar{b}_{oi}) \right\}$$
$$+ \left\{ \sum_{i=1}^{n} \left( + \frac{\bar{\Phi}_i \eta_i}{2 M_i R_i \tau_i} (\tilde{\bar{\Lambda}}_i(0) + 2\bar{b}_{\wp}) - \frac{\varsigma_{ki}}{2 b_{ki}} \tilde{\bar{k}}_{1i}^2 + \frac{\varsigma_{ki}}{2 b_{ki}} \bar{k}_{1i}^2 \right) \right\}$$
$$+ \left\{ \sum_{i=1}^{n} \left( -\frac{\varsigma_{\Xi i}}{2 b_{\Xi i}} \tilde{\bar{\Xi}}_i^2 + \frac{\varsigma_{\Xi i}}{2 b_{\Xi i}} \bar{\Xi}_i^2 - \frac{\varsigma_{Hi}}{2 b_{Hi}} h_{\bar{u}_i} \tilde{\bar{H}}_i^2 + \frac{\varsigma_{Hi}}{2 b_{Hi}} h_{\bar{u}_i} \bar{H}_i^2 \right) \right\} \tag{4.32}$$

Through Lemma 2.1 in Chap. 3, we have

$$-\bar{k}_{1i} J_i(t) \tanh(J_i \mu_i^{-1}) \le 0.2785 \bar{k}_{1i} \mu_i - \bar{k}_{1i} \left| J_i(t) \right| \tag{4.33}$$

By Eqs. (4.33), (4.32) can be derived as

$$\begin{aligned}\dot{\bar{V}} \leq &\left\{\sum_{i=1}^{n}\left(-\bar{\Im}_i J_i^2 - \frac{\varsigma_{ki}}{2b_{ki}}\tilde{\tilde{k}}_{1i}^2 - \frac{\varsigma_{\Xi i}}{2b_{\Xi i}}\tilde{\tilde{\Xi}}_i^2 - \frac{\varsigma_{\mathrm{H}i}}{2b_{\mathrm{H}i}}h_{\bar{u}_i}\tilde{\tilde{\mathrm{H}}}_i^2\right)\right\}\\ &+\left\{\sum_{i=1}^{n}\left(0.2785\bar{k}_{1i}\mu_i + (1-\lambda_{i2})\bar{\Theta}_i\bar{b}_{oi} + \frac{\varsigma_{ki}}{2b_{ki}}\bar{k}_{1i}^2\right)\right\}\\ &+\left\{\sum_{i=1}^{n}\left(\frac{\bar{\Phi}_i\eta_i}{2M_iR_i\tau_i}(\tilde{\tilde{\Lambda}}_i(0)+2\bar{b}_{\wp}) + \frac{\varsigma_{\Xi i}}{2b_{\Xi i}}\bar{\Xi}_i^2 + \frac{\varsigma_{\mathrm{H}i}}{2b_{\mathrm{H}i}}h_{\bar{u}_i}\bar{\mathrm{H}}_i^2\right)\right\}\end{aligned} \tag{4.34}$$

Therefore, for the vehicle $i = 1, 2, \ldots, n$ of closed-loop heterogeneous vehicle platoon system, it can be derived as

$$\dot{\bar{V}} \leq -\hbar_1(\sum_i^n(\bar{V}_i) + \hbar_2 \tag{4.35}$$

where $\hbar_1 = \min\{2\bar{\Im}_i, \min\limits_{1\leq i\leq n}\varsigma_{ki}, \min\limits_{1\leq i\leq n}\varsigma_{\Xi i}, \min\limits_{1\leq i\leq n}\varsigma_{\mathrm{H}i}\}$, $\hbar_2 = \sum\limits_{i=1}^{n}\{0.2785\bar{k}_{1i}\mu_i + (1-$ $\lambda_{i2})\bar{\Theta}_i\bar{b}_{oi} + \frac{\varsigma_{ki}\bar{k}_{1i}^2}{2b_{ki}}\} + \sum\limits_{i=1}^{n}\left\{\frac{\bar{\Phi}_i\eta_i}{2M_iR_i\tau_i}(\tilde{\tilde{\Lambda}}_i(0)+2\bar{b}_{\wp}) + \frac{\varsigma_{\Xi i}\bar{\Xi}_i^2}{2b_{\Xi i}} + \frac{\varsigma_{\mathrm{H}i}h_{\bar{u}_i}\bar{\mathrm{H}}_i^2}{2b_{\mathrm{H}i}}\right\}$.

By multiplying both sides of the Eq. (4.35) by the decayed function $e^{\bar{\lambda}_1 t}$ and integrating it over $[0, t]$, it is obtained that

$$\bar{V} \leq \left(\bar{V}(0) - \frac{\hbar_2}{\hbar_1}\right)e^{-\bar{\lambda}_1 t} + \frac{\hbar_2}{\hbar_1} \leq \bar{V}(0) + \frac{\hbar_2}{\hbar_1} \tag{4.36}$$

The above formula indicates $(1/2)\sum_{i=1}^{n}J_i^2(t) \leq \bar{V}(0) + \hbar_2/\hbar_1$, and $\|J(t)\| \leq \sqrt{2(\bar{V}(0)+\hbar_2/\hbar_1)}$, where $\bar{V}(0) \geq 0$. According to the formula (4.36), under the bounded initial conditions, $\hat{\hat{k}}_{1i}$, $\hat{\hat{\Xi}}_i$, $\hat{\hat{\mathrm{H}}}_i$ are all ultimately uniformly bounded and satisfy$\tilde{\tilde{k}}_{1i}$, $\tilde{\tilde{\Xi}}_i$, $\tilde{\tilde{\mathrm{H}}}_i \in L_\infty$. The convergence accuracy of the spacing error can be improved by reducing the parameter $\hbar_2$ and increasing the parameter $\hbar_1$. Meanwhile, the system state can be quickly converged to a small neighborhood of the origin by selecting suitable design parameters. Meanwhile, the formula (4.13) and (4.14) show that$r_i(t)$, $\bar{e}_i(t)$, $\dot{\bar{e}}_i(t)$are all ultimately uniformly bounded, which proves the reliability and robustness of the designed controller. Based on the uniform boundedness of $\bar{e}_i(t)$ and $\dot{\bar{e}}_i(t)$, if the initial conditions $\bar{e}_i(0) = 0$, $\dot{\bar{e}}_i(0) = 0$are satisfied, it follows from Definition 4.3 that for any given constant $y_1 > 0$, there exists a constant $y_2 > 0$ that makes $\sup\limits_i \max\left\{|\bar{e}_i(0)|, |\dot{\bar{e}}_i(0)|\right\} < y_2 \Rightarrow \sup\limits_i\sup\limits_{t\geq 0}\max\left\{|\bar{e}_i(t)|, |\dot{\bar{e}}_i(t)|\right\} < y_1$ true. Therefore, according to Definition 4.1, the string stability of heterogeneous vehicle platoon system can be ensured.

Furthermore, for proving the finite time stability of heterogeneous vehicle platoon system, the Lyapunov function is considered as

$$\bar{V}^f = \sum_{i=1}^{n-1}(\bar{V}_i^f) + \bar{V}_n^f \tag{4.37}$$

where $\bar{V}_i^f = \frac{1}{2}J_i^2$, $\bar{V}_n^f = \frac{1}{2}J_n^2$. By taking the derivative of $\bar{V}^f(t)$,

$$\begin{aligned}\dot{\bar{V}}^f \le & \sum_{i=1}^{n-1}\left(-\bar{\Phi}_i\left[\frac{\eta_i(1-\lambda_{i2})h_{\bar{u}_i}(\bar{\mathrm{H}}_i-\tilde{\bar{\mathrm{H}}}_i)\bar{u}_i(t)}{M_iR_i\tau_i\hat{\bar{\mathrm{H}}}_i}-\frac{\bar{a}_i(t)}{\tau_i}+\bar{\Xi}_i\right]J_i(t)\right)\\ &+\sum_{i=1}^{n-1}\left(-\bar{\Phi}_i\left[-\frac{1}{M_i\tau_i}\left[\left(\rho_aA_iC_{ai}\left(\frac{\bar{v}_i^2}{2}+\tau_i\bar{v}_i\bar{a}_i\right)\right)\right]+\frac{\eta_i\bar{\Lambda}_i}{M_iR_i\tau_i}\right]J_i(t)\right)\\ &+\sum_{i=1}^{n-1}((1-\lambda_{i2})\bar{\Theta}_i\left|\bar{\Gamma}_iJ_i\right|)+J_n\dot{J}_n\end{aligned} \tag{4.38}$$

Then, using formulas (4.18), (4.19) and (4.33), we can obtain that

$$\begin{aligned}\dot{\bar{V}}^f \le & \left\{\sum_{i=1}^{n}(0.2785\hat{\bar{k}}_{1i}\mu_i-\hat{\bar{k}}_{1i}\left|J_i(t)\right|-\bar{k}_{2i}J_i^2+(1-\lambda_{i2})\bar{\Theta}_i\bar{b}_{oi})\right\}\\ &+\left\{\sum_{i=1}^{n}\left(\frac{\bar{\Phi}_i}{M_i\tau_i}\tilde{\bar{\Xi}}_iJ_i-\frac{\bar{\Phi}_i\eta_i}{M_iR_i\tau_i}\tilde{\bar{\Lambda}}_iJ_i+\bar{\Phi}_i\bar{\mathrm{B}}_ih_{\bar{u}_i}(t)\bar{u}_i(t)\frac{\tilde{\bar{\mathrm{H}}}_i}{\hat{\bar{\mathrm{H}}}_i}J_i\right)\right\}\end{aligned} \tag{4.39}$$

Based on the above $\tilde{\bar{k}}_{1i}$, $\tilde{\bar{\Xi}}_i$, $\tilde{\bar{\mathrm{H}}}_i \in L_\infty$, we can see that, for positive definite constants $\overset{\frown}{\bar{k}}_{1i}$, $\overset{\frown}{\bar{\Xi}}_i$, $\overset{\frown}{\bar{\mathrm{H}}}_i$, there exist $\left|\tilde{\bar{k}}_{1i}\right| \le \overset{\frown}{\bar{k}}_{1i}$, $\left|\tilde{\bar{\Xi}}_i\right| \le \overset{\frown}{\bar{\Xi}}_i$, and $\left|\tilde{\bar{\mathrm{H}}}_i\right| \le \overset{\frown}{\bar{\mathrm{H}}}_i$ hold. Therefore, we know $\left|\frac{\tilde{\bar{\mathrm{H}}}_i}{\hat{\bar{\mathrm{H}}}_i}\right| \le \frac{\overset{\frown}{\bar{\mathrm{H}}}_i}{|\bar{\mathrm{H}}_i|+\overset{\frown}{\bar{\mathrm{H}}}_i}$, and $\bar{k}_{1i}$ is a positive definite constant, $\left|\tilde{\bar{k}}_{1i}\right| \le \overset{\frown}{\bar{k}}_{1i}$, thus $\hat{\bar{k}}_{1i}$ is bounded and $\left|\hat{\bar{k}}_{1i}\right| \le \bar{k}_{1i} + \overset{\frown}{\bar{k}}_{1i}$ is satisfied. By using formulas (4.4) and (4.5), $h_{\bar{u}_i}(t)\bar{u}_i(t)$ is uniformly bounded and $\left|h_{\bar{u}_i}(t)\bar{u}_i(t)\right| \le \overset{\frown}{\bar{U}}$ is satisfied. Based on Theorem 4.1, we can get $\tilde{\bar{\Lambda}}_i \le \sqrt{\tilde{\bar{\Lambda}}_{ii}(0)+2\bar{b}_\wp}$ is bounded. Thus, based on the above conditions, Eq. (4.39) can be transformed into

$$\begin{aligned}\dot{\bar{V}}^f \le & \left\{\sum_{i=1}^{n}(-\bar{k}_{2i}J_i^2+0.2785(\bar{k}_{1i}+\overset{\frown}{\bar{k}}_{1i})\mu_i-(\bar{k}_{1i}+\overset{\frown}{\bar{k}}_{1i})\left|J_i(t)\right|)\right\}\\ &+\left\{\sum_{i=1}^{n}\left((1-\lambda_{i2})\bar{\Theta}_i\bar{b}_{oi}+\frac{\bar{\Phi}_i\tilde{\bar{\Xi}}_iJ_i}{M_i\tau_i}-\frac{\bar{\Phi}_i\eta_i\tilde{\bar{\Lambda}}_iJ_i}{M_iR_i\tau_i}+\frac{\bar{\Phi}_i\bar{\mathrm{B}}_i\overset{\frown}{\bar{U}}\overset{\frown}{\bar{\mathrm{H}}}_i}{|\bar{\mathrm{H}}_i|+\overset{\frown}{\bar{\mathrm{H}}}_i}J_i\right)\right\}\end{aligned} \tag{4.40}$$

Thus, when the following conditions are met,

$$
\begin{aligned}
&(\bar{k}_{1i}+\hat{\bar{k}}_{1i})+\left|\frac{\bar{\Phi}_i\eta_i}{M_iR_i\tau_i}\right|(\sqrt{\tilde{\bar{\Lambda}}_i(0)}+2\bar{b}_{\wp})\\
&\geq(1-\lambda_{i2})\bar{\Theta}_i\bar{b}_{oi}+\left|\frac{\bar{\Phi}_i}{M_i\tau_i}\right|\bar{\Xi}_i+\left|\bar{\Phi}_i\bar{B}_i\right|\frac{\hat{\bar{U}}\hat{\bar{H}}_i}{\left|\bar{H}_i\right|+\hat{\bar{H}}_i}+0.2785(\bar{k}_{1i}+\hat{\bar{k}}_{1i})\mu_i+\hat{\bar{\Upsilon}}_i
\end{aligned}
\tag{4.41}
$$

where $\hat{\bar{\Upsilon}}_i > 0$is constant.Then, formula (4.40) can be derived as

$$
\begin{aligned}
\dot{\bar{V}}^f &\leqslant \sum_{i=1}^{n}\{-\bar{k}_{2i}J_i^2-\hat{\bar{\Upsilon}}_i\,|J_i|\}\\
&\leqslant \sum_{i=1}^{n}\{-\hat{\bar{\Upsilon}}_i\sqrt{2\bar{V}^f}\}\leqslant-\sqrt{2}\hat{\bar{\Upsilon}}_i\sqrt{\bar{V}^f}.
\end{aligned}
\tag{4.42}
$$

Therefore, under the condition $\bar{V}^f(0)\geqslant 0$, $\dot{\bar{V}}^f\leqslant-\sqrt{2}\hat{\bar{\Upsilon}}_i\sqrt{\bar{V}^f}$, $\hat{\bar{\Upsilon}}_i>0$. According to lemma of global finite-time stability, it can be seen that $\bar{V}^f$ is bounded and can converge to zero in finite time, which proves the global finite-time stability of heterogeneous vehicle platoon system. $J_i(t)$ and $\bar{e}_i(t)$ can achieve convergence performance in finite time. Based on the above analysis, the string stability of heterogeneous vehicle platoon system can also be guaranteed within finite time.

Finally, the heterogeneous vehicle platoon system under consideration can reach a stable state within a finite time. When the heterogeneous vehicle platoon system is in a steady state, it can be reasonably assumed that the velocity state of all followers in the system is consistent, that is $\bar{v}_i=\bar{v}_0$. It can be seen from the formula (4.7) that the decayed function $\bar{\Pi}_i(t)$ is also in a zero state at steady state. Therefore, when heterogeneous vehicle platoon system is in a steady state, the impact of the improved exponential spacing policy on traffic flow can be characterized by the relationship between traffic flow rate and traffic flow density. Based on $\hat{\bar{e}}_i\ (t)$ in the Eq. (4.7), the ideal spacing distance $\hat{\Gamma}_{de}$ in the steady state period can be expressed as

$$
\hat{\Gamma}_{de}=L+\bar{\varphi}_d \tag{4.43}
$$

Thus, the traffic flow rate $F(\rho)$ of heterogeneous vehicle platoon system can be characterized as

$$
F(\rho)=\rho\bar{v}_0=\frac{\bar{v}_0}{L+\bar{\varphi}_d} \tag{4.44}
$$

where the steady state traffic flow density is $\rho=1/(L+\bar{\varphi}_d)$.

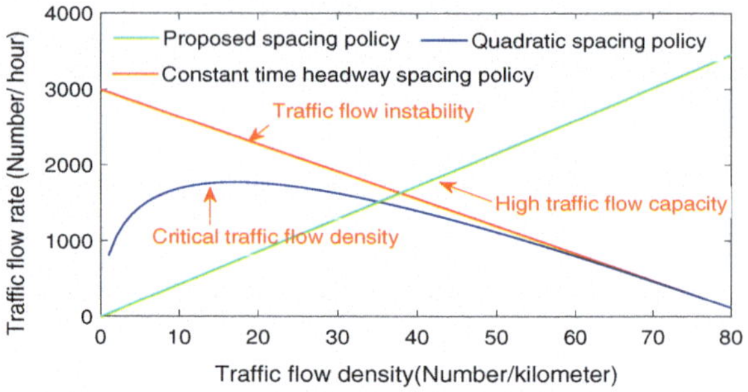


**Fig. 4.2** Traffic flow rate under different spacing policies

Based on the above, it can be seen that according to Definition 4.2, the improved exponential spacing policy can satisfy the stability condition $\partial F/\partial \rho > 0$ of traffic flow, the stability of traffic flow of heterogeneous vehicle platoon system can be ensured. Proof complete. □

**Remark 4.5** Through the proposed control scheme, the inter-vehicle spacing distance is reduced under the condition of high-speed driving of vehicle platoon system, so that the steady-state inter-vehicle spacing distance error in Eq. (4.7) can achieve high traffic flow rate and traffic flow density. It is similar to the traditional constant policy. However, the traditional constant policy can not guarantee the string stability effectively. Compared with the traditional constant policy, constant time headway spacing policy and quadratic spacing policy (as mentioned in Chap. 1), in this chapter, the vehicle velocity information, exponential decay function, fault upper boundary factor and velocity information of the leader vehicle are introduced to design the spacing policy. The robustness and stability of the closed-loop control system can be improved while the traffic flow capacity can be ensured effectively. Meanwhile, the string stability of the system can also be ensured.

**Remark 4.6** According to the traffic flow rate under different spacing policies, as shown in Fig. 4.2, the existing constant time headway spacing policy [13] cannot guarantee the stability of traffic flow. Meanwhile, the existing quadratic spacing policy [6] can ensure the string stability and maximum traffic flow capacity under the condition of critical traffic flow density. Different from the quadratic spacing policy above, the proposed exponential spacing policy can always maintain the high traffic flow capacity without affecting the string stability by introducing exponential function, upper bound of fault factor and the velocity of the proceeding vehicle.

## 4.4 Method Verification and Analysis

For evaluating the proposed control scheme, in Case 1, the numerical simulation experiment was conducted for a vehicle platoon system with 7 heterogeneous vehicles ($i = 0, 1, \ldots, 6$ ). The dynamic parameters of the heterogeneous vehicles are shown in Chap. 3. In the meantime, for comprehensively evaluating the control performance of the proposed method, the comparative experiments are conducted in Case 2, in which the actuator fault of the vehicles in the vehicle platoon system is specifically considered. In addition, in both cases, the input saturation and external disturbances problems of all vehicles are considered. The external disturbances considered is $\bar{d}_i(t) = 0.01\sin(2t + 1)$. For the vehicle 0, its initial position, velocity, and acceleration are shown in Chap. 3.

Based on the proposed improved exponential spacing policy, when the initial position, velocity and acceleration of the follower vehicle are selected as shown in Chap. 3, the non-zero initial spacing errors $\bar{e}_{xi}(0)$ and $\bar{e}_i(0)$ can always be maintained at zero, $i = 1, 2, \ldots, 6$. For the vehicle $i = 1, 2, \ldots, 6$, the initial values $\hat{\bar{k}}_{1i}(0)$, $\hat{\bar{\mathrm{H}}}_i(0)$, and $\hat{\bar{\Xi}}_i(0)$ of the adaptive compensation laws are selected as $0.1, 0.1, 0.35$, respectively. The nonlinear observer gain is $l_i$=1. The relevant parameters $\varepsilon$, $\phi$, $\bar{\kappa}$, $\bar{\zeta}_i$, and $\bar{\varphi}_d$ of the improved exponential spacing policy designed are selected as $0.2, 7, 0.01, 10, 8$, respectively. In this chapter, the actual measured inter-vehicle spacing distance between two adjacent vehicles is $\bar{\varphi}_i = \bar{x}_{i-1}(t) - \bar{x}_i(t) - L_i$.

The road slope angle function is $\bar{\delta}_i = \begin{cases} 2.1\%, 180 \leq \bar{x}_i < 320 \\ 1.2\%, 320 \leq \bar{x}_i < 1000 \\ 0\%, otherwise \end{cases}$.

The acceleration function of the leader vehicle is $\bar{a}_0(t) = \begin{cases} 2.5\,\mathrm{m/s^2}, 5s \leq t < 10s \\ -1.5\,\mathrm{m/s^2}, 20s \leq t < 25s \\ 0\,\mathrm{m/s^2}, otherwise \end{cases}$.

### 4.4.1 Case 1: Actuator Input Saturation and External Disturbances

In the process of this case experiment, considering that all follower vehicles in the heterogeneous vehicle platoon system have the actuator input saturation and external disturbances. For the $i$th follower vehicle, the constraint threshold of input saturation is $\bar{u}_{bi}$=80, where $i = 1, 2, \ldots, 6$. The physical fault is not considered in this case. The experimental results of this case are shown in Figs. 4.3, 4.4, 4.5, and 4.6, which proves the robustness and reliability of the proposed scheme.

As shown in Fig. 4.3a, for the given initial state of the vehicle $i = 1, 2, \ldots, 6$, as shown in Chap. 3, all the $\overset{\frown}{\bar{e}}_i(t)$ are in non-zero states, while the initial state of inter-vehicle spacing errors $\bar{e}_i(t)$ can remain at zero, which proves that the proposed spacing policy can overcome the assumed condition of zero initial spacing error. The maximum vehicle spacing error is 0.157m, and the inter-vehicle spacing error $\bar{e}_i(t)$ of the $i$th follower vehicle is smaller than the spacing error of the proceeding vehicle, that is $\|\bar{e}_6\| \leq \|\bar{e}_5\| \leq \cdots \leq \|\bar{e}_1\|$. All spacing errors can converge to zero within a finite time ($T \leq 30s$). It verifies that the proposed scheme can ensure the string stability of heterogeneous vehicle platoon system in a finite time. The position trajectory

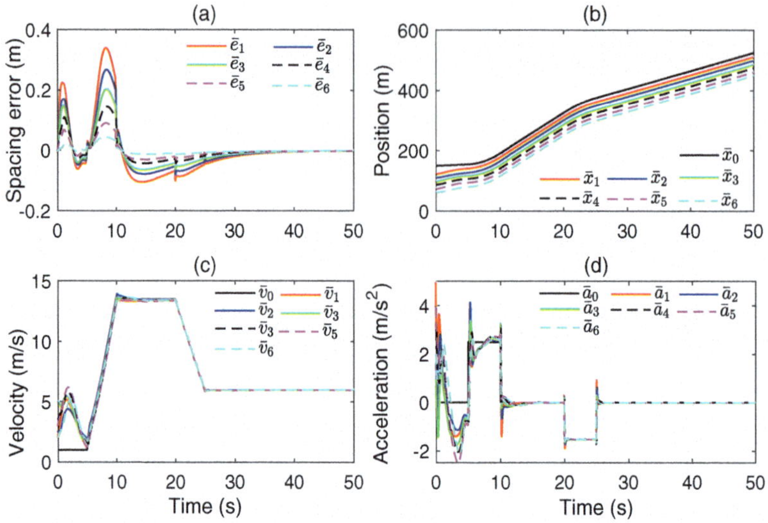


**Fig. 4.3** Experimental results of the proposed scheme in Case 1. **a** Spacing error $\bar{e}_i(t)$, **b** Position $\bar{x}_i(t)$, **c** Velocity $\bar{v}_i(t)$, **d** Acceleration $\bar{a}_i(t)$

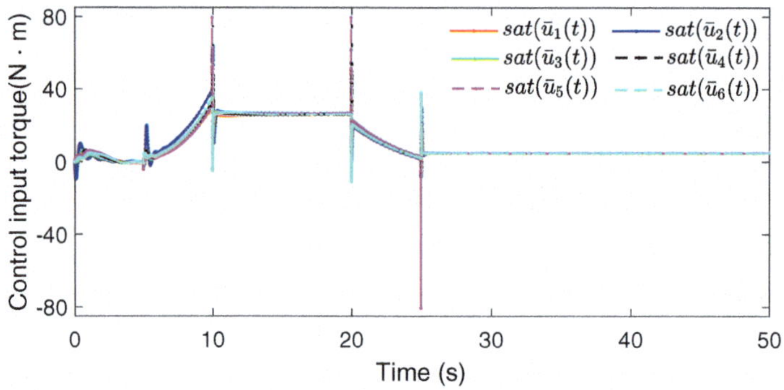


**Fig. 4.4** Control input torque $sat(\bar{u}_i(t))$ under proposed scheme in Case 1

of the vehicle shown in Fig. 4.3b indicates that collision and overtaking problems are avoided in both transient and steady-state processes. As shown in Fig. 4.3c, d, in the acceleration stage ($5s \leq t < 10s$) and deceleration stage ($20s \leq t < 25s$), the velocity and acceleration of all the follower vehicles can be consistent with the velocity and acceleration of the leader vehicle within a finite time. Figure 4.4 describes the saturation control torque input $sat(\bar{u}_i(t))$ of the follower vehicle. Figure 4.5a shows the coupled sliding mode surface $J_i(t)$, and Fig. 4.5b, c, d show the convergence of the adaptive compensation law $\hat{\bar{k}}_{1i}$, $\hat{\bar{\mathrm{H}}}_{\mathrm{i}}$, $\hat{\bar{\Xi}}_i$, respectively. The redundancy estimation $\hat{\bar{\Lambda}}_i$ of the proposed nonlinear observer is shown in Fig. 4.6.

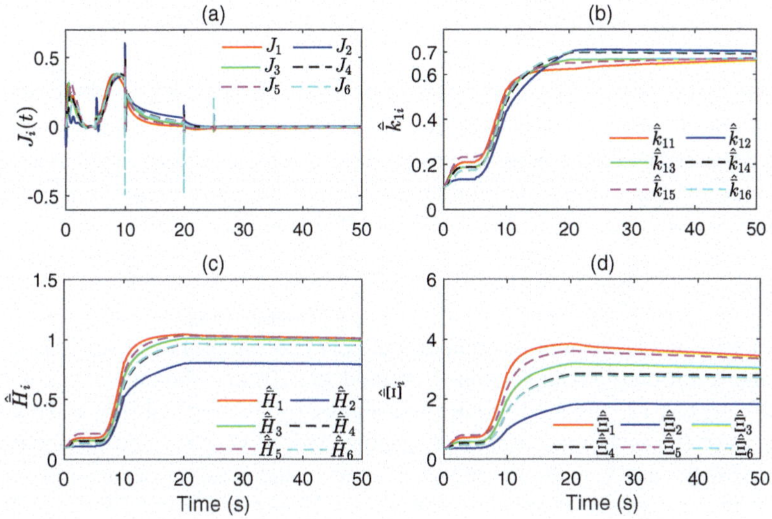


**Fig. 4.5** Experimental results of the proposed scheme in Case 1. **a** Coupled sliding mode surface $J_i(t)$, **b** Adaptive law $\hat{\hat{k}}_{1i}$, **c** Adaptive law $\hat{\hat{\mathrm{H}}}_{\mathrm{i}}$, **d** Adaptive law $\hat{\hat{\Xi}}_i(t)$

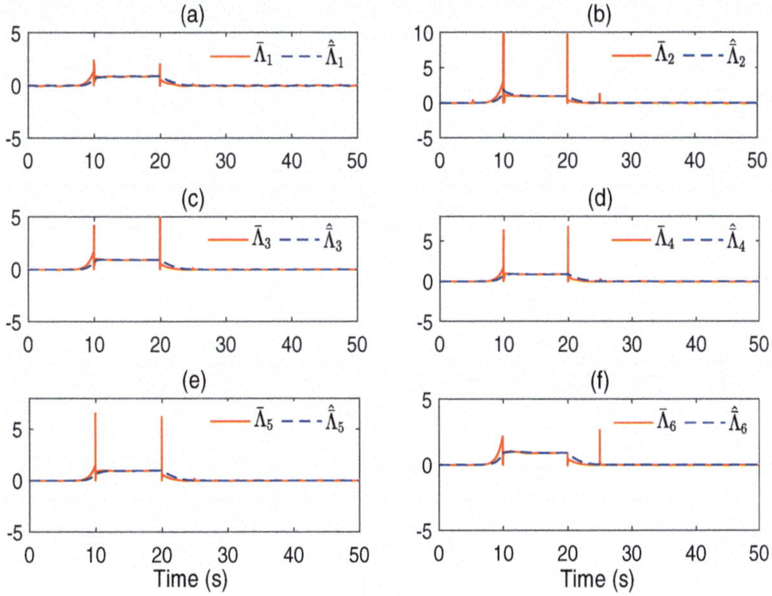


**Fig. 4.6** Redundancy term $\bar{\Lambda}_i$ of nonlinear observer and its estimation $\hat{\hat{\Lambda}}_i$ under proposed scheme in Case 1

## 4.4.2 Case 2: Actuator Fault, Saturation and Disturbances

Based on case 1, the actuator fault suffered by the vehicle $i = 6$ at the rear of heterogeneous vehicle platoon system is considered. The problems of actuator input saturation and external disturbances are involved in all vehicles of heterogeneous vehicle platoon system. The input saturation threshold constraint considered is $\bar{u}_{bi} = 52$. For the vehicles $i = 6$, $\sigma_6 = 1$, $\Delta_6(t) = 0.05\cos(5t)$, and $\bar{\Theta}_6 = 1.54$.

The partial actuator fault function is $\lambda_6 = \begin{cases} 0.05, 15 \leq t < 20 \\ 0.35, 35 \leq t < 40 \\ 0, otherwise \end{cases}$.

The experimental results of comparison between the proposed scheme in Case 2 and the traditional existing control method [9, 10] are shown in Figs. 4.7, 4.8, 4.9, 4.10, 4.11, 4.12, and 4.13. Figure 4.7a shows good string stability and a smaller spacing error (the maximum transient spacing error of 0.145 m). At the mean time, compared with case 1, when the physical faults of the actuator occur in the vehicle $i = 6$ at the end of the platoon and in stages $15 \leq t < 20$ and $35 \leq t < 40$, the string stability in Fig. 4.7a can still be ensured. Compared with Figs. 4.7a, and 4.11a shows a larger chattering and a larger transient spacing error. Figures 4.7b and 4.11b show the position tracking trajectory of the follower vehicle. Under acceleration and deceleration conditions, Fig. 4.11c shows a longer convergence time and a larger velocity tracking error than Fig. 4.7c. When the vehicle at the end of the platoon has the actuator fault, compared with Figs. 4.7d, and 4.11d has a larger transient

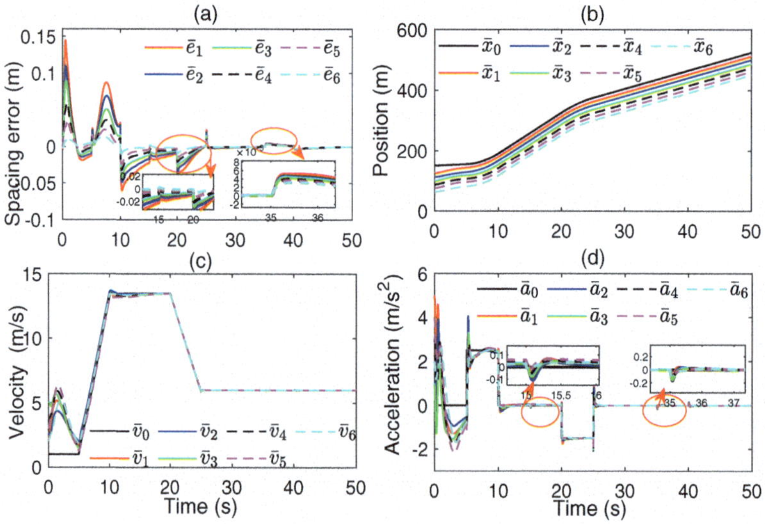

**Fig. 4.7** Experimental results of the proposed scheme in Case 2. **a** Spacing error $\bar{e}_i(t)$, **b** Position $\bar{x}_i(t)$, **c** Velocity $\bar{v}_i(t)$, **d** Acceleration $\bar{a}_i(t)$

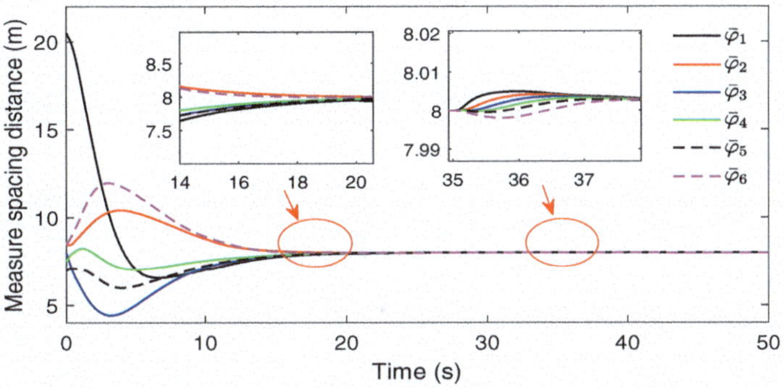


**Fig. 4.8** Actual measure spacing distance $\bar{\varphi}_i(t)$ under the proposed scheme in Case 2

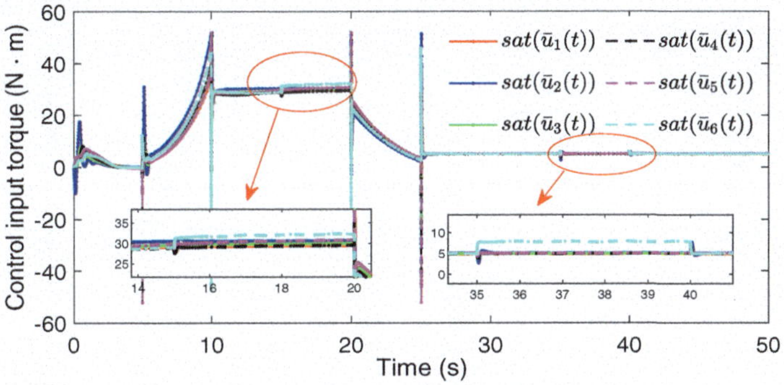


**Fig. 4.9** Control input torque $sat(\bar{u}_i(t))$ under the proposed scheme in Case 2

acceleration error, and its maximum transient spacing error is $0.1\,\mathrm{m/s^2}$in the stage $15 \leq t < 20$, and the maximum transient spacing error is $0.2\,\mathrm{m/s^2}$ in the stage $35 \leq t < 40$.

The actual measure inter-vehicle spacing distance $\bar{\varphi}_i$ of the vehicle platoon system under the method proposed and the traditional control method are shown in Figs. 4.8 and 4.12, respectively. When the vehicle $i = 6$ at the rear of the heterogeneous vehicle platoon system suffers from the actuator fault, the actual measure inter-vehicle spacing distance in Fig. 4.8 can converge to the desired inter-vehicle spacing distance constant within a finite time, which has faster convergence speed and higher accuracy compared with Fig. 4.12. The above analysis shows that Theorem 4.2 of the proposed method has stronger robustness.

The redundant estimation $\hat{\hat{\Lambda}}_i$ of the nonlinear observer under proposed scheme in case 2 is shown in Fig. 4.10, which indicates that the observer can still achieve better estimation performance even if the actuator fault occurs in the rear vehicle. Figures 4.9 and 4.13 show the control saturation input $sat(\bar{u}_i(t))$ under proposed scheme and the traditional control method, respectively.

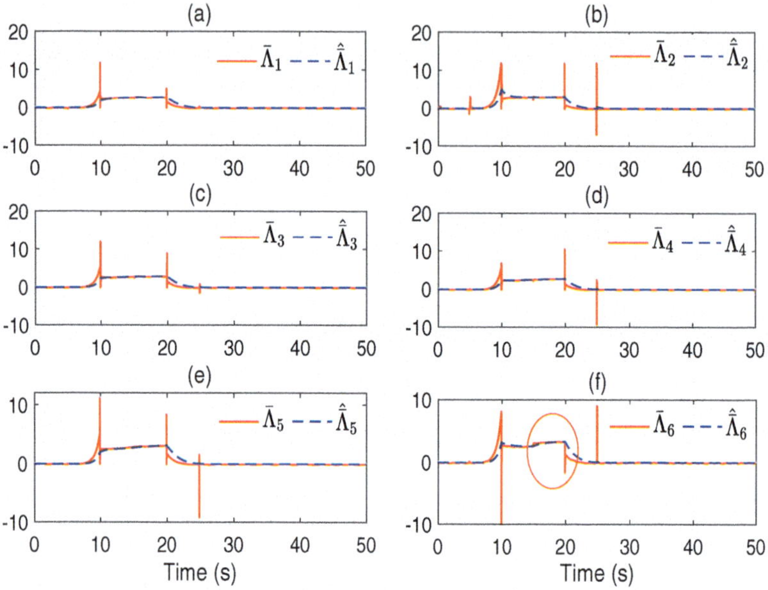


**Fig. 4.10** Redundancy term $\bar{\Lambda}_i$ of nonlinear observer and its estimation $\hat{\bar{\Lambda}}_i$ under proposed scheme in Case 2

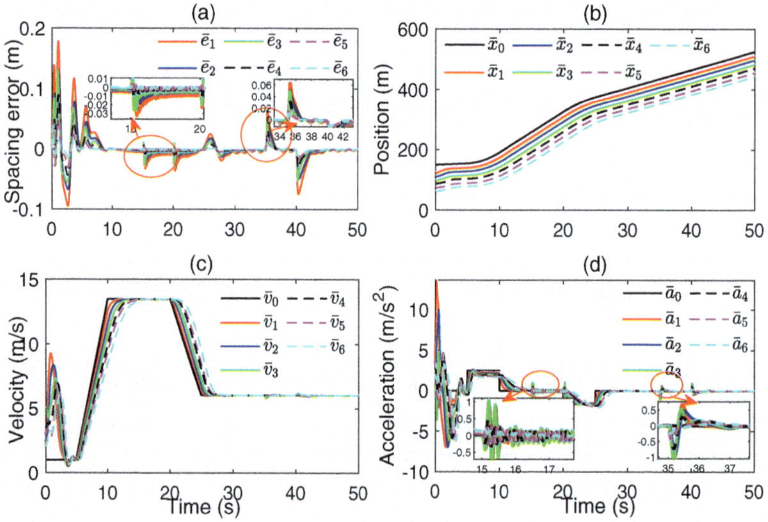


**Fig. 4.11** Experimental results of the traditional control scheme in Case 2. **a** Spacing error $\bar{e}_i(t)$, **b** Position $\bar{x}_i(t)$, **c** Velocity $\bar{v}_i(t)$, **d** Acceleration $\bar{a}_i(t)$

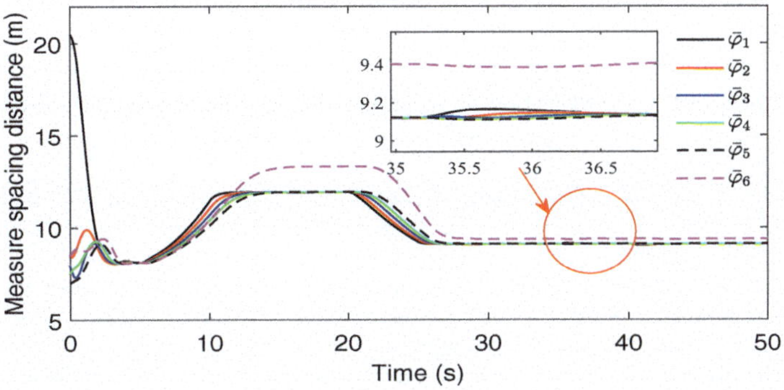


**Fig. 4.12** Actual measure spacing distance $\bar{\varphi}_i(t)$ under the traditional control scheme in Case 2

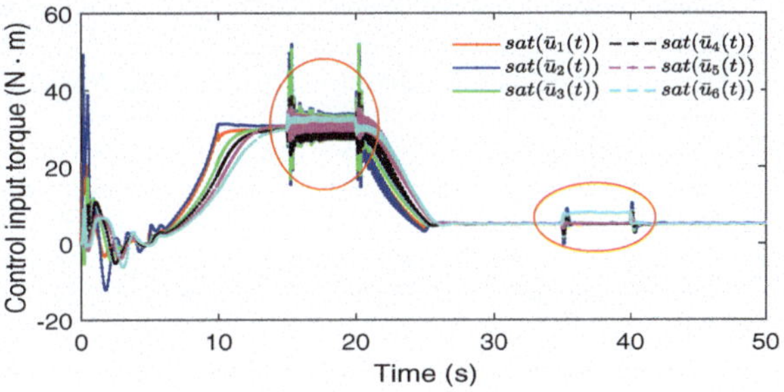


**Fig. 4.13** Control input torque $sat(\bar{u}_i(t))$ under the traditional control scheme in Case 2

## 4.5 Summary

For the problems of actuator fault, input saturation and external disturbances of heterogeneous vehicle platoon system, in this chapter, a distributed adaptive platoon finite time resilient control strategy based on improved exponential spacing strategy is proposed. Firstly, on the basis of Chap. 3, the improved exponential spacing error strategy is designed by introducing the upper bound of fault factors and the velocity information of the leader vehicle. The time headway constant design limitation of the traditional quadratic spacing policy is removed, the traffic flow capacity is improved, and the string stability of the system is ensured. Based on the error compensator, a nonlinear observer is designed to compensate the saturation constraint approximation error, bias fault and external disturbances. Then, a distributed adaptive platoon finite time resilient control strategy is designed to achieve a more ideal safety spacing distance. Based on the unknown parameters, the adaptive law compensations are designed for eliminating the influence of actuator fault, saturation and disturbances. It is proved that the improved exponential spacing policy can effectively improve the traffic flow capacity of the system and ensure the string stability of the system. Meanwhile, the theory proves that the proposed control strategy can guarantee the global finite-time stability of the closed-loop system. Finally, the effectiveness of the proposed method is verified by the comparative experimental cases.

## References


1. Gao Z, Zhang Y, Guo G (2022) Fixed-time prescribed performance adaptive fixed-time sliding mode control for vehicular platoons with actuator saturation. *IEEE Transactions on Intelligent Transportation Systems*, 23(12): 24176–24189.
2. Pan C, Chen Y, Liu Y, et al (2022) Distributed finite-time fault-tolerant control for heterogeneous vehicular platoon with saturation. *IEEE Transactions on Intelligent Transportation Systems*, 23(11): 21259–21273.
3. Guo G, and Li D (2019) Adaptive sliding mode control of vehicular platoons with prescribed tracking performance. *IEEE Transactions on Vehicular Technology*, 68(8): 7511–7520.
4. Wang J, Luo X, Wang L, et al (2020) Integral sliding mode control using a disturbance observer for vehicle platoons. *IEEE Transactions on Industrial Electronics*, 67(8): 6639–6648.
5. Xiao W, Cassandras C (2021) Decentralized optimal merging control for connected and automated vehicles with safety constraint guarantees. *Automatica*, 123: 109333.
6. Chen C, Xiao T, Qiu T, et al (2020) Smart-contract-based economical platooning in blockchain-enabled urban internet of vehicles. *IEEE Transactions on Industrial Informatics*, 16(6): 4122–4133.
7. Peters A, Middleton R, Mason O (2014) Leader tracking in homogeneous vehicle platoons with broadcast delays. *Automatica*, 50(1): 64–74.
8. Zhou J, and Peng H (2005) Range policy of adaptive cruise control vehicles for improved flow stability and string stability. *IEEE Transactions on Intelligent Transportation Systems*, 6(2): 229–237.
9. Kwon J, and Chwa D (2014) Adaptive bidirectional platoon control using a coupled sliding mode control method. *IEEE Transactions on Intelligent Transportation Systems*, 15(5): 2040–2048.
10. Guo X, Wang J, Liao F, et al (2017) Distributed adaptive sliding mode control strategy for vehicle-following systems with nonlinear acceleration uncertainties. *IEEE Transactions on Vehicular Technology*, 66(2): 981–991.
11. Sungu H, Inoue M, and Imura J (2015) Nonlinear spacing policy based vehicle platoon control for local string stability and global traffic flow stability. *2015 European Control Conference*, 3396–3401.
12. Ploeg J, Kazerooni E. S, Lijster G, et al (2015) Graceful degradation of cooperative adaptive cruise control. *IEEE Transactions on Intelligent Transportation Systems*, 16(1): 488–497.
13. Liu A, Li T, Gu Y, et al (2021) Cooperative extended state observer based control of vehicle platoons with arbitrarily small time headway. *Automatica*, 129: 109678.
14. Herman I, Knorn S, Ahlén A (2017) Disturbance scaling in bidirectional vehicle platoons with different asymmetry in position and velocity coupling. *Automatica*, 82: 13–20.
15. Besselink B, and Johansson K (2017) String stability and a delay-based spacing policy for vehicle platoons subject to disturbances. *IEEE Transactions on Automatic Control*, 62(9): 4376–4391.
16. Xu L, Zhuang W, Yin G, et al (2019) Modeling and robust control of heterogeneous vehicle platoons on curved roads subject to disturbances and delays. *IEEE Transactions on Vehicular Technology*, 68(12): 11551–11564.
17. Warnick S, and Rodriguez A (2000) A systematic anti-windup strategy and the longitudinal control of a platoon of vehicles with control saturation. *IEEE Transactions on Vehicular Technology*, 49(3): 1006–1016.
18. Guo X, Wang J, Liao F, et al (2018) CNN-based distributed adaptive control for vehicle-following platoon with input saturation. *IEEE Transactions on Intelligent Transportation Systems*, 19(10): 3121–3132.
19. Zhang D, Liu G, Zhou H, et al (2018) Adaptive sliding mode fault-tolerant coordination control for four-wheel independently driven electric vehicles. *IEEE Transactions on Industrial Electronics*, 65(11): 9090–9100.

20. Li Y, Yang T, and Tong S (2020) Adaptive neural networks finite-time optimal control for a class of nonlinear systems. *IEEE Transactions on Neural Networks and Learning Systems*, 31(11): 4451–4460.
21. Li Y, Li K, and Tong S (2020) Adaptive neural network finite-time control for multi-input and multi-output nonlinear systems with positive powers of odd rational numbers. *IEEE Transactions on Neural Networks and Learning Systems*, 31(7): 2532–2543.
22. Guo J, Luo Y, Li K (2017) Adaptive fuzzy sliding mode control for coordinated longitudinal and lateral motions of multiple autonomous vehicles in a platoon. *Science China Technological Sciences*, 60: 576–586.
23. Gao F, Hu X, Li S, et al (2018) Distributed adaptive sliding mode control of vehicular platoon with uncertain interaction topology. *IEEE Transactions on Industrial Electronics*, 65(8): 6352–6361.
24. Ma Y, Li Z, Malekian R, et al (2019) Hierarchical fuzzy logic-based variable structure control for vehicles platooning. *IEEE Transactions on Intelligent Transportation Systems*, 20(4): 1329–1340.

# Chapter 5
# Distributed Prescribed Performance Resilient Control for Heterogeneous Intelligent Connected Vehicle Platoon Systems with Compound Cyber Attacks

For the network communication layer of the vehicle platoon systems, VANETs based on the new generation of network communication C-V2X technology can form the mobile networks with high-speed mobile vehicles and road infrastructure as nodes. Thus, the V2V and V2I communication interaction and perception are achieved. However, the credibility, availability and integrity of the interactive data in the vehicle platoon systems at high speed are vulnerable to cyber attacks due to the openness and vulnerability of vehicle wireless communication network and the complexity of traffic environment. The malicious cyber attack is one of the main reasons that cause the communication faults of the network cyber physical systems (which mentioned in Chap. 1). Malicious attackers can pretend to be mobile vehicles or roadside communication facilities, and simultaneously launch cyber attacks on the V2V communication interaction and the communication channel of control layer of the vehicle platoon system in their range. The multiple communication faults induced by the compound cyber attacks can achieve the control performance loss of the platoon system, or even crash, and induce traffic collisions. In addition, in the most of malicious cyber attacks, deception attacks directly tamper with the system communication interaction data by monitoring the system communication transmission and using the obtained system information to destroy its integrity and credibility. Thus, the deception attacks can cause the communication faults and have stronger destructive power to the platoon control system. Therefore, how to design a distributed adaptive resilient control strategy to enhance the robustness and fast stability of the vehicle platoon systems is of great significance and great challenge for the multiple communication faults caused by malicious compound deception attacks.

On the other hand, unlike the physical faults that originate from the defects or faults of the internal components of the vehicle platoon subsystem, the communication faults caused by deliberate artificial cyber attacks are based on the existing resource allocation of malicious attackers. Usually, the state observability of the vehicle system can be destroyed, and the damage to the ICVPS control system is

 
C. Pan and Y. Chen, *Resilient Control of Intelligent Connected Vehicle Platoon Systems*,
https://doi.org/10.1007/978-981-97-8850-7_5

more intelligent and targeted (as shown in Chap. 1). From the perspective of secure control, how to achieve collision avoidance and mobile communication maintenance at high velocity are important research topics. Moreover, how to constrain the spacing error range to maintain the string stability and transient/steady state performance under the compound cyber attacks conditions of the vehicle platoons also have the positive practical significance. For achieving the collision avoidance and network communication maintenance under the compound cyber attacks, the prescribed performance control method has great advantages. However, compared with the existing prescribed performance control research [1, 2], the research on how to ensure the string stability and closed-loop fast finite time stability of the vehicle platoon system with the compound cyber attacks under the predefined asymmetric constraint performance is still insufficient, which has important research value.

Inspired by the above analysis, in this chapter, the distributed adaptive prescribed performance finite time resilient control scheme for the HICVPS with the compound cyber attacks [3]. The main contributions are as follows. Firstly, the fault of V2V communication interaction and controller-actuator communication in control layer caused by the compound deception attacks initiated by multiple malicious nodes are considered. Different from the feedback channel communication faults [4, 5], the V2V communication interaction faults are induced based on man-in-the-middle attacks. Considering the unknown nonlinear and external disturbances of the HICVPS, the nonlinear state compensator is designed to eliminate the influence of multiple communication faults induced by the compound cyber attacks on the system state information and ensure the boundability of the system state signal. The adaptive compensation laws are designed to update the unknown fuzzy weight and the upper bound of the unknown cyber attacks. Secondly, based on the improved exponential spacing policy proposed in Chap. 4, the string stability and scalability of the HICVPS under the compound cyber attacks are guaranteed. The asymmetric spacing error domain constraint is constructed to ensure collision avoidance and the maintenance of mobile communication at high velocity.

Finally, the hyperbolic arctangent asymmetric Barrier Lyapunov function is constructed, and based on the leader-bidirectional information flow topology, the distributed adaptive prescribed performance fast finite time resilient control scheme is designed to overcome the influence of the compound cyber attacks and disturbances on the control performance. The resilient control scheme proposed in this chapter can not only ensure the string stability under the prescribed performance, but also realize the fast finite time stability of the closed-loop system. The transient/steady state performance of the system under the compound cyber attacks conditions are also ensured. The proposed method is different from the existing prescribed asymmetric performance string control method in the sense of the global finite time [6] and the existing prescribed asymmetric performance string control method in the sense of the uniformly bounded stability [7].

## 5.1 Introduction

Intelligent connected vehicle platoons and vehicle-to-everything (V2X) collaborative perception enable the interactions between drivers, vehicles and road infrastructure in a new and efficient way. The vehicle platoon systems is said to be the effective technical mean to solve the three major problems in the international traffic field, such as traffic safety, traffic congestion and environmental pollution [8]. The platoon control of intelligent connected vehicles is mainly divided into longitudinal control and lateral control [9]. The platoon longitudinal control consists of vehicle node dynamic model, the spacing policy, the controller design, the communication network topology, and string stability performance [10]. The string stability ensures that the cyber attacks influence on the state of the preceding vehicle node will not cause amplified fluctuations in the states of the following vehicle nodes. In this chapter, the platoon longitudinal control of the vehicles is considered to keep the desired inter-vehicle distance and the string stability under the compound cyber attacks.

As a large-scale cyber-physical system, the vehicle platoon systems can closely integrates the vehicle node entities with the wireless communication networks and control units. As shown in Fig. 5.1, the considered platoon system is a three-layer architecture, including the physical layer, communication layer, and control layer [8, 11]. From the perspective of the wireless network communication layer of the platoon system, VANETs take high-speed driving vehicles in the platoon system and road infrastructure as nodes to form mobile networks. VANETs can realize the V2V and V2I communication interaction. However, the whole platoon system is easy to crash even when only one vehicle node is attacked by the malicious nodes due to the network connection characteristic of the platoon [12]. Malicious nodes often can masquerade as the moving vehicle node or infrastructure node to launch attacks. Under the malicious nodes, the communication channel on the control layer of the vehicle platoon system within their range will be break, the information exchanged in V2V communication also will be tampered with [13]. Ultimately, the hazardous traffic accident occur .

Actually, the vehicle platoon system is known to be vulnerable to a variety of malicious cyber attack nodes due to the open wireless network characteristic and physical limitations. In recent works, a lot of control methods have been developed to deal with the threat of malicious cyber attacks and ensure the robustness of the platoon tracking system. In [14], for the sensor attacks issues, the distributed resilient control and estimation methods was studied for the vehicle platoon system. In [15], to deal with the denial-of-service attacks problems of the heterogeneous multi-agent systems, the distributed output-feedback resilient tracking method was developed. For the HICVPS under the hybrid attacks of false-data injection and message delay attacks, the distributed robust platoon controller was proposed to ensure the inner-vehicle stability and string stability in [16]. However, how to design the distributed platoon resilient control methods and enhance the resilience performance of the

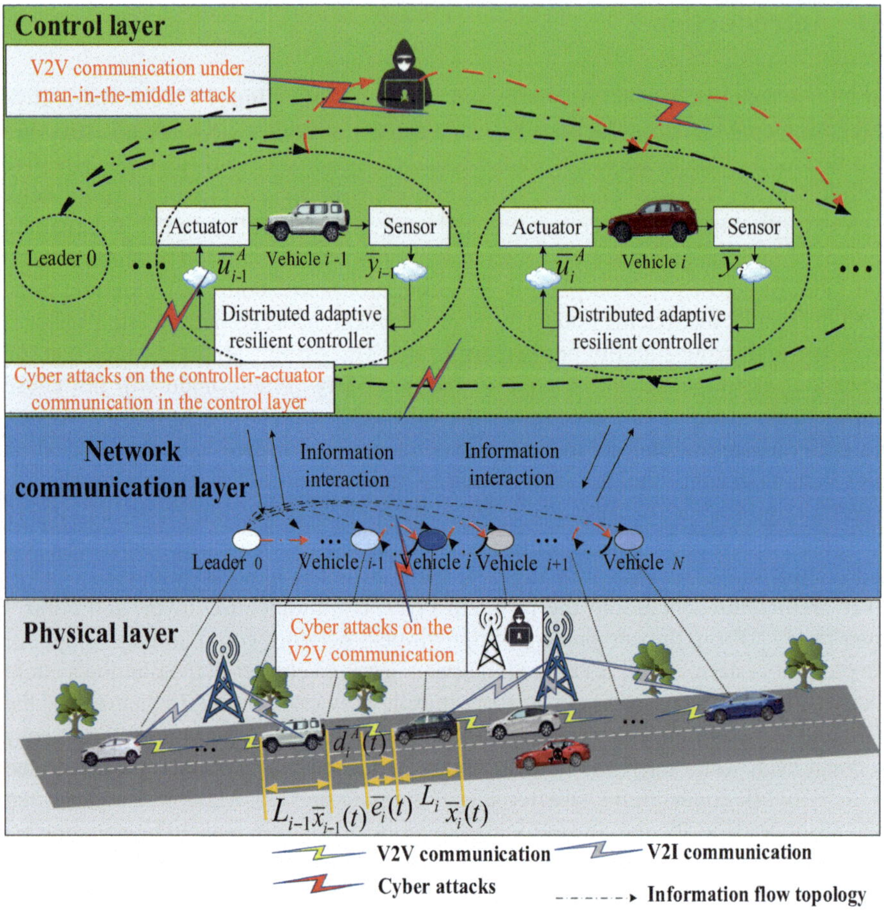


**Fig. 5.1** Three-layer control structure of the HICVPS under compound cyber attacks

vehicle platoon system for the issues of compound cyber attacks on the V2V communication layer and the communication channel of the control layer are still an open and meaningful problem.

From the point of view of resilient security control, the transient and steady-state performances of the spacing error are the important characteristic problems for the longitudinal control of the vehicle platoon system. Due to the safety requirements of collision avoidance and network communication connection maintenance in the practical traffic system and application, the inter-vehicle spacing error usually cannot exceed a certain threshold [17]. Prescribed performance control can guarantee both transient and steady-state performances and has been studied by many scholars. Therefore, prescribed performance control is a good way to cope with the issues of collision avoidance and safety control requirements of the vehicle platoon system under malicious cyber attacks. In [18], the lane control of the autonomous vehicles scheme was proposed to maintain lane-keeping errors within the prescribed performance boundaries under the issues of the rollover prevention and control input saturation. In [19], the adaptive prescribed performance fuzzy tracking control method was also proposed for the nonlinear switched stochastic systems with unknown control directions. In [20], the disturbance observer based prescribed performance adaptive fuzzy containment control was studied to ensure the local neighborhood containment errors can converge into the prescribed boundary. In [21], the adaptive fuzzy output-feedback control was developed to achieve the prescribed performance error constraints for the nonlinear systems. However, it is still an open research topic for how to guarantee the string stability of the platoon in prescribed asymmetric constraints performance and fast finite time stability for the vehicle platoon system with compound cyber attacks.

## 5.2 Problem Statement

In this chapter, the three-layer three-dimensional structure of the HICVPS under the compound cyber attacks is shown in Fig. 5.1, which mainly includes distributed adaptive platoon finite time resilient control layer, physical layer of the HICVPS, and the network communication layer. In the network communication layer, not only the sensor-controller-actuator communication interaction in the closed-loop control loop is included, but also the V2V communication interaction of the VANETs based on the leader bi-directional information flow topology is also involved. However, the open wireless mobile communication network VANETs may make the HICVPS vulnerable to cyber attacks from the malicious nodes in the road traffic system. Deception attacks jointly launched by the multiple malicious nodes within the same range can not only cause the information transmission faults in the controller-actuator communication channel of the control layer, but also destroy the V2V communication from VANETs of the network communication layer as shown in Fig. 5.1. Thus, the interaction states information from the vehicle $i-1$ are also destroyed. The security and stability of the vehicle platoon system control will also be seriously threatened.

### *5.2.1 Modeling of Intelligent Connected Vehicle Platoon Systems*

Consider the HICVPS consisting of $N+1$ intelligent connected vehicles, in which the nonlinear third-order longitudinal dynamics model [6, 7] of $N$ follower vehicles is

$$
\begin{aligned}
\dot{\bar{x}}_i(t) &= \bar{v}_i(t) \\
\dot{\bar{v}}_i(t) &= \bar{a}_i(t) \\
\dot{\bar{a}}_i(t) &= \bar{f}_i(\bar{X}_i) + u_i(t)\tau_i^{-1} + \bar{w}_i(t) \\
\bar{y}_i(t) &= \bar{x}_i(t)
\end{aligned}
\tag{5.1}
$$

where $i = 1, 2, \ldots N$ represents $N$ follower vehicles. $\bar{x}_i(t)$, $\bar{v}_i(t)$, $\bar{a}_i(t)$ are respectively the position, velocity, and acceleration of the vehicle $i$. $\bar{X}_i = \left[\, \bar{v}_i \; \bar{a}_i \,\right]^T$, $\bar{y}_i(t)$ is the output of the platoon system. $u_i(t) = \bar{F}_i(t)/M_i$ represents the control input torque of the vehicle $i$ at time $t$. $\bar{F}_i(t)$ is the output torque of throttle/brake. $\tau_i$ is the engine inertia delay; $M_i$ represents the mass of the vehicle. In engineering applications, due to the limitations of physical equipment and measurement means, the actual dynamics parameters of the vehicles in road traffic environment are often difficult to accurately obtain, such as mechanical damping, air resistance coefficient, etc. Therefore, the smooth nonlinear functions $\bar{f}_i(\bar{X}_i)$ are introduced to represent the bounded unmodeled dynamics. $\bar{w}_i(t)$ describes the external environmental disturbances, including road with slope, strong wind and sand, where $\bar{w}_i(t)$ met the condition $|\bar{w}_i(t)| \le \bar{\bar{w}}_i$, $\bar{\bar{w}}_i \in \mathrm{R}^+$. The dynamics of the leader vehicle in the platoon system is expressed as

$$
\begin{aligned}
\dot{\bar{x}}_0(t) &= \bar{v}_0(t) \\
\dot{\bar{v}}_0(t) &= \bar{a}_0(t)
\end{aligned}
\tag{5.2}
$$

where $\bar{x}_0(t)$, $\bar{v}_0(t)$, and $\bar{a}_0(t)$ are respectively the position, velocity and acceleration of the vehicle 0. In this chapter, the states information of the leader vehicle 0 is considered to be known.

### *5.2.2 Modeling of Malicious Compound Cyber Attacks*

First, considering the deception attacks launched by the malicious nodes within the range, which leads to the information transmission faults of the controller-actuator communication channel of the control layer, the tampered control signal of the HICVPS can be modeled as [22, 23],

$$
u_i^A(t) = u_i(t) + \bar{\varepsilon}_i(t)\bar{\Psi}_i(\bar{X}_i) \tag{5.3}
$$

where $u_i^A(t)$ is the control input torque under the condition of cyber attacks. $\bar{\varepsilon}_i(t)\bar{\Psi}_i(\bar{X}_i)$ is the communication attack function of the control layer, which originates from the attack injection of malicious nodes in the network communication layer and acts on the physical devices of the control layer actuators. $\bar{\varepsilon}_i(t)$ is the unknown time-varying attack weight and meets the condition that $|\bar{\varepsilon}_i(t)| \le \bar{\bar{\varepsilon}}_i$, where $\bar{\bar{\varepsilon}}_i > 0$ is the unknown positive definite constant. $\bar{\Psi}_i(\bar{X}_i)$, $t \ge 0$ represents the cyber attack function of the control layer suffered by the vehicle $i$. $\|\bar{\Psi}_i(\bar{X}_i)\| \leq \bar{Z}_i(\bar{X}_i)$, $\bar{Z}_i(\bar{X}_i) > 0$ is a known bounded function.

Second, different from the feedback channel communication attack, as shown in Fig. 5.1, the state interaction information of V2V communication from the vehicle $i-1$ will be tampered with and replayed. The malicious nodes may launch deception attacks on VANETs through the mobile wireless network of the network communication layer of the HICVPS (which may be induced by man-in-the-middle attacks). Then the impact of the cyber attacks on the actual measured inter-vehicle distance $\bar{d}_i^A(t)$ of the vehicle $i$ [24] can be modeled as follows:

$$\bar{d}_i^A(t) = \bar{x}_{i-1}(t) + \widehat{\Gamma}_i\,(\bar{x}_{i-1}(t)) - \bar{x}_i(t) - L_i \tag{5.4}$$

where $\bar{d}_i^A(t)$ is the actual measurement of inter-vehicle distance of the vehicle $i$ under the influence of cyber attacks, $i = 1, 2, \ldots N$. $L_i$ is the length of the vehicle $i$. $\widehat{\Gamma}_i\,(\bar{x}_{i-1}(t))$ is the considered V2V communication attack model. Due to the physical constraints and concealment requirements of the attack signal, it can be reasonably assumed to be a bounded signal $\widehat{\Gamma}_i\,(\bar{x}_{i-1}(t))$, which can be parameterized as $\Gamma_i(t))\bar{x}_{i-1}(t)$. $\Gamma_i(t) > -1$ is the time-varying bounded weight function, $|\Gamma_i(t)| \leq \bar{\Gamma}_i$ and $\bar{\Gamma}_i$ is bounded. When $\Gamma_i(t) \equiv 0$, then $\bar{d}_i(t) = \bar{x}_{i-1}(t) - \bar{x}_i(t) - L_i$, it means that no communication attacks has occurred in VANETs. The vehicle $i$ is able to have a complete and trusted information exchange with the neighbor vehicle.

**Remark 5.1** By monitoring the communication transmission of the vehicle platoon system, malicious attack nodes launch the deception attacks on the vehicle platoon targets within the range. The obtained system interaction information was utilized to directly tamper with the system communication interaction data and destroy its integrity and credibility. However, due to physical constraints such as communication capacity, energy and resource allocation in high-speed traffic environment, as well as abnormal signal detection, the requirements of covert attack, and the amplitude and information capacity of attack signals are usually bounded [4, 25]. It is therefore reasonable to assume that $\bar{\varepsilon}_i(t)$, $\bar{\Psi}_i(\bar{X}_i)$, and $\widehat{\Gamma}_i\,(\bar{x}_{i-1}(t))$ are bounded.

For realizing the distributed adaptive platoon resilient control, the lemmas and assumptions given are as follows:

**Lemma 5.1** ([2, 19, 21]) *Considering $\bar{f}(\bar{x})$ as an unknown continuous nonlinear function defined on a closed set $\bar{\Omega} \in \mathrm{R}^N$, there exists a fuzzy logic system $\bar{W}^{\mathrm{T}}\bar{\Phi}(\bar{x})$ that makes $\bar{f}(\bar{x}) = \bar{W}^{\mathrm{T}}\bar{\Phi}(\bar{x}) + \bar{\mathrm{E}}(\bar{x})$, $|\bar{\mathrm{E}}(\bar{x})| \leq \bar{\bar{\mathrm{E}}}$, in which $\bar{\mathrm{E}}(\bar{x})$ is the function approximation error, $\bar{\bar{\mathrm{E}}}>0$ is an arbitrarily small positive definite constant. $\bar{x} = [\bar{x}_1, \ldots, \bar{x}_N]^T \in \mathrm{R}^N$ is the input variable, and $\bar{W} = [\bar{W}_1, \ldots, \bar{W}_N]^T$ is the optimal weight vector, $\bar{W} = \arg\min_{\hat{\bar{W}} \in \mathrm{R}^n} \{\sup_{\bar{x} \in \bar{\Omega}} \left|\bar{f}(\bar{x}) - \hat{\bar{W}}^T\bar{\Phi}(\bar{x})\right|\}$, where $\hat{\bar{W}}$ is the estimate of $\bar{W}$. $\bar{\Phi}(\bar{x}) = [\bar{\Phi}_1(\bar{x}), \ldots, \bar{\Phi}_N(\bar{x})]^T$ is the fuzzy basis function vector. $\bar{\Phi}_l(\bar{x}) = (\prod_{i=1}^N \bar{\kappa}_{F_i^l}(\bar{x}_i)) \times \sum_{l=1}^M (\prod_{i=1}^N \bar{\kappa}_{F_i^l}(\bar{x}_i))^{-1}$ where $\bar{\kappa}_{F_i^l}(\bar{x}_i)$ is the membership function of the fuzzy set $F_i^l$, $i = 1, 2, \ldots, N$, $l = 1, 2 \ldots, M$.*

**Lemma 5.2** ([26]) *For any given positive definite constant* $\bar{G} > 0$*, if there is a time-varying function* $\bar{e}(t)$ *satisfying the condition* $|\bar{e}(t)| < \bar{G}$*, then an inequality is* $artanh(\frac{\bar{e}^2(t)}{\bar{G}^2}) < \frac{\bar{e}^2(t)}{\bar{G}^2-\bar{e}^2(t)}$ *established, where* $artanh(\frac{\bar{e}^2(t)}{\bar{G}^2}) = \frac{1}{2}[\ln(1+\frac{\bar{e}^2(t)}{\bar{G}^2}) - \ln(1-\frac{\bar{e}^2(t)}{\bar{G}^2})]$.

**Lemma 5.3** ([27]) *(Fast finite time stability) Considering a nonlinear system* $\dot{\bar{x}} = \bar{G}(\bar{x}, t)$*,* $\bar{x}(0) = 0$*, if there is a positive definite Lyapunov function* $V(\bar{x}) > 0$ *that makes the inequality* $\dot{V}(\bar{x})+\bar{\eta}V(\bar{x})+\bar{\chi}V^r(\bar{x}) \leq 0$ *true,* $\bar{\eta}$*,* $\bar{\chi} > 0$*, and,* $0 < r < 1$*. Then, the system* $\dot{\bar{x}} = \bar{G}(\bar{x}, t)$ *is fast finite-time stable, and the convergence time is satisfied that* $T \leq (\bar{\eta}(1-r))^{-1}\ln(\bar{\eta}V^{1-r}(\bar{x}(0))+\chi)\chi^{-1}$.

**Assumption 5.1** ([17]) $\forall \bar{X}_1, \bar{X}_2 \in R^i$, there exists the constant $\bar{b}_i \in R^+$ that meets the condition that $\left|f_i(\bar{X}_1) - f_i(\bar{X}_2)\right| \leq \bar{b}_i \left\|\bar{X}_1 - \bar{X}_2\right\|, i = 1, 2, \ldots n$, where $f_i(\bar{X}_1)$, $f_i(\bar{X}_2)$ are nonlinear function, $\|\cdot\|$ is the binary norm of vector.

**Control objective** For the HICVPS with compound cyber attacks and external disturbances, the control objectives are as follows:

1. The stability of each follower vehicle in the HICVPS can be guaranteed. The velocity and acceleration of each follower vehicle can quickly be consistent with that of the leader vehicle 0. That is to say, in the process of the HICVPS control, that is $\lim_{t\to\infty} |\bar{e}_i(t)| = 0$, $\lim_{t\to\infty} |\bar{v}_{i-1}(t) - \bar{v}_i(t)| = 0$, and $\lim_{t\to\infty} |\bar{a}_{i-1}(t) - \bar{a}_i(t)| = 0$.
2. The fast finite-time stability of the HICVPS closed-loop under the prescribed asymmetric constraints can be ensured, that is, meet Lemma 5.3 fast finite-time stability. The impact of communication cyber attacks and disturbances on the system is reduced. On the other hand, the string stability of the HICVPS can be satisfied.

## 5.3 Resilient Control Design

For eliminating the impact of compounds cyber attacks and disturbances on the stability of the HICVPS, the distributed adaptive fast finite time platoon resilient control scheme is designed. The control structure is shown in Fig. 5.1. Firstly, the nonlinear state compensator is constructed to ensure the boundness of state signals under compounds cyber attacks. Secondly, the distributed adaptive platoon fast finite time resilient controller with prescribe performance is designed to ensure the string stability and fast finite time stability of closed-loop system under the prescribed asymmetric constraints.

### 5.3.1 Designe of Nonlinear State Compensator

The unknown disturbances and malicious attack nodes may destroy the information transmission of the control layer, and VANETs-based V2V communication attacks will also destroy the control performance and the string stability of the HICVPS, increasing the risk of rear-end collision. Therefore, it is necessary to construct a nonlinear state compensator for eliminating the influence of compound cyber attacks on the state information and ensuring the boundness of the state signals. The vehicle longitudinal dynamics model (5.1) with the communication attacks of the control layer (5.3) can be deformed into:

$$\dot{\bar{a}}_i(t) = \bar{f}_i(\hat{\bar{X}}_i) + \Delta \bar{f}_i + \bar{w}_i(t) + u_i^A(t)\tau_i^{-1} \tag{5.5}$$

where $i = 1, 2, \ldots N$, $\dot{\bar{x}}_i(t) = \bar{v}_i(t)$, $\dot{\bar{v}}_i(t) = \bar{a}_i(t)$, $\hat{\bar{v}}_i(t)$, $\hat{\bar{a}}_i(t)$ are the states estimation of $\bar{v}_i(t)$, $\bar{a}_i(t)$, respectively. To further simplify the derivation, we define that $\hat{\bar{X}}_i = \left[\, \hat{\bar{v}}_i \; \hat{\bar{a}}_i \,\right]^T$, $\Delta \bar{f}_i = \bar{f}_i(\bar{X}_i) - \bar{f}_i(\hat{\bar{X}}_i)$. According to Lemma 5.1, $\hat{\bar{f}}_i(\hat{\bar{X}}_i \,|\, \hat{\bar{W}}_i) = \hat{\bar{W}}_i^T \bar{\Phi}_i(\hat{\bar{X}}_i)$is defined, in which $\hat{\bar{W}}_i$ is the estimation of the unknown optimal weight $\bar{W}_i$, and meets the condition that $\bar{W}_i = \arg \min\limits_{\hat{\bar{W}} \in \bar{\Omega}_i} \{ \sup\limits_{(\hat{\bar{v}}_i, \hat{\bar{a}}_i) \in U_i} \left| \hat{\bar{f}}_i(\hat{\bar{X}}_i \,|\, \hat{\bar{W}}_i) - \bar{f}_i(\hat{\bar{X}}_i) \right| \}$. $\bar{\Omega}_i$ is the closed set belonging to $\hat{\bar{W}}_i$, and $U_i$ is a closed set belonging to $\hat{\bar{v}}_i(t)$ and $\hat{\bar{a}}_i(t)$. $\bar{\mathrm{E}}_i(t)$ is the minimum fuzzy approximation error, and $\bar{\mathrm{E}}_i(t) = \bar{f}_i(\hat{\bar{X}}_i) - \hat{\bar{f}}_i(\hat{\bar{X}}_i \,|\, \bar{W}_i)$, $\left| \bar{\mathrm{E}}_i(t) \right| \le \bar{\bar{\mathrm{E}}}_i$, $\bar{\bar{\mathrm{E}}}_i$ is an unknown positive definite constant. According to the above design, the formula (5.5) is rewritten as

$$\dot{\bar{a}}_i(t) = \bar{W}_i^T \bar{\Phi}_i(\hat{\bar{X}}_i) + \Delta \bar{f}_i + \bar{w}_i(t) + \bar{\mathrm{E}}_i(t) + u_i^A(t)\tau_i^{-1} \tag{5.6}$$

where $\hat{\bar{x}}_i(t)$ is the estimation of $\bar{x}_i(t)$. We define that $\bar{e}_{xi} = \bar{x}_i - \hat{\bar{x}}_i$, $\bar{e}_{vi} = \bar{v}_i - \hat{\bar{v}}_i$, and$\bar{e}_{ai} = \bar{a}_i - \hat{\bar{a}}_i$ are the estimation errors. Using the formula (5.6), the nonlinear state compensator is constructed as

$$\dot{\hat{\bar{a}}}_i(t) = \hat{\bar{W}}_i^T \bar{\Phi}_i(\hat{\bar{X}}_i) + k_3(\bar{y}_i - \hat{\bar{x}}_i) + (u_i(t) + \hat{\bar{\bar{\varepsilon}}}_i \bar{\mathrm{Z}}_i(\hat{\bar{X}}_i))\tau_i^{-1} \tag{5.7}$$

where $\dot{\hat{\bar{x}}}_i(t) = \hat{\bar{v}}_i(t) + k_1(\bar{y}_i - \hat{\bar{x}}_i)$, $\dot{\hat{\bar{v}}}_i(t) = \hat{\bar{a}}_i(t) + k_2(\bar{y}_i - \hat{\bar{x}}_i)$, $k_1$, $k_2$, $k_3$ are the designed compensator gain. $\hat{\bar{\bar{\varepsilon}}}_i$ is the estimate of the attack parameters $\bar{\bar{\varepsilon}}_i$, which will be designed in the next section. According to formula (5.6) and formula (5.7), the compensator error system (5.8) can be obtained as

$$\dot{\bar{e}}_{ai}(t) = \tilde{\bar{W}}_i^T \bar{\Phi}_i(\hat{\bar{X}}_i) - k_3\bar{e}_{xi} + \bar{w}_i(t) + \Delta \bar{f}_i + \bar{\mathrm{E}}_i(t) + \bar{\varepsilon}_i(t)\bar{\Psi}_i(\bar{X}_i)\tau_i^{-1} - \hat{\bar{\bar{\varepsilon}}}_i \bar{\mathrm{Z}}_i(\hat{\bar{X}}_i)\tau_i^{-1} \tag{5.8}$$

where $\dot{\bar{e}}_{xi}(t) = \bar{e}_{vi} - k_1\bar{e}_{xi}$, $\dot{\bar{e}}_{vi}(t) = \bar{e}_{ai} - k_2\bar{e}_{xi}$, $\tilde{\bar{W}}_i = \bar{W}_i - \hat{\bar{W}}_i$.

Let $\bar{e}_i^o = [\bar{e}_{xi}(t), \bar{e}_{vi}(t), \bar{e}_{ai}(t)]^T$, the compensator error system (5.8) is simplified to

$$\dot{\bar{e}}_i^o = A\bar{e}_i^o + \tilde{\bar{F}} + \bar{w} + \bar{\mathrm{E}} + \Delta\bar{f} + \tilde{\bar{\Psi}} \tag{5.9}$$

where $\Delta\bar{f} = \begin{bmatrix} 0 \\ 0 \\ \Delta\bar{f}_i \end{bmatrix}$, $\tilde{\bar{F}} = \begin{bmatrix} 0 \\ 0 \\ \tilde{\bar{W}}_i^T \bar{\Phi}_i(\hat{\bar{X}}_i) \end{bmatrix}$, $\bar{w}(t) = \begin{bmatrix} 0 \\ 0 \\ \bar{w}_i(t) \end{bmatrix}$, $\bar{\mathrm{E}} = \begin{bmatrix} 0 \\ 0 \\ \bar{\mathrm{E}}_i(t) \end{bmatrix}$,
$A = \begin{bmatrix} -k_1 & 1 & 0 \\ -k_2 & 0 & 1 \\ -k_3 & 0 & 0 \end{bmatrix}$, and $\tilde{\bar{\Psi}} = \begin{bmatrix} 0 & 0 & \frac{1}{\tau_i}(\bar{\varepsilon}_i(t)\bar{\Psi}_i(\bar{X}_i) - \hat{\bar{\varepsilon}}_i\bar{Z}_i(\hat{\bar{X}}_i)) \end{bmatrix}^T$.

By selecting the appropriate compensator gain parameters$k_1$, $k_2$, $k_3$, the matrix $A$is made to be a strict Hurwitz matrix. So for any given positive definite symmetric matrix $\bar{Q}^T = \bar{Q} > 0$, there is a positive definite matrix $\bar{P}^T = \bar{P} > 0$ that satisfies the condition $A^T\bar{P} + \bar{P}A = -\bar{Q}$.

***Proof*** For verifying the validity of the proposed nonlinear state compensator (5.7), it is necessary to prove the boundedness of the compensator error system (5.9) by selecting Lyapunov function .We define the Lyapunov function as $\bar{V}_i^0 = \bar{e}_i^{o^T}\bar{P}\bar{e}_i^o$. Based on the formula (5.9), the time derivative with respect to $\bar{V}_i^0$ is obtained

$$\dot{\bar{V}}_i^0 = -\bar{e}_{i^o}^T\bar{Q}\bar{e}_i^o + 2\bar{e}_{i^o}^T\bar{P}(\tilde{\bar{F}} + \bar{w} + \bar{\mathrm{E}} + \Delta\bar{f} + \tilde{\bar{\Psi}}) \tag{5.10}$$

Based on Assumption 4.1, and the Gaussian function properties $0 < \bar{\Phi}_i^T(\hat{\bar{X}}_i(t)) \times \bar{\Phi}_i(\hat{\bar{X}}_i(t)) \leq 1$ in fuzzy logic systems, we have that

$$2\bar{e}_{i^o}^T\bar{P}\bar{w} \leq \left\|\bar{e}_i^o\right\|^2 + \left\|\bar{P}\right\|^2 \bar{\bar{w}}_i^2 \tag{5.11}$$

$$2\bar{e}_{i^o}^T\bar{P}\bar{\mathrm{E}} \leq \left\|\bar{e}_i^o\right\|^2 + \left\|\bar{P}\right\|^2 \bar{\bar{\mathrm{E}}}_i^2 \tag{5.12}$$

$$2\bar{e}_{i^o}^T\bar{P}\Delta\bar{f} \leq \left\|\bar{e}_i^o\right\|^2 + \left\|\bar{P}\right\|^2 \bar{b}_{i1}^2 \left\|\bar{e}_i^o\right\|^2 \tag{5.13}$$

$$2\bar{e}_{i^o}^T\bar{P}\tilde{\bar{F}} \leq \left\|\bar{e}_i^o\right\|^2 + \left\|\bar{P}\right\|^2 \tilde{\bar{W}}_i^T\tilde{\bar{W}}_i \tag{5.14}$$

$$2\bar{e}_{i^o}^T\bar{P}\tilde{\bar{\Psi}} \leq \left\|\bar{e}_i^o\right\|^2 + \tau_i^{-2}\left\|\bar{P}\right\|^2 (\bar{\bar{\varepsilon}}_i^2\bar{Z}_i^2(\bar{X}_i) + \hat{\bar{\varepsilon}}_i^2\bar{Z}_i^2(\hat{\bar{X}}_i)) \tag{5.15}$$

By substituting formula (5.11) to (5.15) into formula (5.10), we get

$$\dot{\bar{V}}_i^0 \leq -\lambda_0\left\|\bar{e}_i^o\right\|^2 + \left\|\bar{P}\right\|^2 \tilde{\bar{W}}_i^T\tilde{\bar{W}}_i + \bar{p}_0 + \tau_i^{-2}\left\|\bar{P}\right\|^2 \hat{\bar{\varepsilon}}_i^2\bar{Z}_i^2(\hat{\bar{X}}_i) \tag{5.16}$$

where $\lambda_{\min}(\bar{Q})$ is the smallest eigenvalue of the matrix $\bar{Q}$, $\bar{b}_{i1}$ is a positive definite constant, $\lambda_0 = \lambda_{\min}(\bar{Q}) - 4 - \|\bar{P}\|^2 \bar{b}_{i1}^2$, $\bar{p}_0 = \|\bar{P}\|^2 \bar{\bar{w}}_i^2 + \|\bar{P}\|^2 \bar{\bar{\mathrm{E}}}_i^2 + \tau_i^{-2} \|\bar{P}\|^2 \bar{\bar{\varepsilon}}_i^2 \bar{\mathrm{Z}}_i^2(\bar{X}_i)$. □

It should be pointed out here that if $\lambda_0 > 0$, $\|\bar{P}\|^2 \tilde{\bar{W}}_i^T \tilde{\bar{W}}_i$, and $\|\bar{P}\|^2 \hat{\bar{\varepsilon}}_i^2 \bar{\mathrm{Z}}_i^2(\hat{\bar{X}}_i)$ are bounded, then the compensator error $\bar{e}_i^o$ is bounded, Thus, it can be seen that the compensator error system (5.9) is stable. Therefore, the boundedness of the function terms $\|\bar{P}\|^2 \tilde{\bar{W}}_i^T \tilde{\bar{W}}_i$, and $\|\bar{P}\|^2 \hat{\bar{\varepsilon}}_i^2 \bar{\mathrm{Z}}_i^2(\hat{\bar{X}}_i)$ will be guaranteed by designing suitable adaptive laws in the subsequent closed-loop stability control design of the system.

### 5.3.2 *Design of Distributed Prescribed Performance Resilient Controller*

In this section, for the problems of the compound cyber-attacks and disturbances of the HICVPS, the distributed adaptive prescribed performance fast finite time resilient control scheme is proposed under the framework of adaptive sliding mode control technology. Then, the fast finite-time stability and the string stability of the closed-loop system under the prescribed asymmetric constraint performance can be ensured.

For controlling the actual measured inter-vehicle distance to the desired inter-vehicle distance while ensuring the string stability under the prescribed asymmetric constraint performance, the inter-vehicle distance error is introduced. In this section, the improved exponential spacing policy proposed in Chap. 4 is used. The desired inter-vehicle spacing $\bar{d}_{si}(t)$ is

$$\bar{d}_{si}(t) = \bar{\delta}_d + \frac{\bar{\eta}_i(\hat{\bar{v}}_i^2 - \bar{v}_0^2)}{2\bar{\varphi}} + \bar{\phi}\left(1 - \exp\left(-\frac{\bar{\bar{v}}_i}{\bar{\omega}}\right)\right) \tag{5.17}$$

where $\bar{\delta}_d$ is the ideal inter-vehicle distance constant in the static state; $\bar{\eta}_i$ represents the safety factor based on road environment and weather conditions, and $\bar{\varphi} > 0$ is the maximum deceleration of the vehicles. $\bar{\bar{v}}_i = \hat{\bar{v}}_i - \bar{v}_0$, $\bar{\phi}$, and $\bar{\omega}$ are the designed positive definite constants.

The utilized improved exponential spacing policy is

$$\begin{aligned}
&\bar{e}_i(t) = \bar{\bar{e}}_i(t) - \bar{\Lambda}_i(t)\\
&\bar{\bar{e}}_i(t) = \bar{\bar{d}}_i^A(t) - \bar{d}_{si}(t)\\
&\bar{\Lambda}_i(t) = [\bar{\bar{e}}_i(0) + \bar{E}_{1i}(t) + \bar{E}_{2i}(t)]\exp(-\bar{\varpi}_i t)
\end{aligned} \tag{5.18}$$

where $\bar{\bar{d}}_i^A(t) = (1 + \Gamma_i(t))\hat{\bar{x}}_{i-1}(t) - \hat{\bar{x}}_i(t) - L_i$, $\bar{x}_0(t) = \hat{\bar{x}}_0(t)$, $\bar{v}_0(t) = \hat{\bar{v}}_0(t)$, $\bar{a}_0(t) = \hat{\bar{a}}_0(t)$. $\bar{E}_{1i}(t) = (\bar{\varpi}_i \bar{\bar{e}}_i(0) + \dot{\bar{\bar{e}}}_i(0))t$, $\bar{E}_{2i}(t) = 0.5(\bar{\varpi}_i^2 \bar{\bar{e}}_i(0) + 2\bar{\varpi}_i \dot{\bar{\bar{e}}}_i(0) +$

$\dddot{\bar{e}}_i(0))t^2$. $\bar{e}_i(t)$ is the spacing error of two adjacent vehicles of the vehicle $i$ and the vehicle $i$–1, $\bar{\varpi}_i$ is the positive definite constant by design. It can be seen from the Eq. (5.18), $\bar{e}_i(0) = 0$, $\dot{\bar{e}}_i(0) = 0$, and $\ddot{\bar{e}}_i(0) = 0$, which show that the initial spacing error $\bar{\bar{e}}_i(t)$ can always be maintained as zero for any initial value. Assuming that no attacks occur when $t = 0$, then, $u_i^A(0) = u_i(0)$ and $\Gamma_i(0) = 0$.

Due to the limited communication range and information perception capability based on VANETs in the vehicle platoon system, such as dedicated Short range Communication Standard (DSRC), it can usually achieve the bilateral communication of high-speed moving targets within tens of meters. Also, for the V2V communication and V2I communication, the maximum communication range can reach 300 m [28].

On the other hand, there usually exists the minimum communication distance for avoiding the collision accidents such as rear-end collision in the process of high-speed movement of the vehicle platoon system. Based on the above analysis, it is of practical significance to consider the inter-vehicle distance error constraint in this chapter. Thus, the actual measured inter-vehicle distance $\bar{\bar{d}}_i^A(t)$ and the desired inter-vehicle distance $\bar{d}_{si}(t)$ are defined to meet the following constraints

$$0 < \bar{h}_{\min} < \bar{\bar{d}}_i^A(t) < \bar{h}_{\max} \tag{5.19}$$

$$0 < \bar{h}_{\min} < \bar{d}_{si}(t) < \bar{h}_{\max} \tag{5.20}$$

where $\bar{h}_{\min}$ represents the minimum safe inter-vehicle distance for avoiding collision. $\bar{h}_{\max}$ represents the maximum effective communication distance for maintaining the tight structure of the vehicle platoon and achieving the limited communication perception capability in reality. In combination with formula (5.19) and (5.20), the prescribed constraint of inter-vehicle spacing error $\bar{e}_i(t)$ can be expressed as

$$\bar{\bar{h}}_{\min} < \bar{e}_i(t) < \bar{\bar{h}}_{\max} \tag{5.21}$$

where $\bar{\bar{h}}_{\min} = \bar{h}_{\min} - \bar{d}_{si\min} + \bar{\varsigma}_{\min,i}$, $\bar{\bar{h}}_{\max} = \bar{h}_{\max} - \bar{d}_{si\min} + \bar{\varsigma}_{\max,i}$, $\bar{d}_{si\min}$ is the minimum value of $\bar{d}_{si}(t)$, and it is satisfied $\bar{h}_{\min} < \bar{d}_{si\min}(t) < \bar{h}_{\max}$. $\bar{\varsigma}_{\min,i}$, and $\bar{\varsigma}_{\max,i}$ are the minimal positive definite constants. When $\left|\bar{\bar{h}}_{\min}\right| \neq \left|\bar{\bar{h}}_{\max}\right|$, the prescribed constraint of inter-vehicle spacing error $\bar{e}_i(t)$ is asymmetric.

For ensuring the steady performance of the prescribed constraint of inter-vehicle spacing error and the compound cyber attacks, the sliding mode surface is defined for the vehicle $i = 1, 2, \ldots N$,

$$\bar{s}_i(t) = \dot{\bar{e}}_i(t) + \sigma_{1i}\bar{e}_i(t) + \int_0^t \sigma_{2i} sgn^{r_1}(\bar{e}_i(t))d\tau \tag{5.22}$$

where $\sigma_{1i}$, $\sigma_{2i}$ are positive definite constants, $r_1 > 1$, $sgn^{r_1}(\bar{e}_i) = |\bar{e}_i|^{r_1} sign(\bar{e}_i)$.

Second, for realizing the string stability of the HICVPS, the coupling sliding mode surface is introduced as

$$\bar{S}_i(t) = \partial_1 \bar{s}_i(t) - \partial_2 \bar{s}_{i+1}(t),\ i = 1, 2, \ldots, N \tag{5.23}$$

where $\partial_1$ and $\partial_2$ are designed positive definite constants. $0 < \partial_1/\partial_2 \le 1$ is the coupling strength between $\bar{s}_i(t)$ and $\bar{s}_{i+1}(t)$. As can be seen from Eq. (5.23), $\bar{S}_i(t)$ and $\bar{s}_i(t)$ have the consistent convergence characteristics. For the vehicle $i = N$, $\bar{s}_{N+1}(t) = \dot{\bar{s}}_{N+1}(t) = 0$.

In addition, for realizing the prescribed performance control, the inter-vehicle spacing error is strictly kept within the prescribed asymmetric constraint domain (5.21). Constructing the hyperbolic arctangent asymmetric barrier Lyapunov function as

$$\bar{V}_{bi}(t) = \frac{1}{2}\overset{\frown S}{\xi}_i\ artanh\left(\frac{\bar{e}_i^2(t)}{\bar{\bar{h}}_{\max}^2}\right) + \frac{1}{2}\breve{\xi}_i^S artanh\left(\frac{\bar{e}_i^2(t)}{\bar{\bar{h}}_{\min}^2}\right) \tag{5.24}$$

where $\overset{\frown S}{\xi}_i = \xi_i^{-1}(1 + sgn(\bar{e}_i(t)))$, $\breve{\xi}_i^S = \xi_i^{-1}(1 - sgn(\bar{e}_i(t)))$, $\xi_i$ is the positive definite constant. According to formula (5.24), it can be seen that $\bar{V}_{bi}(t)$ is positive definite and continuously differentiable in the prescribed asymmetric constraint domain $\bar{\bar{h}}_{\min} < \bar{e}_i(t) < \bar{\bar{h}}_{\max}$.

Therefore, for the vehicle $i = 1, 2, \ldots, N-1$ of the HICVPS, the distributed adaptive prescribed performance fast finite time resilient controller is designed as

$$\begin{aligned} u_i(t) = &-\frac{\bar{\varphi}\tau_i}{\bar{\eta}_i\hat{\bar{\upsilon}}_i}\{\frac{-\breve{k}_i\bar{S}_i}{\partial_1} - \frac{H_i}{\partial_1} - \frac{1}{\partial_1}\frac{\mathfrak{S}_i^2\bar{S}_i(t)}{\left|\mathfrak{S}_i\bar{S}_i(t)\right| + g_{\mathfrak{S}i}} - (1 + \hat{\bar{\Gamma}}_i)\left|\bar{B}_{i-1}\right|\} \\ &-\frac{\bar{\varphi}\tau_i}{\bar{\eta}_i\hat{\bar{\upsilon}}_i}\{\frac{\bar{\eta}_i\hat{\upsilon}_i}{\bar{\varphi}}\hat{\bar{W}}_i^T\bar{\Phi}_i(\hat{\bar{X}}_i) - \frac{\bar{\eta}_i\bar{\bar{\upsilon}}_i}{\bar{\varphi}\tau_i}\hat{\bar{\bar{\varepsilon}}}_i\bar{Z}_i(\hat{\bar{X}}_i) + \frac{\bar{\eta}_i\hat{\bar{\upsilon}}_i}{\bar{\varphi}}k_3\bar{e}_{xi}\} \end{aligned} \tag{5.25}$$

where $\mathfrak{S}_i = \partial_1[-(\hat{\bar{a}}_i(t) + (k_2 - k_1^2)\bar{e}_{xi} + k_1\bar{e}_{\upsilon i}) - \bar{\eta}_i(\hat{\bar{a}}_i(t) + k_2\bar{e}_{xi})\bar{\varphi}^{-1}\hat{\bar{a}}_i(t) - \bar{\eta}_i(\hat{\bar{a}}_i(t) + k_2\bar{e}_{xi})\bar{\varphi}^{-1}k_2\bar{e}_{xi} - \bar{\eta}_i\hat{\bar{\upsilon}}_i\bar{\varphi}^{-1}k_2(\bar{e}_{\upsilon i} - k_1\bar{e}_{xi}) + \bar{\eta}_i\bar{a}_0^2\bar{\varphi}^{-1} + \frac{\bar{\phi}\dot{\bar{\bar{\upsilon}}}_i^2}{\bar{\omega}^2}\exp(-\frac{\bar{\bar{\upsilon}}_i}{\bar{\omega}}) - \ddot{\bar{\Lambda}}_i + \sigma_{1i}\dot{\bar{e}}_i(t) + \sigma_{2i}\mathrm{sgn}^{r_1}(\bar{e}_i(t))] - \partial_2\dot{\bar{s}}_{i+1}(t)$, $H_i = \overset{\frown S}{\xi}_i[\frac{\bar{\bar{h}}_{\max}^2\bar{e}_i(t)}{\bar{\bar{h}}_{\max}^4 - \bar{e}_i^4(t)}(\frac{1}{\partial_1} + \frac{\partial_2}{\partial_1}\frac{\bar{s}_{i+1}^2(t)\bar{S}_i}{\left|\bar{s}_{i+1}\bar{S}_i\right| + \iota_{hi}})] + \breve{\xi}_i^S[\frac{\bar{\bar{h}}_{\min}^2\bar{e}_i(t)}{\bar{\bar{h}}_{\min}^4 - \bar{e}_i^4(t)}(\frac{1}{\partial_1} + \frac{\partial_2}{\partial_1}\frac{\bar{s}_{i+1}^2(t)\bar{S}_i}{\left|\bar{s}_{i+1}\bar{S}_i\right| + \iota_{hi}})]$ $\bar{B}_{i-1} = (\hat{\bar{a}}_{i-1}(t) + (k_2 - k_1^2)\bar{e}_{x(i-1)} + k_1\bar{e}_{\upsilon(i-1)})$, $\bar{e}_{xi} = \bar{y}_i - \hat{\bar{x}}_i$, $\bar{e}_{\upsilon i} = \dot{\bar{y}}_i - \hat{\bar{\upsilon}}_i$, $g_{\mathfrak{S}i}$, $\iota_{hi}$ are all the small positive definite constants, $\breve{k}_i$ is the positive definite constant, and meets that $\breve{k}_i > 4^{-1}\bar{\partial}_1\bar{\eta}_i\bar{\bar{\upsilon}}_i\bar{\varphi}^{-1}$, $\bar{\bar{\upsilon}}_i = \sup\left|\hat{\bar{\upsilon}}_i\right|$.

For the vehicle $i = N$ of the HICVPS, the distributed adaptive prescribed performance fast finite time resilient controller is designed as

$$u_N(t) = -\frac{\bar{\varphi}\tau_N}{\bar{\eta}_N\hat{\upsilon}_N}\{\frac{-\breve{k}_N\bar{S}_N}{\partial_1} - \frac{H_N}{\partial_1} - \frac{1}{\partial_1}\frac{\mathfrak{S}_N^2\bar{S}_N(t)}{\left|\mathfrak{S}_N\bar{S}_N(t)\right| + g_{\mathfrak{S}N}} - \frac{\bar{\eta}_N\bar{\bar{\upsilon}}_N}{\bar{\varphi}\tau_N}\hat{\bar{\bar{\varepsilon}}}_N(t)\bar{Z}_N(\hat{\bar{X}}_N)\}$$

$$-\frac{\bar{\varphi}\tau_N}{\bar{\eta}_N\hat{\upsilon}_N}\{-(1+\hat{\bar{\Gamma}}_N(t))\left|\bar{B}_{N-1}\right| + \frac{\bar{\eta}_N\hat{\bar{\upsilon}}_N}{\bar{\varphi}}\hat{\bar{W}}_N^T\bar{\Phi}_N(\hat{\bar{X}}_N) + \frac{\bar{\eta}_i\hat{\bar{\upsilon}}_i}{\bar{\varphi}}k_3\bar{e}_{xi}\} \tag{5.26}$$

where $\bar{S}_N(t) = \partial_1\bar{s}_N(t), \bar{s}_{N+1}(t) = \dot{\bar{s}}_{N+1}(t) = 0$, $H_N = \frac{\overset{S}{\widehat{\bar{\xi}}}_N}{\partial_1}\frac{\bar{\bar{h}}_{\max}^2\bar{e}_N(t)}{\bar{\bar{h}}_{\max}^4 - \bar{e}_N^4(t)} + \frac{\overset{S}{\breve{\bar{\xi}}}_N}{\partial_1}\frac{\bar{\bar{h}}_{\min}^2\bar{e}_N(t)}{\bar{\bar{h}}_{\min}^4 - \bar{e}_N^4(t)}$.

The constructed adaptive update laws ($i = 1, 2, \ldots N$) are

$$\dot{\hat{\bar{W}}}_i = \alpha_{wi}\partial_1\frac{\bar{\eta}_i\hat{\bar{\upsilon}}_i}{\bar{\varphi}}\bar{\Phi}_i(\hat{\bar{X}}_i)\bar{S}_i(t) - \wp_{wi}\hat{\bar{W}}_i \tag{5.27}$$

$$\dot{\hat{\bar{\bar{\varepsilon}}}}_i = \alpha_{\varepsilon i}\partial_1\frac{\bar{\eta}_i\bar{\bar{\upsilon}}_i}{\bar{\varphi}\tau_i}\bar{Z}_i(\hat{\bar{X}}_i)\bar{S}_i(t) - \wp_{\varepsilon i}\hat{\bar{\bar{\varepsilon}}}_i \tag{5.28}$$

$$\dot{\hat{\bar{\Gamma}}}_i = \alpha_{\Gamma i}\partial_1\left|\bar{B}_{i-1}\right|\bar{S}_i(t) - \wp_{\Gamma i}\hat{\bar{\Gamma}}_i \tag{5.29}$$

where $\hat{\bar{W}}_i, \hat{\bar{\Gamma}}_i, \hat{\bar{\bar{\varepsilon}}}_i$ are the estimation of $\bar{W}_i, \bar{\Gamma}_i, \bar{\bar{\varepsilon}}_i$, respectively. The estimated errors are respectively defined as $\tilde{\bar{W}}_i = \bar{W}_i - \hat{\bar{W}}_i$, $\tilde{\bar{\Gamma}}_i = \bar{\Gamma}_i - \hat{\bar{\Gamma}}_i$, $\tilde{\bar{\bar{\varepsilon}}}_i = \bar{\bar{\varepsilon}}_i - \hat{\bar{\bar{\varepsilon}}}_i$. $\alpha_{wi}$, $\alpha_{\Gamma i}$, and$\alpha_{\varepsilon i}$are positive constants, $\hat{\bar{W}}_i(0) \geq 0$, $\hat{\bar{\Gamma}}_i(0) \geq 0$, $\hat{\bar{\bar{\varepsilon}}}_i(0) \geq 0$. $\wp_{\varepsilon i}$, $\wp_{\Gamma i}$ and $\wp_{wi}$ are the minimal positive definite constants and satisfy the condition $\wp_{wi}(2\alpha_{wi})^{-1} > \partial_1\bar{\eta}_i\bar{\bar{\upsilon}}_i\bar{\varphi}^{-1} + \left\|\bar{P}\right\|^2 > 0$, and $\wp_{\varepsilon i}(2\alpha_{\varepsilon i})^{-1} > (\left\|\bar{P}\right\|^2\bar{Z}_i^2(\hat{\bar{X}}_i) + 1)\tau_i^{-2}\left\|\bar{P}\right\|^2\bar{Z}_i^2(\hat{\bar{X}}_i) > 0$.

### 5.3.3 Analysis of System Stability

**Theorem 5.1** *Considering the problems that the HICVPS (5.1) is subjected to the controller-actuator communication attacks of the control layer (5.3) and the V2V communication attacks based on VANETs (5.4). Based on the nonlinear state compensator (5.7), hyperbolic arctangent asymmetric barrier Lyapunov function (5.24), and the adaptive compensation and updated laws (5.27) to (5.29), the distributed adaptive prescribed performance fast finite time resilient controllers (5.25)(5.26) are designed. The fast finite time stability of closed-loop systems of the inter-vehicle spacing errors (5.18) is achieved under the prescribed asymmetric constraints (5.21). The string stability is also guaranteed under the cyber attacks conditions, and the convergence time of the system is* $T \leqslant \breve{k}_i^{-1}\ln(2\sqrt{2}\breve{k}_i\bar{\bar{V}}^{0.5}(0) + 2\widehat{\Psi}_i)(2\widehat{\Psi}_{ii})^{-1}$

***Proof*** First, for verifying the ultimately uniformly bounded stability and string stability of the heterogeneous closed-loop vehicle platoon system, the Lyapunov function is selected as

$$\bar{V}(t) = \sum_{i=1}^{N} (\bar{V}_{si}(t) + \bar{V}_{ei}(t) + \bar{V}_{bi}(t) + \bar{V}_i^0(t)) \tag{5.30}$$

where $\bar{V}_{si}(t) = (1/2)\bar{S}_i^2(t)$, $\bar{V}_{ei}(t) = \bar{V}_{wi}(t) + \bar{V}_{\Gamma i}(t) + \bar{V}_{\varepsilon i}(t)$, $\bar{V}_{wi}(t) = (1/2\alpha_{wi})$ $\tilde{\bar{W}}_i^T \tilde{\bar{W}}_i$, $\bar{V}_{\Gamma i}(t) = (1/2\alpha_{\Gamma i})\tilde{\bar{\Gamma}}_i^2$, and $\bar{V}_{\varepsilon i}(t) = (1/2\alpha_{\varepsilon i})\tilde{\bar{\varepsilon}}_i^2$.

Taking the time derivative of $\bar{S}_i(t)$, $i = 1, 2, \ldots, N-1$ and substituting it into Eqs. (5.18), (5.22) and (5.23) to obtain,

$$\dot{\bar{S}}_i(t) = \partial_1 \dot{\bar{s}}_i(t) - \partial_2 \dot{\bar{s}}_{i+1}(t) \leq \partial_1[(1 + \Gamma_i(t))\bar{B}_{i-1} - \bar{\eta}_i \hat{\bar{\upsilon}}_i \bar{\varphi}^{-1} \dot{\hat{\bar{a}}}_i] + \mathsf{S}_i \tag{5.31}$$

For the vehicle$i = N$, it can be seen that $\bar{S}_N(t) = \partial_1 \bar{s}_N(t)$. Taking the derivative of $\bar{S}_N(t)$ and combining the formula (5.23), we can obtain that

$$\dot{\bar{S}}_N(t) = \partial_1 \dot{\bar{s}}_N(t) \leq \partial_1[(1 + \Gamma_N(t))\bar{B}_{N-1} - \bar{\eta}_N \hat{\bar{\upsilon}}_N \bar{\varphi}^{-1} \dot{\hat{\bar{a}}}_N] + \mathsf{S}_N \tag{5.32}$$

where $\mathsf{S}_i$, $i = 1, 2, \ldots, N$ are defined in Eqs. (5.25) and (5.26).

Further, by substituting Eqs. (5.31) and (5.32) into $\dot{\bar{V}}_{si}(t)$ , and combining with Eqs. (5.3) and (5.7), we get

$$\begin{aligned} \dot{\bar{V}}_{si}(t) \leqslant & -\partial_1 \bar{S}_i(t) \frac{\bar{\eta}_i \hat{\bar{\upsilon}}_i}{\bar{\varphi}} [\hat{\bar{W}}_i^T \bar{\Phi}_i(\hat{\bar{X}}_i) + k_3 \bar{e}_{xi} + \frac{1}{\tau_i} \bar{\varepsilon}_i(t) \bar{\Psi}_i(\hat{\bar{X}}_i) + \frac{u_i(t)}{\tau_i}] \\ & + \mathfrak{S}_i \bar{S}_i(t) + \partial_1(1 + \Gamma_i(t))\bar{B}_{i-1}\bar{S}_i(t) \end{aligned} \tag{5.33}$$

According to the above analysis, the conditions of $|\bar{\varepsilon}_i(t)| \leq \bar{\bar{\varepsilon}}_i$, $|\Gamma_i(t)| \leq \bar{\Gamma}_i$, $\left\| \bar{\Psi}_i(\hat{\bar{X}}_i) \right\| \leq \bar{\mathsf{Z}}_i(\hat{\bar{X}}_i)$ are obtained, and the (5.33) can be derived as

$$\begin{aligned} \dot{\bar{V}}_{si}(t) \leqslant & \partial_1\{(1 + \bar{\Gamma}_i)\left|\bar{B}_{i-1}\right| - \frac{\bar{\eta}_i \hat{\bar{\upsilon}}_i}{\bar{\varphi}} \hat{\bar{W}}_i^T \bar{\Phi}_i(\hat{\bar{X}}_i) - \frac{\bar{\eta}_i \hat{\bar{\upsilon}}_i}{\bar{\varphi}\tau_i} u_i(t)\}\bar{S}_i(t) \\ & + \partial_1\{\frac{\bar{\eta}_i \hat{\bar{\upsilon}}_i}{\bar{\varphi}\tau_i} \bar{\bar{\varepsilon}}_i \bar{Z}_i(\hat{\bar{X}}_i) - \frac{\bar{\eta}_i \hat{\bar{\upsilon}}_i}{\bar{\varphi}} k_3 \bar{e}_{xi}\}\bar{S}_i(t) + \left|\mathfrak{S}_i \bar{S}_i(t)\right| \end{aligned} \tag{5.34}$$

Combining with the proposed distributed adaptive resilient controller (5.25) and (5.26), the formula (5.34) can be expressed as

$$\dot{\bar{V}}_{si}(t) \leqslant -\breve{k}_i \bar{S}_i^2 + g_{\mathfrak{S}i} + \partial_1 \tilde{\bar{\Gamma}}_i \bar{B}_{i-1} \bar{S}_i(t) + \partial_1 \frac{\bar{\eta}_i \bar{\bar{\upsilon}}_i}{\bar{\varphi}\tau_i} \tilde{\bar{\bar{\varepsilon}}}_i \bar{Z}_i(\hat{\bar{X}}_i)\bar{S}_i(t) - H_i \bar{S}_i(t) \tag{5.35}$$

where $-\mathfrak{S}_i^2 \bar{S}_i^2(t)(\left|\mathfrak{S}_i \bar{S}_i(t)\right| + g_{\mathfrak{S}i})^{-1} \leqslant -\left|\mathfrak{S}_i \bar{S}_i(t)\right| + g_{\mathfrak{S}i}$.

Taking the time derivative of $\bar{V}_{bi}(t)$ and combining the Eqs. (5.22) and (5.23) to get

$$
\begin{aligned}
\dot{V}_{bi}(t) &= [\widehat{\xi}_i^S \frac{\bar{\bar{h}}_{\max}^2 \bar{e}_i(t)}{\bar{\bar{h}}_{\max}^4 - \bar{e}_i^4(t)} + \breve{\xi}_i^S \frac{\bar{\bar{h}}_{\min}^2 \bar{e}_i(t)}{\bar{\bar{h}}_{\min}^4 - \bar{e}_i^4(t)}]\dot{\bar{e}}_i(t) \\
&\leqslant [\widehat{\xi}_i^S \frac{\bar{\bar{h}}_{\max}^2 \bar{e}_i(t)}{\bar{\bar{h}}_{\max}^4 - \bar{e}_i^4(t)} + \breve{\xi}_i^S \frac{\bar{\bar{h}}_{\min}^2 \bar{e}_i(t)}{\bar{\bar{h}}_{\min}^4 - \bar{e}_i^4(t)}]\times \\
&\quad (\frac{\bar{S}_i(t)}{\partial_1} + \partial_2 \left|\bar{s}_{i+1}(t)\bar{S}_i(t)\right| (\partial_1)^{-1} - \sigma_{1i}\bar{e}_i(t))
\end{aligned} \tag{5.36}
$$

Meanwhile, according to Lemma 5.2, the following conditions can be obtained,

$$
\begin{aligned}
&\widehat{\xi}_i^S [\frac{\bar{\bar{h}}_{\max}^2 \bar{e}_i(t)}{\bar{\bar{h}}_{\max}^4 - \bar{e}_i^4(t)} (\partial_1^{-1}\bar{S}_i(t) + \partial_2\partial_1^{-1}\left|\bar{s}_{i+1}(t)\bar{S}_i(t)\right| - \sigma_{1i}\bar{e}_i(t))] \\
\leqslant\ &\widehat{\xi}_i^S [\frac{\bar{\bar{h}}_{\max}^2 \bar{e}_i(t)}{\bar{\bar{h}}_{\max}^4 - \bar{e}_i^4(t)} (\partial_1^{-1}\bar{S}_i(t) + \partial_2\partial_1^{-1}\left|\bar{s}_{i+1}(t)\bar{S}_i(t)\right|) - \frac{\sigma_{1i}\bar{\bar{h}}_{\max}^2}{\bar{\bar{h}}_{\max}^2 + e_i^2(t)} (\frac{\bar{e}_i^2(t)}{\bar{\bar{h}}_{\max}^2 - \bar{e}_i^2(t)})] \\
\leqslant\ &\widehat{\xi}_i^S [\frac{\bar{\bar{h}}_{\max}^2 \bar{e}_i(t)}{\bar{\bar{h}}_{\max}^4 - \bar{e}_i^4(t)} (\partial_1^{-1}\bar{S}_i(t) + \partial_2\partial_1^{-1}\left|\bar{s}_{i+1}(t)\bar{S}_i(t)\right|) - \frac{\sigma_{1i}\bar{\bar{h}}_{\max}^2}{\bar{\bar{h}}_{\max}^2 + \bar{e}_i^2(t)} \operatorname{artanh}(\frac{\bar{e}_i^2(t)}{\bar{\bar{h}}_{\max}^2})]
\end{aligned} \tag{5.37}
$$

Similar to formula (5.37), based on Lemma 5.2, can be obtained

$$
\begin{aligned}
&\breve{\xi}_i^S [\frac{\bar{\bar{h}}_{\min}^2 \bar{e}_i(t)}{\bar{\bar{h}}_{\min}^4 - \bar{e}_i^4(t)} (\partial_1^{-1}\bar{S}_i(t) + \partial_2\partial_1^{-1}\left|\bar{s}_{i+1}(t)\bar{S}_i(t)\right| - \sigma_{1i}\bar{e}_i(t))] \\
\leqslant\ &\breve{\xi}_i^S [\frac{\bar{\bar{h}}_{\min}^2 \bar{e}_i(t)}{\bar{\bar{h}}_{\min}^4 - \bar{e}_i^4(t)} (\partial_1^{-1}\bar{S}_i(t) + \partial_2\partial_1^{-1}\left|\bar{s}_{i+1}(t)\bar{S}_i(t)\right|) - \frac{\sigma_{1i}\bar{\bar{h}}_{\min}^2}{\bar{\bar{h}}_{\min}^2 + \bar{e}_i^2(t)} \operatorname{artanh}(\frac{\bar{e}_i^2(t)}{\bar{\bar{h}}_{\min}^2})]
\end{aligned} \tag{5.38}
$$

Based on the formula (5.37) and the formula (5.38), the formula (5.36) can be transformed into

$$
\begin{aligned}
\dot{V}_{bi}(t) \leqslant\ &\widehat{\xi}_i^S [\frac{\bar{\bar{h}}_{\max}^2 \bar{e}_i(t)}{\bar{\bar{h}}_{\max}^4 - \bar{e}_i^4(t)} (\partial_1^{-1}\bar{S}_i(t) + \partial_2\partial_1^{-1}\left|\bar{s}_{i+1}(t)\bar{S}_i(t)\right|)] \\
&- \widehat{\xi}_i^S \frac{\sigma_{1i}\bar{\bar{h}}_{\max}^2}{\bar{\bar{h}}_{\max}^2 + \bar{e}_i^2(t)} \operatorname{artanh}(\frac{\bar{e}_i^2(t)}{\bar{\bar{h}}_{\max}^2}) - \breve{\xi}_i^S \frac{\sigma_{1i}\bar{\bar{h}}_{\min}^2}{\bar{\bar{h}}_{\min}^2 + \bar{e}_i^2(t)} \operatorname{artanh}(\frac{\bar{e}_i^2(t)}{\bar{\bar{h}}_{\min}^2}) \\
&+ \breve{\xi}_i^S [\frac{\bar{\bar{h}}_{\min}^2 \bar{e}_i(t)}{\bar{\bar{h}}_{\min}^4 - \bar{e}_i^4(t)} (\partial_1^{-1}\bar{S}_i(t) + \partial_2\partial_1^{-1}\left|\bar{s}_{i+1}(t)\bar{S}_i(t)\right|)]
\end{aligned} \tag{5.39}
$$

Furthermore, the time derivative of $\bar{V}_{ei}(t)$ is

$$\dot{\bar{V}}_{ei}(t) = -\frac{1}{\alpha_{wi}}\tilde{\bar{W}}_i^T\dot{\hat{\bar{W}}}_i - \frac{1}{\alpha_{\Gamma i}}\tilde{\bar{\Gamma}}_i\dot{\hat{\bar{\Gamma}}}_i - \frac{1}{\alpha_{\varepsilon i}}\tilde{\bar{\bar{\varepsilon}}}_i\dot{\hat{\bar{\bar{\varepsilon}}}}_i \tag{5.40}$$

By substituting the adaptive laws formula (5.27), formula (5.28) and formula (5.29) into the above formula, we get

$$\begin{aligned}\dot{\bar{V}}_{ei}(t) = &-\eth_1\frac{\bar{\eta}_i\hat{\bar{\upsilon}}_i}{\bar{\varphi}}\tilde{\bar{W}}_i^T\bar{\Phi}_i(\hat{\bar{X}}_i)\bar{S}_i(t) + \frac{\wp_{wi}}{\alpha_{wi}}\tilde{\bar{W}}_i^T\hat{\bar{W}}_i + \frac{\wp_{\varepsilon i}}{\alpha_{\varepsilon i}}\tilde{\bar{\bar{\varepsilon}}}_i\hat{\bar{\bar{\varepsilon}}}_i\\ &-\eth_1\frac{\bar{\eta}_i\hat{\bar{\upsilon}}_i}{\bar{\varphi}\tau_i}\tilde{\bar{\varepsilon}}_i Z_i(\hat{\bar{X}}_i)\bar{S}_i(t)_i - \eth_1\tilde{\bar{\Gamma}}_i\left|B_{i-1}\right|\bar{S}_i(t) + \frac{\wp_{\Gamma i}}{\alpha_{\Gamma i}}\tilde{\bar{\Gamma}}_i\hat{\bar{\Gamma}}_i\end{aligned} \tag{5.41}$$

By formula (5.35), and (5.41), we can obtain that

$$\begin{aligned}\dot{\bar{V}}_{si}(t) + \dot{\bar{V}}_{ei}(t) \leqslant &-\breve{k}_i\bar{S}_i^2 + g_{\mathfrak{S}i} - H_i\bar{S}_i(t) + \frac{\wp_{wi}}{\alpha_{wi}}\tilde{\bar{W}}_i^T\hat{\bar{W}}_i\\ &-\eth_1\frac{\bar{\eta}_i\hat{\bar{\upsilon}}_i}{\bar{\varphi}}\tilde{\bar{W}}_i^T\bar{\Phi}_i(\hat{\bar{X}}_i)\bar{S}_i(t) + \frac{\wp_{\varepsilon i}}{\alpha_{\varepsilon i}}\tilde{\bar{\bar{\varepsilon}}}_i\hat{\bar{\bar{\varepsilon}}}_i + \frac{\wp_{\Gamma i}}{\alpha_{\Gamma i}}\tilde{\bar{\Gamma}}_i\hat{\bar{\Gamma}}_i\end{aligned} \tag{5.42}$$

In conjunction with (5.42), and $\dot{\bar{V}}_{bi}(t)$ (5.39), we can further derive,

$$\begin{aligned}&\dot{\bar{V}}_{si}(t) + \dot{\bar{V}}_{ei}(t) + \dot{\bar{V}}_{bi}(t)\\ &\leqslant -\breve{k}_i\bar{S}_i^2 + g_{\mathfrak{S}i} + \frac{\wp_{wi}}{\alpha_{wi}}\tilde{\bar{W}}_i^T\hat{\bar{W}}_i - \eth_1\frac{\bar{\eta}_i\hat{\bar{\upsilon}}_i}{\bar{\varphi}}\tilde{\bar{W}}_i^T\bar{\Phi}_i(\hat{\bar{X}}_i)\bar{S}_i(t)\\ &+\frac{\wp_{\varepsilon i}}{\alpha_{\varepsilon i}}\tilde{\bar{\bar{\varepsilon}}}_i\hat{\bar{\bar{\varepsilon}}}_i + \frac{\wp_{\Gamma i}}{\alpha_{\Gamma i}}\tilde{\bar{\Gamma}}_i\hat{\bar{\Gamma}}_i + \breve{\xi}_i^S\Big[\frac{\bar{\bar{h}}_{\max}^2\bar{e}_i(t)\iota_{hi}}{\bar{\bar{h}}_{\max}^4 - \bar{e}_i^4(t)}\frac{\eth_2}{\eth_1} - \frac{\sigma_{1i}\bar{\bar{h}}_{\max}^2}{\bar{\bar{h}}_{\max}^2 + \bar{e}_i^2(t)}\operatorname{artanh}\Big(\frac{\bar{e}_i^2(t)}{\bar{\bar{h}}_{\max}^2}\Big)\Big]\\ &+\breve{\xi}_i^S\Big[\frac{\bar{\bar{h}}_{\min}^2\bar{e}_i(t)\iota_{hi}}{\bar{\bar{h}}_{\min}^4 - \bar{e}_i^4(t)}\frac{\eth_2}{\eth_1} - \frac{\sigma_{1i}\bar{\bar{h}}_{\min}^2}{\bar{\bar{h}}_{\min}^2 + \bar{e}_i^2(t)}\operatorname{artanh}\Big(\frac{\bar{e}_i^2(t)}{\bar{\bar{h}}_{\min}^2}\Big)\Big]\end{aligned} \tag{5.43}$$

where $-\bar{s}_{i+1}^2(t)\bar{S}_i^2(\left|\bar{s}_{i+1}\bar{S}_i\right| + \iota_{hi})^{-1} \leqslant -\left|\bar{s}_{i+1}(t)\bar{S}_i(t)\right| + \iota_{hi}$.

Secondly, formula (5.43) can be further expressed as

$$\begin{aligned}&\dot{\bar{V}}_{si}(t) + \dot{\bar{V}}_{ei}(t) + \dot{\bar{V}}_{bi}(t)\\ &\leqslant -\breve{k}_i\bar{S}_i^2 + g_{\mathfrak{S}i} + \frac{\wp_{wi}}{\alpha_{wi}}\tilde{\bar{W}}_i^T\hat{\bar{W}}_i - \frac{\eth_1\bar{\eta}_i\hat{\bar{\upsilon}}_i}{\bar{\varphi}}\tilde{\bar{W}}_i^T\bar{\Phi}_i(\hat{\bar{X}}_i)\bar{S}_i(t)\end{aligned}$$

$$+\frac{\wp_{\varepsilon i}}{\alpha_{\varepsilon i}}\tilde{\bar{\varepsilon}}_i\hat{\bar{\varepsilon}}_i+\frac{\wp_{\Gamma i}}{\alpha_{\Gamma i}}\tilde{\bar{\Gamma}}_i\hat{\bar{\Gamma}}_i+E_i-2\xi_i^a[\frac{\overset{\frown}{\xi}{}_i^S}{2}\operatorname{artanh}(\frac{\bar{e}_i^2(t)}{\bar{\bar{h}}_{\max}^2})+\frac{\overset{\smile}{\xi}{}_i^S}{2}\operatorname{artanh}(\frac{\bar{e}_i^2(t)}{\bar{\bar{h}}_{\min}^2})] \tag{5.44}$$

where $E_i=\max\{\sup(\frac{\bar{\bar{h}}_{\max}^2\iota_{hi}}{\bar{\bar{h}}_{\max}^4-\bar{e}_i^4(t)}\frac{\partial_2}{\partial_1}),\sup(\frac{\bar{\bar{h}}_{\min}^2\iota_{hi}}{\bar{\bar{h}}_{\min}^4-\bar{e}_i^4(t)}\frac{\partial_2}{\partial_1})\}\frac{2|\bar{e}_i(t)|}{\xi_i},\xi_i^a=\min\{\inf(\frac{\sigma_{1i}\bar{\bar{h}}_{\min}^2}{\bar{\bar{h}}_{\min}^2+\bar{e}_i^2(t)}),\inf(\frac{\sigma_{1i}\bar{\bar{h}}_{\max}^2}{\bar{\bar{h}}_{\max}^2+\bar{e}_i^2(t)})\}>0$.

Using Young's inequality, we can see

$$-\tilde{\bar{W}}_i^T\bar{\Phi}_i(\hat{\bar{X}}_i)\bar{S}_i(t)\leqslant\tilde{\bar{W}}_i^T\tilde{\bar{W}}_i+\frac{1}{4}\bar{S}_i^2(t) \tag{5.45}$$

Substituting Eqs. (5.24) and (5.45) into the above Eq. (5.44),

$$\begin{aligned}&\dot{\bar{V}}_{si}(t)+\dot{\bar{V}}_{ei}(t)+\dot{\bar{V}}_{bi}(t)\\&\leqslant-\breve{k}_i\bar{S}_i^2+g_{\mathfrak{S}i}+\frac{\wp_{wi}}{\alpha_{wi}}\tilde{\bar{W}}_i^T\hat{\bar{W}}_i+\frac{\partial_1\bar{\eta}_i\bar{\bar{\upsilon}}_i}{4\bar{\varphi}}\bar{S}_i^2(t)+\frac{\partial_1\bar{\eta}_i\bar{\bar{\upsilon}}_i}{\bar{\varphi}}\tilde{\bar{W}}_i^T\tilde{\bar{W}}_i\\&\quad+\frac{\wp_{\varepsilon i}}{\alpha_{\varepsilon i}}\tilde{\bar{\varepsilon}}_i\hat{\bar{\varepsilon}}_i+\frac{\wp_{\Gamma i}}{\alpha_{\Gamma i}}\tilde{\bar{\Gamma}}_i\hat{\bar{\Gamma}}_i+E_i-2\xi_i^a\bar{V}_{bi}(t)\end{aligned} \tag{5.46}$$

Based on Young's inequality, the following conclusions can be obtained

$$\frac{\wp_{wi}}{\alpha_{wi}}\tilde{\bar{W}}_i^T\hat{\bar{W}}_i\leqslant-\frac{\wp_{wi}}{2\alpha_{wi}}\tilde{\bar{W}}_i^T\tilde{\bar{W}}_i+\frac{\wp_{wi}}{2\alpha_{wi}}\bar{W}_i^T\bar{W}_i \tag{5.47}$$

$$\frac{\wp_{\varepsilon i}}{\alpha_{\varepsilon i}}\tilde{\bar{\varepsilon}}_i\hat{\bar{\varepsilon}}_i\leqslant-\frac{\wp_{\varepsilon i}}{2\alpha_{\varepsilon i}}\tilde{\bar{\varepsilon}}_i^2+\frac{\wp_{\varepsilon i}}{2\alpha_{\varepsilon i}}\bar{\bar{\varepsilon}}_i^2 \tag{5.48}$$

$$\frac{\wp_{\Gamma i}}{\alpha_{\Gamma i}}\tilde{\bar{\Gamma}}_i\hat{\bar{\Gamma}}_i\leqslant-\frac{\wp_{\Gamma i}}{2\alpha_{\Gamma i}}\tilde{\bar{\Gamma}}_i^2+\frac{\wp_{\Gamma i}}{2\alpha_{\Gamma i}}\bar{\Gamma}_i^2 \tag{5.49}$$

By substituting formula (5.47) to (5.49) into formula (5.46), we get

$$\begin{aligned}&\dot{\bar{V}}_{si}(t)+\dot{\bar{V}}_{ei}(t)+\dot{\bar{V}}_{bi}(t)+\dot{\bar{V}}_i^0\\&\leqslant-\hbar_k\bar{S}_i^2+g_{\mathfrak{S}i}-\widehat{\wp}_{wi}\tilde{\bar{W}}_i^T\tilde{\bar{W}}_i+\wp_{wi}(2\alpha_{wi})^{-1}\bar{W}_i^T\bar{W}_i\\&-\widehat{\wp}_{\varepsilon i}\tilde{\bar{\varepsilon}}_i^2+E_i+\widehat{\wp}^1{}_{\varepsilon i}\bar{\bar{\varepsilon}}_i^2-2\lambda_0\bar{V}_i^0+\bar{p}_0-\wp_{\Gamma i}(2\alpha_{\Gamma i})^{-1}\tilde{\bar{\Gamma}}_i^2\\&\quad+\wp_{\Gamma i}(2\alpha_{\Gamma i})^{-1}\bar{\Gamma}_i^2-2\xi_i^a\bar{V}_{bi}(t)\end{aligned} \tag{5.50}$$

where $\overset{1}{\widehat{\wp}}_{\varepsilon i} = \tau_i^{-2} + \tau_i^{-2}\|\bar{P}\|^2\bar{Z}_i^2(\hat{\bar{X}}_i) + \wp_{\varepsilon i}(2\alpha_{\varepsilon i})^{-1}$, $\hbar_k = \breve{k}_i - \eth_1\bar{\eta}_i\bar{\bar{\upsilon}}_i(4\bar{\varphi})^{-1}$, $\widehat{\wp}_{\varepsilon i} = -\tau_i^{-2}\|\bar{P}\|^4\bar{Z}_i^4(\hat{\bar{X}}_i) - \tau_i^{-2}\|\bar{P}\|^2\bar{Z}_i^2(\hat{\bar{X}}_i) + \wp_{\varepsilon i}(2\alpha_{\varepsilon i})^{-1}$, $\widehat{\wp}_{wi}=\wp_{wi}(2\alpha_{wi})^{-1} - \eth_1\bar{\eta}_i\bar{\upsilon}_i\bar{\varphi}^{-1} - \|\bar{P}\|^2$, $\hbar_k, \widehat{\wp}_{wi}, \widehat{\wp}_{\varepsilon i} \in \mathbb{R}^+$.

From formula (5.50), it can be further obtained that,

$$
\begin{aligned}
\dot{\bar{V}}(t) &= \sum_{i=1}^{N}\{\dot{\bar{V}}_{si}(t) + \dot{\bar{V}}_{ei}(t) + \dot{\bar{V}}_{bi}(t) + \dot{\bar{V}}_i^0\} \\
&\leqslant \sum_{i=1}^{N}\{-2\hbar_k\bar{V}_{si}(t) - 2\alpha_{wi}\widehat{\wp}_{wi}\bar{V}_{wi}(t) - 2\alpha_{\varepsilon i}\widehat{\wp}_{\varepsilon i}\bar{V}_{\varepsilon i}(t)\} \\
&+ \sum_{i=1}^{N}\{-\wp_{\Gamma i}\bar{V}_{\Gamma i}(t) - 2\xi_i^a\bar{V}_{bi}(t) + g_{\mathfrak{S}i} + \frac{\wp_{wi}}{2\alpha_{wi}}\bar{W}_i^T\bar{W}_i + \overset{1}{\widehat{\wp}}_{\varepsilon i}\bar{\bar{\varepsilon}}_i^2\} \\
&+ \sum_{i=1}^{N}\{\frac{\wp_{\Gamma i}}{2\alpha_{\Gamma i}}\bar{\Gamma}_i^2 + E_i - 2\lambda_0\bar{V}_i^0 + \bar{p}_0\} \\
&\leqslant -\mathfrak{F}_1\bar{V} + \mathfrak{F}_2
\end{aligned} \tag{5.51}
$$

where $\mathfrak{F}_1$ and $\mathfrak{F}_2$ both are positive constants. According to Formula (5.25) to (5.28), $\mathfrak{F}_1 = \min\{2\hbar_k, \min_{1\leqslant i\leqslant N} 2\alpha_{wi}\widehat{\wp}_{wi}, \min_{1\leqslant i\leqslant N} \wp_{\Gamma i}, \min_{1\leqslant i\leqslant N} 2\alpha_{\varepsilon i}\widehat{\wp}_{\varepsilon i}, 2\xi_i^a, 2\lambda_0\}$, $\mathfrak{F}_2 = \sum_{i=1}^{N}\{g_{\mathfrak{S}i} + \frac{\wp_{wi}}{2\alpha_{wi}}\bar{W}_i^T\bar{W}_i + \frac{\wp_{\Gamma i}}{2\alpha_{\Gamma i}}\bar{\Gamma}_i^2 + \overset{1}{\widehat{\wp}}_{\varepsilon i}\bar{\bar{\varepsilon}}_i^2 + E_i + \bar{p}_0\}$.

Thus, multiply the attenuation function $e^{\mathfrak{F}_1 t}$ on both sides of the Eq. (5.51) and integrate it on the $[0, t]$ to obtain $\bar{V}(t) \leqslant -\mathfrak{F}_1\bar{V} + \mathfrak{F}_2 \leqslant (\bar{V}(0) - \frac{\mathfrak{F}_2}{\mathfrak{F}_1})e^{-\mathfrak{F}_1 t} + \frac{\mathfrak{F}_2}{\mathfrak{F}_1} \leqslant \bar{V}(0) + \frac{\mathfrak{F}_2}{\mathfrak{F}_1}$, Therefore, according to Lyapunov stability theory, $\bar{S}_i(t), \tilde{\bar{W}}_i, \bar{\tilde{\Gamma}}_i(t), \bar{\tilde{\varepsilon}}_i(t)$, and $\bar{e}_i^o$ are bounded. Then, $\|\bar{P}\|^2\tilde{\bar{W}}_i^T\tilde{\bar{W}}_i$, $\|\bar{P}\|^2\hat{\bar{\varepsilon}}_i^2\bar{Z}_i^2(\hat{\bar{X}}_i)$ in Eq. (5.15) are bounded, as well as $\hat{\bar{\Gamma}}_i(t)$, $\hat{\bar{\varepsilon}}_i(t)$, $\hat{\bar{W}}_i$ in closed-loop systems are bounded. Therefore, when the initial condition $\bar{V}(0) > 0$ are met, the vehicle platoon system of closed-loop heterogeneity is ultimately uniformly bounded, which means that there is a normal number large enough $\bar{V}_{\max}$ to satisfy the conditions,

$$
\begin{aligned}
\bar{V}(t) &\leqslant \sum_{i=1}^{N}\{\bar{S}_i^2(0) + \tilde{\bar{W}}_i^T\tilde{\bar{W}}_i(0) + \bar{\tilde{\varepsilon}}_i^2(0) + \bar{\tilde{\Gamma}}_i^2(0) + \bar{e}_i^o(0)\} \\
&+ \sum_{i=1}^{N}\{(1 + \operatorname{sgn}(\bar{e}_i(0)))\operatorname{artanh}(\frac{\bar{e}_i^2(0)}{\bar{\bar{h}}_{\max}^2}) + (1 - \operatorname{sgn}(\bar{e}_i(0)))\operatorname{artanh}(\frac{\bar{e}_i^2(t)}{\bar{\bar{h}}_{\min}^2})\} \\
&\leqslant 2\bar{V}_{\max}\widehat{\mu}^{-1}
\end{aligned} \tag{5.52}
$$

where $\widehat{\mu}= \max\{1, \min\limits_{1\leq i\leq N} \alpha_{wi}, \min\limits_{1\leq i\leq N} \alpha_{\Gamma i}, \min\limits_{1\leq i\leq N} \alpha_{\varepsilon i}, \min\limits_{1\leq i\leq N} \xi_i^a\}$. In addition, the condition $\lim\limits_{t\to\infty} \sum\limits_{i=1}^{N} \frac{1}{2}\bar{S}_i^2(t) \leqslant \bar{V}(0) + \frac{\mathfrak{F}_2}{\mathfrak{F}_1}$ is also met. Therefore, $\bar{V}_{bi}(t)$ and $\bar{S}_i(t)$ are bounded, and can converge to the neighborhood of the origin. Using formula (5.21) and (5.22), $\bar{s}_i(t)$ and $\bar{e}_i(t)$ can also converge to the neighborhood of the origin. It can be seen from formula (5.18) that, $\bar{e}_i(0)$=0, and $\bar{V}_{bi}(0)$ is positive definite, the prescribed asymmetric constraint domain of the initial condition $\bar{\bar{h}}_{\min} < \bar{e}_i(0) < \bar{\bar{h}}_{\max}$ can be satisfied, then it can converge to the prescribed asymmetric constraint domain $\bar{\bar{h}}_{\min} < \bar{e}_i(t) < \bar{\bar{h}}_{\max}$ by increasing $\mathtt{F}_1$ and decreasing $\mathtt{F}_2$ at the same time. At the same time, according to Definition 3.1 of string stability, since $\bar{e}_i(t)$ is uniformly bounded, for a given arbitrary $\phi > 0$, $\exists\varphi > 0$, then, $\sup\limits_i \max\{|\bar{e}_i(0)|, |\dot{\bar{e}}_i(0)|\} < \phi \Rightarrow \sup\limits_i \sup\limits_{t\geq 0} \max\{|\bar{e}_i(t)|, |\dot{\bar{e}}_i(t)|\} < \varphi$. Therefore, the stability of the HICVPS can be guaranteed.

Secondly, for verifying the fast finite-time stability of the closed-loop HICVPS, the Lyapunov function is selected as

$$\bar{\bar{V}}(t) = \sum_{i=1}^{N} \bar{V}_{si}(t) \tag{5.53}$$

Taking the time derivative of formula (5.52), and we get

$$\begin{aligned}\dot{\bar{\bar{V}}}(t) \leqslant &\sum_{i=1}^{N}\{-\breve{k}_i\bar{S}_i^2 + g_{\mathfrak{S}i} + \eth_1\tilde{\bar{\Gamma}}_i\left|\bar{B}_{i-1}\right|\bar{S}_i(t)\}\\ &+\sum_{i=1}^{N}\{\eth_1\frac{\bar{\eta}_i\bar{\upsilon}_i}{\varphi\tau_i}\bar{\bar{\varepsilon}}_i Z_i(\hat{\bar{X}}_i)\bar{S}_i(t) - \breve{\xi}_i^S\frac{\bar{\bar{h}}_{\max}^2\bar{e}_i(t)}{\bar{\bar{h}}_{\max}^4 - \bar{e}_i^4(t)}(\frac{\eth_2}{\eth_1}\frac{\bar{s}_{i+1}^2(t)\bar{S}_i^2}{\left|\bar{s}_{i+1}\bar{S}_i\right| + \iota_i} + \frac{\bar{S}_i(t)}{\eth_1})\}\\ &-\sum_{i=1}^{N}\{\breve{\xi}_i^S\frac{\bar{\bar{h}}_{\min}^2\bar{e}_i(t)}{\bar{\bar{h}}_{\min}^4 - \bar{e}_i^4(t)}(\frac{\bar{S}_i(t)}{\eth_1} + \frac{\eth_2}{\eth_1}\frac{\bar{s}_{i+1}^2(t)\bar{S}_i^2}{\left|\bar{s}_{i+1}\bar{S}_i\right| + \iota_i})\}\end{aligned} \tag{5.54}$$

where $\bar{s}_{N+1}(t) = \dot{\bar{s}}_{N+1}(t) = 0$. Further, the Eq. (5.53) can be simplified to

$$\begin{aligned}\dot{\bar{\bar{V}}}(t) \leqslant &\sum_{i=1}^{N}\{-\breve{k}_i\bar{S}_i^2 + g_{\mathfrak{S}i} + \eth_1\tilde{\bar{\Gamma}}_i\left|\bar{B}_{i-1}\right|\bar{S}_i(t)\}\\ &+\sum_{i=1}^{N}\{\eth_1\frac{\bar{\eta}_i\bar{\upsilon}_i}{\varphi\tau_i}\bar{\bar{\varepsilon}}_i Z_i(\hat{\bar{X}}_i)\bar{S}_i(t) - K_{1i}(\frac{\bar{S}_i(t)}{\eth_1} + \frac{\eth_2}{\eth_1}\left|\bar{s}_{i+1}\bar{S}_i\right| - \frac{\eth_2}{\eth_1}\iota_i)\}\end{aligned} \tag{5.55}$$

where $K_{1i} = \frac{2|\bar{e}_i(t)|}{\xi_i} \min\{\inf(\frac{\bar{h}^2_{\max}}{\bar{h}^4_{\max} - \bar{e}^4_i(t)}), \inf(\frac{\bar{h}^2_{\min}}{\bar{h}^4_{\min} - \bar{e}^4_i(t)})\}$, $-\frac{\bar{s}^2_{i+1}(t)\bar{S}^2_i}{|\bar{s}_{i+1}\bar{S}_i| + \iota_i} \leq -|\bar{s}_{i+1}\bar{S}_i| + \iota_i$.

Based on the above analysis of the ultimately uniformly bounded stability of the closed loop system, we can know $\tilde{\bar{\Gamma}}_i(t)$ and $\tilde{\bar{\varepsilon}}_i(t)$ are bounded. Therefore, there are two positive definite constants $\breve{\bar{\Gamma}}_i$ and $\breve{\bar{\varepsilon}}_i$ that satisfy the conditions $\left|\tilde{\bar{\Gamma}}_i\right| \leq \breve{\bar{\Gamma}}_i$ and $\left|\tilde{\bar{\varepsilon}}_i\right| \leq \breve{\bar{\varepsilon}}_i$. Then we have

$$-\widehat{\Psi}_i \geqslant \partial_1 \breve{\bar{\Gamma}}_i \left|\bar{B}_{i-1}\right| + g_{\Theta i} + \partial_1 \frac{\bar{\eta}_i \bar{\bar{v}}_i}{\bar{\varphi}\tau_i} \breve{\bar{\varepsilon}}_i \bar{Z}_i(\hat{\bar{X}}_i) - K_{1i}\frac{1}{\partial_1} - K_{1i}\frac{\partial_2}{\partial_1}|\bar{s}_{i+1}| + K_{1i}\frac{\partial_2}{\partial_1}\iota_i \tag{5.56}$$

When the above condition formula (5.55) is satisfied, $\widehat{\Psi}_i \in \mathrm{R}^+$, the formula (5.54) can be further derived as

$$\begin{aligned}\dot{\bar{\bar{V}}}(t) &\leqslant \sum_{i=1}^{N}\{-\breve{k}_i\bar{S}^2_i - \widehat{\Psi}_i\left|\bar{S}_i\right|\} \\ &\leqslant -2\breve{k}_i\bar{\bar{V}}(t) - \sqrt{2}\widehat{\Psi}_i\bar{\bar{V}}^{0.5}(t)\end{aligned} \tag{5.57}$$

Based on the fast finite time stability Lemma 5.3, it can be seen that the closed-loop HICVPS is fast finite time stable. $\bar{e}_i(t)$ can converge to the prescribed asymmetric domain $\bar{\bar{h}}_{\min} < \bar{e}_i(t) < \bar{\bar{h}}_{\max}$ in finite time $T$, $T \leq \frac{1}{\breve{k}_i}\ln\frac{2\sqrt{2}\breve{k}_i\bar{\bar{V}}^{0.5}(0) + 2\widehat{\Psi}_i}{2\widehat{\Psi}_{ii}}$. Proof completed. □

## 5.4 Method Verification and Analysis

In this section, the validity of the distributed adaptive prescribed performance fast finite time resilient control scheme in Theorem 5.1 will be verified by two sets of comparative experimental cases. The control model in the experimental case is a heterogeneous platoon composed of seven heterogeneous vehicles, $i = 0, 1, 2, \ldots 6$, and $N = 6$. The considered the nonlinear function in the longitudinal dynamics model as $f_i(\bar{v}_i(t), \bar{a}_i(t)) = -(\bar{a}_i + \iota_i\bar{v}^2_i M^{-1}_i + \bar{d}_{ri}M^{-1}_i)\tau^{-1}_i - 2\iota_i\bar{v}_i\bar{a}_i M^{-1}_i$, where $\iota_i$ denotes the air resistance coefficient, $\bar{d}_{ri}$ is the mechanical resistance of the vehicle, and $\tau_i$ denote the inertia delay of the engine. The external disturbances function is $w_i(t) = 0.05\sin(2t)$, the control input torque is bounded, $\|u_i(t)\| \leq 50N\cdot m$. For the exponential spacing policy, we select $\bar{\phi} = 0.11$, $\bar{\omega} = 1.38$, and $\bar{\varpi}_i = 2.3$. For the longitudinal dynamics parameters of the vehicle, we select $\iota_i = 0.31$, $\bar{d}_{ri} = 0.22N$, the desired inter-vehicle distance $\bar{\delta}_d = 8\,\mathrm{m}$, when the vehicles are stationary, the safety factor is $\bar{\eta}_i = 0.21$, and $\bar{\varphi} = 7.1\,\mathrm{m/s^2}$. The dynamic parameters of the

**Table 5.1** The dynamic parameters of the HICVPS

| Vehicle $i = 1, 2, \ldots, 6$ | 1 | 2 | 3 | 4 | 5 | 6 |
|---|---|---|---|---|---|---|
| $\bar{M}_i(\times 10^3\ \text{kg})$ | 1.8 | 1.85 | 1.8 | 1.9 | 1.65 | 1.4 |
| $\tau_i\,(s)$ | 0.4 | 0.85 | 0.5 | 0.55 | 0.45 | 0.6 |
| $\bar{\eta}_i$ | 3.5 | 4.5 | 5 | 5 | 4 | 4.5 |
| $L_i\,(m)$ | 1.8 | 1.85 | 1.8 | 1.9 | 1.65 | 1.4 |

HICVPS are shown in Table 5.1. The acceleration of the vehicle $i = 0$ is $\bar{a}_0\ (t) = \begin{cases} 0.48t\ \text{m/s}^2,\ 1s \le t < 5s \\ 2.4\ \text{m/s}^2,\ 5s \le t < 10s \\ -0.48t + 7.2\ \text{m/s}^2,\ 10s \le t < 15s \\ 0\,\text{m/s}^2,\ otherwise \end{cases}$.

For the nonlinear state compensator, firstly, the fuzzy membership function is selected as $\mu_{\gamma_j^r}(S_{i,j}) = \exp(-0.5(S_{i,j} + 2r - 6)^2)$, where $r = 1, 2, 3, 4, 5$, $j = 1, 2$, $S_{i,1} = \hat{\bar{v}}_i(t)$, $S_{i,2} = \hat{\bar{a}}_i(t)$. The fuzzy basis function is defined as $\bar{\Phi}_{ir}(\hat{\bar{X}}_i) = \frac{\mu_{\gamma_1^r}(S_{i,1})\mu_{\gamma_2^r}(S_{i,2})}{\sum_{r=1}^{5}(\mu_{\gamma_1^r}(S_{i,1})\mu_{\gamma_2^r}(S_{i,2}))}$, where $\bar{\Phi}_i(\hat{\bar{X}}_i) = [\bar{\Phi}_{i1}(\hat{\bar{X}}_i), \bar{\Phi}_{i2}(\hat{\bar{X}}_i), \ldots, \bar{\Phi}_{i5}(\hat{\bar{X}}_i)]^T$, $\hat{\bar{W}}_i^T = [\hat{\bar{W}}_{i1}, \hat{\bar{W}}_{i2}, \ldots, \hat{\bar{W}}_{i5}]$. We select $\bar{Q}=$ diag$\{4.5, 4.5, 4.5\}$, the compensator gains $k_1$, $k_2$, $k_3$ are 2, 4, 6, respectively. In additional, let matrix $A$ strictly be Hurwitz matrix.

Considering the controller-actuator communication cyber attacks in the control layer, the attack functions are $\bar{\varepsilon}_i(t) = 0.2\cos(t)$, $\bar{\Psi}_i(\bar{X}_i) = 0.2(\bar{a}_i^2 + \bar{v}_i^2 + 0.08)\sin(t)$, $20\,s \le t < 21\,s$, $i = 4$. Considering the V2V communication attacks problem based on VANETs, the attack function is $\widehat{\Gamma}_i\ (\bar{x}_{i-1}(t)) = 0.001\sin(2t)\bar{x}_{i-1}(t)$, $30\,s \le t < 31\,s$, $i = 4$.

By calculating the root mean square error (RMSE) of the spacing error, the comparison of control accuracy of the control performance in Case 1 and Case 2 is shown in Table 5.2, which indicates that the proposed method has higher convergence accuracy.

### 5.4.1 Case 1

For the HICVPS that suffers from multiple compound cyber attacks and external disturbances, we consider the minimum safe distance is $\bar{h}_{\min} = 7.5$ m, and the maximum effective communication distance is $\bar{h}_{\max} = 8.6$ m under the prescribed asymmetric constraint domain. Then, the closed-loop fast finite time stability and string stability of the system in Theorem 5.1 are verifyied, the safety requirements of collision avoidance is ensured and the network communication is maintained. $\bar{\varsigma}_{\max,i} = 0.4105$ and $\bar{\varsigma}_{\min,i} = 0.0105$ are used to compensate for the initial preset spacing error. In

**Table 5.2** Comparison of control performance indicators

| Vehicle $i = 1, 2, \ldots, 6$ | 1 | 2 | 3 | 4 | 5 | 6 |
|---|---|---|---|---|---|---|
| Performance metrics (RMSE) in Case 1 | | | | | | |
| Proposed method | 0.1374 | 0.0909 | 0.0334 | 0.0816 | 0.0155 | 0.0107 |
| Existing traditional method | 0.3255 | 0.2387 | 0.1621 | 0.1443 | 0.0769 | 0.0539 |
| Performance metrics (RMSE) in Case 2 | | | | | | |
| Proposed method | 0.0816 | 0.0419 | 0.0255 | 0.0801 | 0.0138 | 0.0102 |
| Existing traditional method | 0.7613 | 0.5770 | 0.3982 | 0.4020 | 0.1943 | 0.1380 |

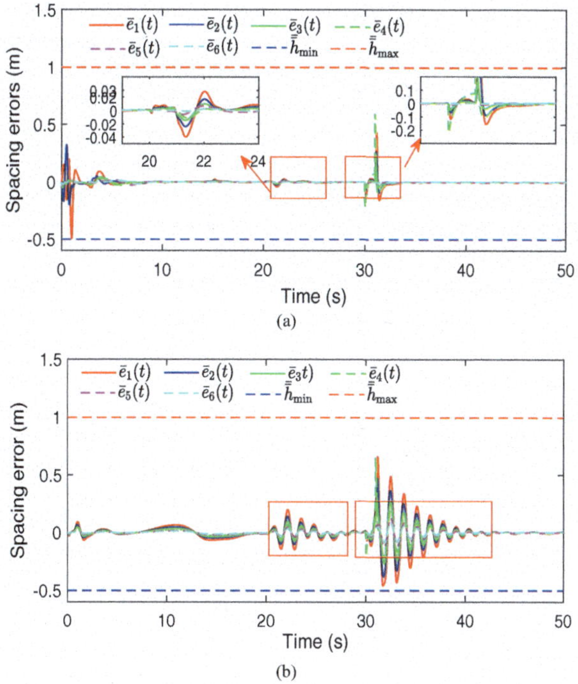


**Fig. 5.2** Experimental results of spacing error under prescribed asymmetric constraints in Case 1. **a** $\bar{e}_i(t)$ with proposed method, **b** $\bar{e}_i(t)$ with existing control method

Case 1, the experimental results of comparison between the proposed Theorem 5.1 method and the existing traditional control method [6, 7] are shown in Figs. 5.2, 5.3, 5.4, 5.5 and 5.6.

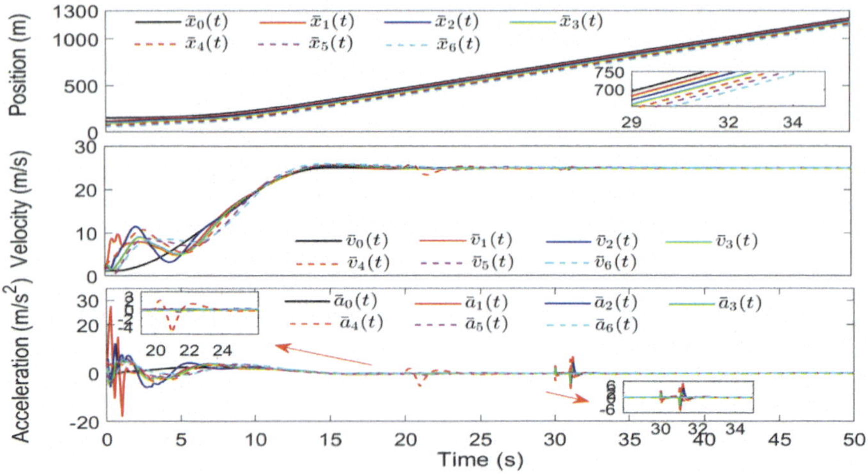


(a)

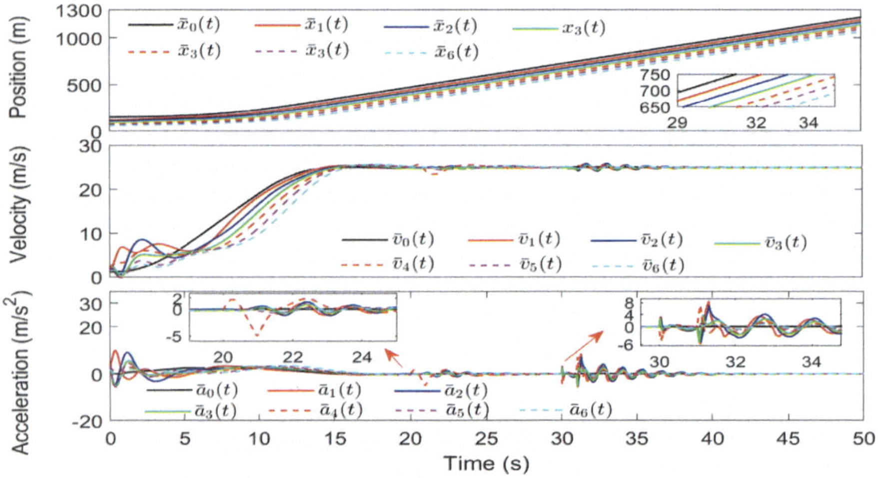


(b)

**Fig. 5.3** Experimental results under prescribed asymmetric constraints in Case 1. **a** System states with proposed method, **b** System states with existing control method

In the HICVPS, the control layer of vehicle $i = 4$suffers from the cyber attacks of controller-actuator communication during $t = 20s \sim 21s$, and the states information from the vehicle $i = 3$ is maliciously tampered during $t = 30s \sim 31s$. As a result, the communication interaction is broken between VANETs-based the vehicle $i = 3$ and the vehicle $i = 4$, as shown in Fig. 5.2a. Compared with Fig. 5.2b (the convergence time $T \leq 20s$), the inter-vehicle spacing error under the proposed method achieves a faster convergence rate in finite time ($T \leq 10s$).

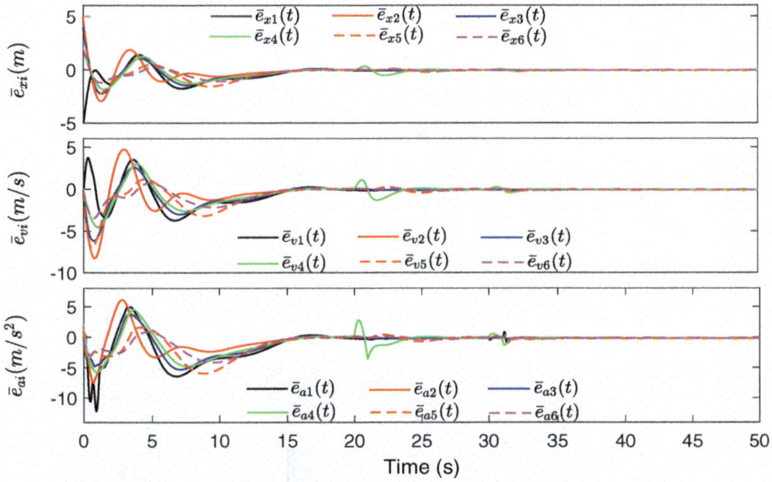


**Fig. 5.4** Compensator errors under proposed method in Case 1

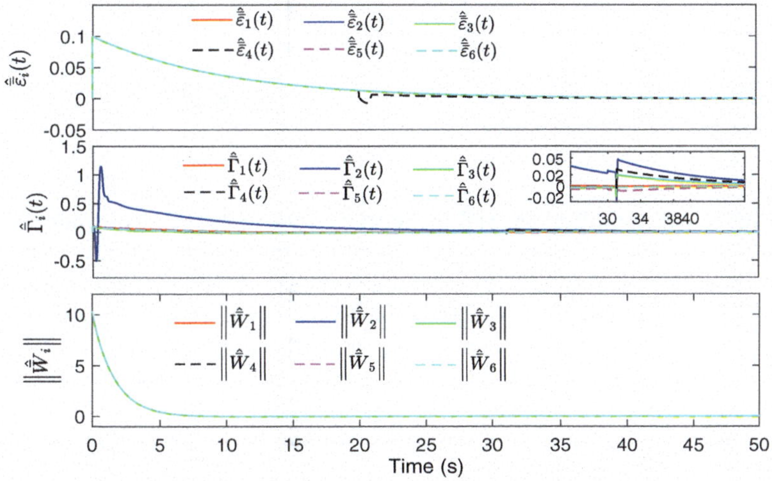


**Fig. 5.5** Adaptive compensation and updated laws under proposed method in Case 1

Meanwhile, under prescribed asymmetric constraints, where $\bar{\bar{h}}_{\min} = -0.5\,\text{m}$, $\bar{\bar{h}}_{\max} = 1\,\text{m}$. Compared with Fig. 5.2a, b shows a larger transient spacing error under communication attacks, which proves that the proposed method can effectively ensure the string stability of the HICVPS. The impact of multiple attacks and disturbances on the system is also eliminated. Figure 5.3a, b show the state of the system under the proposed method and the existing method, such as position, velocity and acceleration. Compared to Fig. 5.3b, a shows better states tracking accuracy and less

**Fig. 5.6** Experimental results of control inputs under prescribed asymmetric constraints in Case 1. **a** Proposed method, **b** Existing control method

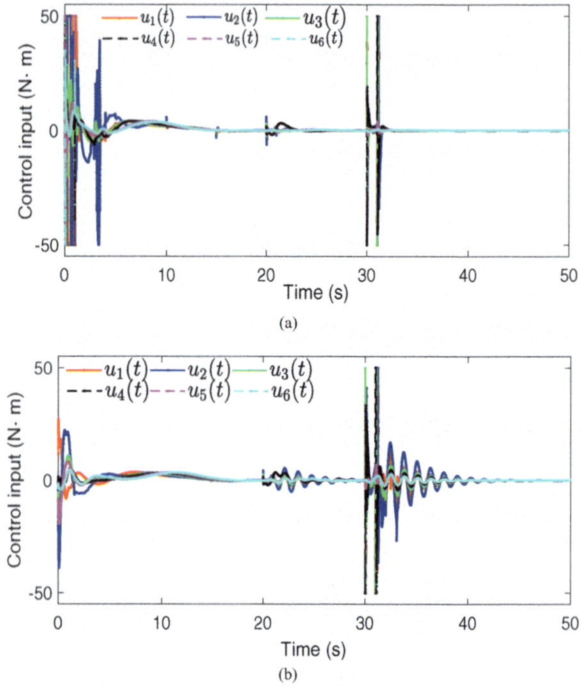


chattering for the vehicles, as well as better system robustness for multiple communication attacks. The observation errors of the designed nonlinear compensator are shown in Fig. 5.4, which verifies the effectiveness of the designed compensator. Figure 5.5 shows the adaptive compensation estimation of unknown parameters and weights. Figure 5.6 shows the control input under the proposed control method and the existing method. In summary, compared with the results of existing control methods, the proposed method enhances the robustness and resilience performances of the HICVPS to multiple communication attacks and disturbances, and the stability of the string under prescribed asymmetric constraints is also ensured.

### 5.4.2 *Case 2*

Based on Case 1, the prescribed asymmetric constraint range is further reduced in Case 2 to further verify the superiority of the proposed control method. We selected the minimum safe distance $\bar{h}_{\min} = 7.7\,\mathrm{m}$, the maximum effective communication distance is $\bar{h}_{\max} = 8.6\,\mathrm{m}$, $\bar{\varsigma}_{\max,i} = 0.2105$, and $\bar{\varsigma}_{\min,i} = 0.0105$. Thus, we can see that $\bar{\bar{h}}_{\min} = -0.3\,\mathrm{m}$, $\bar{\bar{h}}_{\max} = 0.8\,\mathrm{m}$. The comparison experimental results between the proposed method in Case 2 and the existing control method are shown in Figs. 5.7, 5.8, 5.9, 5.10 and 5.11.

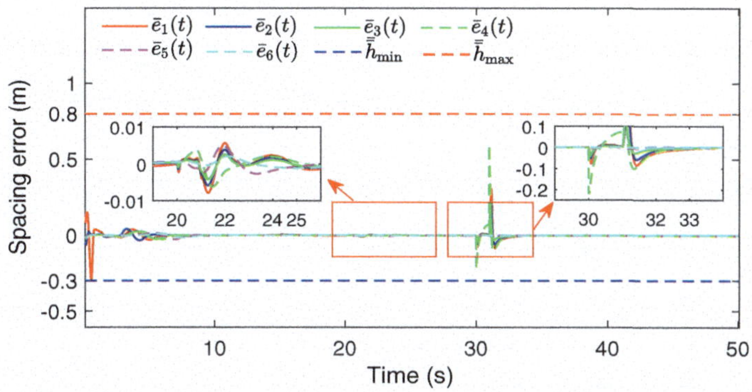


(a)

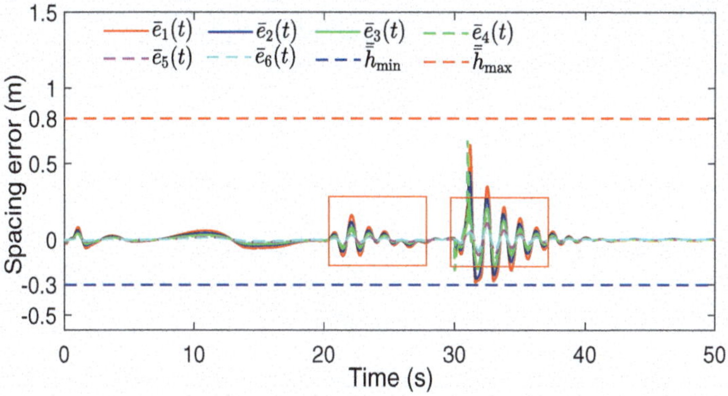


(b)

**Fig. 5.7** Experimental results of spacing error under prescribed asymmetric constraints in Case 2. **a** $\bar{e}_i(t)$ with proposed method, **b** $\bar{e}_i(t)$ with existing control method

As shown in Fig. 5.7, compared with Case 1, in the case of the smaller range of prescribed asymmetric constraints in Case 2, the proposed method can still effectively ensure the closed-loop prescribed performance of the HICVPS under the condition of multiple communication attacks, and achieve the fast finited-time stability and string stability. Compared with the existing control method in Fig. 5.7b (the convergence time $T \leqslant 20s$), Fig. 5.7a shows a faster finite convergence time ($T \leqslant 9s$) and better transient and stable performance under attack conditions. Figure 5.8 shows the heterogeneous state tracking information such as position, speed and acceleration of the vehicle platoon system under the proposed method and the existing control method, which indicates that no collision problems such as rear-end collision and overtaking occur in the whole closed-loop control process. At the same time, when the communication attacks occur, the system states show in Fig. 5.8a can recover to the steady state at a faster convergence rate and maintain higher tracking accuracy than that in Fig. 5.8b.

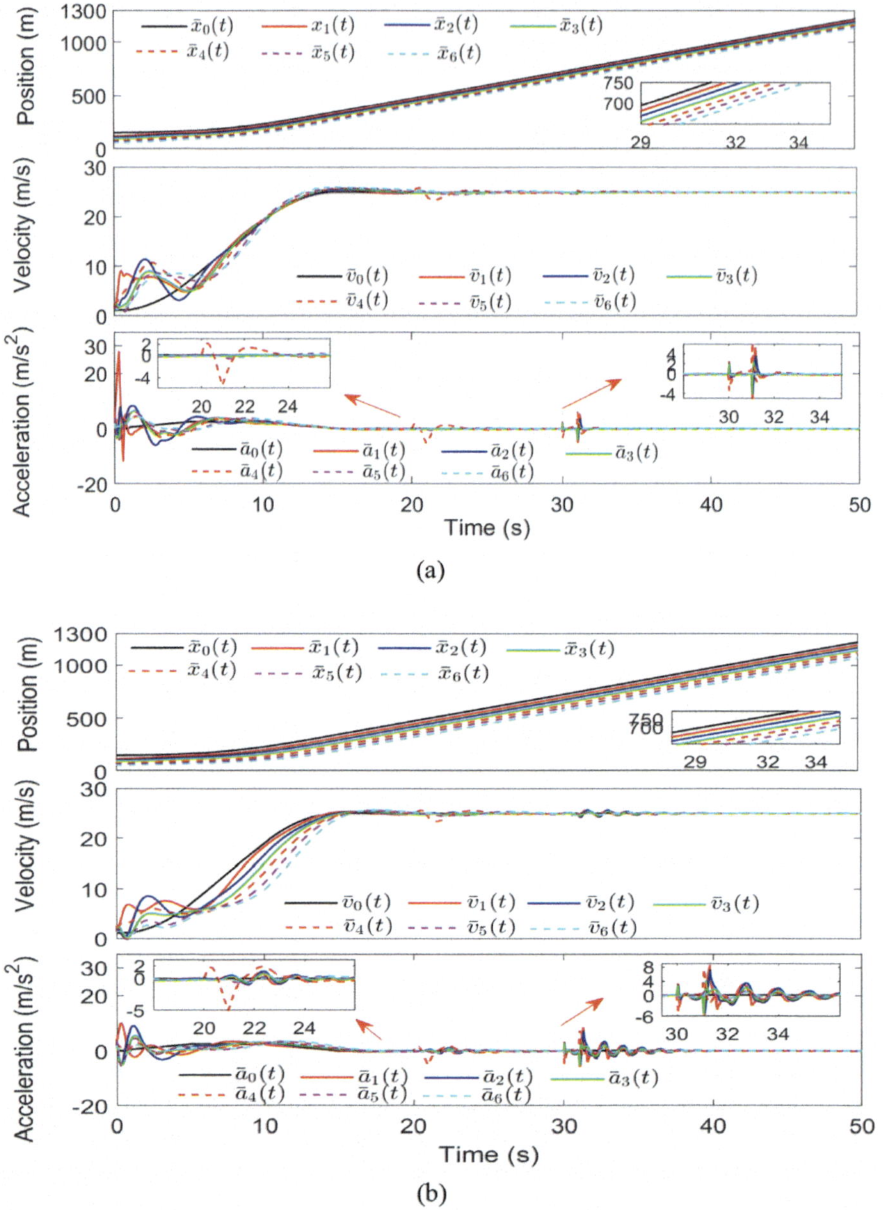


**Fig. 5.8** Experimental results under prescribed asymmetric constraints in Case 2. **a** System states with proposed method, **b** System states with existing control method

Figure 5.9a, b depicts the control inputs under the proposed method and the existing control method, respectively. Figure 5.10 shows the adaptive compensation and updated laws of unknown parameters and weights. The compensated error trajectory in Fig. 5.11 verifies the effectiveness of the designed nonlinear state compensator.

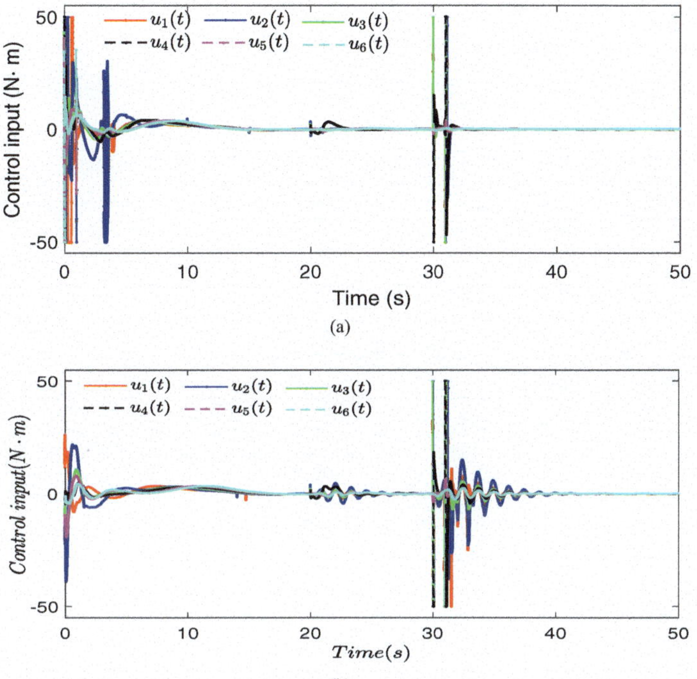


**Fig. 5.9** Experimental results of control inputs under prescribed asymmetric constraints in Case 1. **a** Proposed method, **b** Existing control method

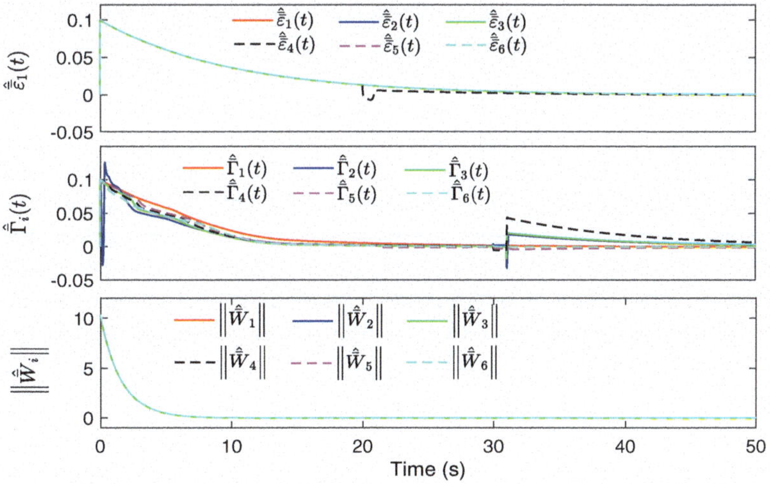


**Fig. 5.10** Adaptive compensation and updated laws under proposed method in Case 2

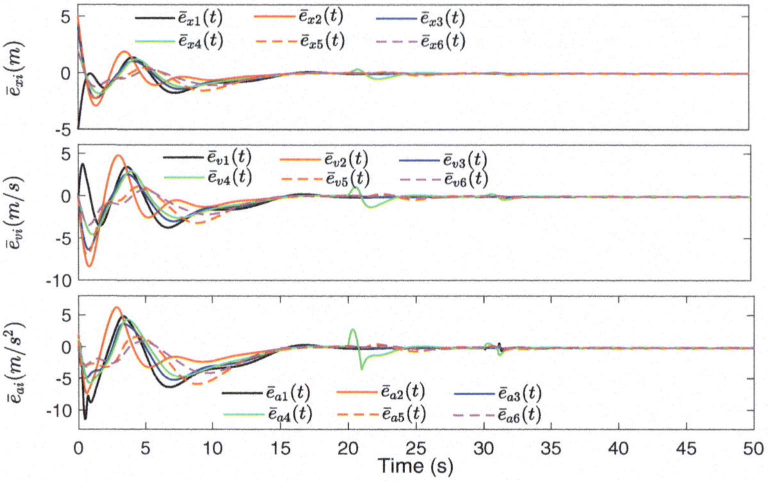


**Fig. 5.11** Compensator errors under proposed method in Case 2

## 5.5 Summary

In this chapter, the distributed adaptive prescribed performance fast finite time resilient control problem of the vehicle platoon system with compound communication attacks and disturbances is considered. Firstly, based on the approximation characteristic of fuzzy logic system, the nonlinear state compensator is constructed to eliminate the influence of communication attacks on the system state information and ensure the boundness of the system state signal. Secondly, the adaptive updated laws are designed to realize the effective estimation of the unknown fuzzy weight and the upper bound of the unknown communication attacks. Based on the improved exponential spacing policy and the hyperbolic arctangent asymmetric barrier Lyapunov function, the distributed adaptive fast finite time resilient scheme with prescribed performance is designed to ensure the string stability, overcome the influence of communication attacks and disturbances on control performance, and achieve fast finite time stability of the closed-loop system. Also, the transient/steady state performance of the HICVPS under multiple communication attacks conditions is ensured. Finally, the comparative experimental results also prove the effectiveness and superiority of the proposed scheme.

## References


1. Stamouli C J, Bechlioulis C P, Kyriakopoulos K J(2022) Robust dynamic average consensus with prescribed transient and steady state performance. *Automatica*, 144: 110503.
2. Shi W(2021) Adaptive fuzzy output-feedback control for nonaffine MIMO nonlinear systems with prescribed performance. *IEEE Transactions on Fuzzy Systems*, 29(5): 1107-1120.

3. Pan C, Chen Y, Chen Q, et al(2023) Distributed fuzzy resilient control for vehicular platooning system with multiple nodes attacks in prescribed performance. *IEEE Transactions on Vehicular Technology*, https://doi.org/10.1109/TVT.2023.3337356.
4. Góes R, Marchand H, Lafortune S, et al(2023) Dealing with sensor and actuator deception attacks in supervisory control. *Automatica*, 147: 110736.
5. Shoukry Y, Tabuada P(2016) Event-triggered state observers for sparse sensor noise/attacks. *IEEE Transactions on Automatic Control*, 61(8): 2079-2091.
6. Guo X, Xu W, Wang J, et al(2021) Distributed neuroadaptive fault-tolerant sliding-mode control for 2-D plane vehicular platoon systems with spacing constraints and unknown direction faults. *Automatica*, 129: 109675.
7. Guo X, Xu W, Wang J, et al(2022) BLF-Based neuroadaptive fault-tolerant control for non-linear vehicular platoon with time-varying fault directions and distance restrictions. *IEEE Transactions on Intelligent Transportation Systems*, 23(8): 12388-12398.
8. Jia D, Lu K, Wang J, et al(2016) A survey on platoon-based vehicular cyber-physical systems. *IEEE Communications Surveys & Tutorials*, 18(1): 263-284.
9. Zhao C, Cai L, and Cheng P(2021) Stability analysis of vehicle platooning with limited communication range and random packet losses. *IEEE Internet of Things Journal*, 8(1): 262-277.
10. Guo X, Xu W, J Wang, et al(2021) Distributed neuroadaptive fault-tolerant sliding-mode control for 2-D plane vehicular platoon systems with spacing constraints and unknown direction faults. *Automatica*, 129: 109675.
11. Besselink B, et al(2016) Cyber-physical control of road freight transport. *Proceedings of the IEEE*, 104(5): 1128-1141.
12. Zhang D, Shen Y P, Zhou Q, et al(2021) Distributed secure platoon control of connected vehicles subject to DoS attack: theory and application. *IEEE Transactions on Systems, Man, and Cybernetics: Systems*, 51(11): 7269-7278.
13. Dutta R G, Hu Y, Yu F, et al(2022) Design and analysis of secure distributed estimator for vehicular platooning in adversarial environment. *IEEE Transactions on Intelligent Transportation Systems*, 23(4): 3418-3429.
14. He X, Hashemi E, Johansson K H(2021) Distributed control under compromised measurements: resilient estimation, attack detection, and vehicle platooning. *Automatica*, 134: 109953.
15. Xiao S, Dong J(2023) Distributed output-feedback resilient fault-tolerant tracking control of uncertain heterogeneous linear MASs under directed topologies and DoS attacks. *Applied Mathematics and Computation*, 443: 127802.
16. Liu Y, Xu L, Cai G, et al(2023) Distributed Robust Platooning Control for Heterogeneous Vehicle Group under Parametric Uncertainty and Hybrid Attacks. *IEEE Transactions on Vehicular Technology*, 72(5): 5677-5689.
17. Liu Y, Yu H, Shi P, et al(2019) Formation control and collision avoidance for a class of multi-agent systems. *Journal of the Franklin Institute*, 356(10): 5395-5420.
18. Hu C, Wang Z, Qin Y, et al(2020) Lane keeping control of autonomous vehicles with prescribed performance considering the rollover prevention and input saturation. *IEEE Transactions on Intelligent Transportation Systems*, 21(7): 3091-3103.
19. Liu Y, and Ma H(2020) Adaptive fuzzy tracking control of nonlinear switched stochastic systems with prescribed performance and unknown control directions. *IEEE Transactions on Systems, Man, and Cybernetics: Systems*, 50(2): 590-599.
20. Wang W, Liang H, Pan Y, et al(2020) Prescribed performance adaptive fuzzy containment control for nonlinear multi-agent systems using disturbance observer. *IEEE Transactions on Cybernetics*, 50(9): 3879-3891.
21. Shi W(2021) Adaptive fuzzy output-feedback control for nonaffine MIMO nonlinear systems with prescribed performance. *IEEE Transactions on Fuzzy Systems*, 29(5): 1107-1120.
22. Cao Z, Niu Y, Song J(2020) Finite-time sliding-mode control of markovian jump cyber-physical systems against randomly occurring injection attacks. *IEEE Transactions on Automatic Control*, 65(3): 1264-1271.

23. Huang X, Dong J(2022) Robust Dynamic actuator failure compensation control of nonlinear systems via cooperative interaction design. *IEEE Transactions on Cybernetics*, 52(7): 6013-6023.
24. Li X, Wen C, Chen C, et al(2022) Adaptive resilient secondary control for microgrids with communication faults. *IEEE Transactions on Cybernetics*, 52(8): 8493-8503.
25. Jin X, Haddad W M, Yucelen T(2017) An adaptive control architecture for mitigating sensor and actuator attacks in cyber-physical systems. *IEEE Transactions on Automatic Control*, 62(11): 6058-6064.
26. Zou Y, W. Huo(2017) Nonlinear robust controller for miniature helicopters without singularity. *IEEE Transactions on Aerospace and Electronic Systems*, 53(3): 1402-1411.
27. Yu S, Yu X, Shirinzadehc B, et al(2005) Continuous finite-time control for robotic manipulators with terminal sliding mode. *Automatica*, 41: 1957-1964.
28. Ali I, Chen Y, Pan C(2022) ECCHSC: Computationally and bandwidth efficient ECC-based hybrid signcryption protocol for secure heterogeneous vehicle-to-infrastructure communications. *IEEE Internet of Things Journal*, 9(6): 4435-4450.

# Chapter 6
# Distributed Fixed Time Resilient Control for Intelligent Connected Vehicle Platoon Systems Under the Communication Faults and Disturbances

Due to physical structure constraints, open wireless network environment and other characteristics, under high-speed driving conditions, the intelligent connected vehicle platoon is sensitive and vulnerable to the external environmental disturbances, such as slippery road surface, strong wind, strong sand, etc., and the communication faults induced by malicious attacks. It is easy to cause the desired inter-vehicle distance to increase or shorten, which leads to system instability and damages the string stability. From the perspective of control, how to ensure the collision avoidance, improve the ICVPS's fault recovery ability and strong resilience under the conditions of communication faults and disturbances, and ensure closed-loop control stability and string stability have important engineering value. The existing finite-time stability control method must rely on initial conditions and system dynamic parameters, and cannot estimate the convergence time in advance. When the initial state of the system is far from the equilibrium point, the convergence speed of the system will be greatly reduced (which can be seen in Chap. 1). Therefore, it cannot fully meet the ICVPS's requirements for robustness and convergence speed under the attack induced communication faults and disturbances. As an extension of finite time control, practical fixed time stability control can provide more robust and resilience performance against faults and disturbances, and can estimate the convergence time of the system in advance without initial state information, which has become a hot spot in the field of control engineering. However, the current research on the design of heterogeneous distributed adaptive fixed-time resilient control scheme for the vehicle platoon with unknown communication faults induced by attacks and disturbances is still incomplete, and it is worth further research.

From the engineering point of view, computing burden and network communication bandwidth seriously restrict the development of the ICVPS, and this contradiction will be more prominent when the scale of the ICVPS expands. Although

 
C. Pan and Y. Chen, *Resilient Control of Intelligent Connected Vehicle Platoon Systems*,
https://doi.org/10.1007/978-981-97-8850-7_6

the traditional periodic data sampling method can alleviate the problem of computing burden, there is still a large waste of resources. In recent years, since data is sampled and transmitted only when predefined events occur, the event-triggered control can provide a more efficient solution to improve communication efficiency and save control energy consumption while maintaining system stability (which is described in detail in Chap. 1). However, how to further combine the event-triggered control method with the distributed adaptive fixed-time resilient control architecture of HICVPS to extend the event triggering interval and save more control energy on the basis of achieving strong robustness and fast convergence control performance under unknown communication faults and disturbances conditions is a challenging problem.

By the above investigations, under the exponential spacing policy and the predecessor-leader following (PLF) information flow topology, in this Chapter, an event-triggered based distributed adaptive fixed-time resilient control scheme for the HICVPS is studied to solve the problem of actuator communication faults induced by deception attacks and external disturbances [1]. The main contributions are listed as follows. Firstly, Different from the traditional disturbance observer based on the sense of asymptotic stability [2, 3], in this Chapter, the sliding mode fixed time disturbance observer is constructed to realize the fixed time fast estimation of the unknown external disturbance of the ICVPS under any initial conditions. The nonlinear compensation of the unknown communication faults and fast estimation of unknown disturbances in HICVPS under high-speed conditions are achieved. Secondly, based on exponential spacing policy and PLF information flow topology, the distributed adaptive fixed-time resilient control scheme is designed to ensure the closed-loop control stability and string stability of the HICVPS in the sense fixed-time stability. Thirdly, different from the existing event triggered mechanism [4, 5], the improved event triggered mechanism with a larger trigger threshold is constructed by introducing the error decaying function, and the inter-event triggering interval is extended by adjusting parameters to avoid the occurrence of Zeno or singularity phenomena. Finally, the effectiveness of the proposed scheme is verified by comparative experiment analysis and SUMO-MATLAB joint simulation platform.

## 6.1 Introduction

In the new generation of advanced wireless communication technology such as vehicle-to-vehicle (V2V) communication, and vehicle-to-infrastructure (V2I) communication, the vehicular platoon system not only can dramatically maintain the desired safety and longitudinal inter-vehicle spacing, but also can track the leader vehicle with the same velocities and accelerations. Then, the all-round communication, interaction, perception, and control are also realized [6]. However, the security and closed-loop stability performance of the vehicular platoon is extremely vulnerable and sensitive to the network action induced communication faults issues, such as cyberattacks, and the external environment disturbances, such as a slippery road slope, strong wind, and strong dust, etc. [7]. Thereupon, its of important to study the resilient control scheme to enhance the resilient recovery capability and ensure the secure and stable operation of the platoon system under the conditions of the external disturbances and the communication faults induced by attacks.

With the communication faults induced by cyber attacks and the external disturbances, the desired inter-vehicle distance will be destroyed, the closed-loop system instability and rear-end collision will occur, the string stability also will be broken [8]. From the control performance viewpoint, how to guarantee the closed-loop system stability and string stability of the ICVPS under the communication faults induced by cyber-attacks and disturbances have crucial engineering practical value. In [9], the resilient control scheme in the sense of the asymptotic stability was studied for avoiding the inter-vehicle collisions and ensure closed-loop system stability under the actuator physical faults and communication attacks. The distributed control schemes were presented in the sense of the global asymptotic stability for the homogeneous vehicles platoon with the communication faults induced by denial of service attacks in [10, 11]. The above works guarantee the closed-loop system stability and collision avoidance. However, the achieved resilience and robustness performance is limited for the ICVPS against the communication faults induced by cyberattacks and disturbances in sense of asymptotic stability. The closed-loop system convergence may be within the infinite time and the string stability can also not be ensured.

For obtaining the finite time convergence performance and improving the resilience and the transient response performance, numerous existing works have been done on the finite time stability control of the ICVPS. In [12], the nonlinear consensus-based finite time stability control method was developed for the ICVPS. In [13], the distributed finite-time tracking control scheme is studied for the HICVPS. It is should be noted that the aforementioned works on finite time platoon control must rely on the system initial conditions and the convergence time can not be estimated in advance. Usually, the communication faults induced by cyberattacks and disturbances may cause more severe results on the closed-loop stability and the string stability of the ICVPS due to the cascaded structure constraints of the platoon. With the traditional finite-time control method, the resilience and secure performance of the ICVPS cannot be fully met. The practical fixed time control method was used in many diverse systems, such as the hypersonic vehicles tracking control in [14], the multiple mechanical systems control in [15], and so on. It becomes a hot topic in the field of control engineering. The practical fixed time control method is the extension of the finite time control, and can support the strong resilience performance. Meanwhile, the convergence time can be estimated in advance without the initial state information. Therefore, it is worthy of exploring the distributed adaptive fixed-time resilient control method for the heterogeneous vehicular platoon with the communication faults induced by cyberattacks and disturbances.

On the other assistance, as far as the network resource is concerned, the computation burden and limited network bandwidth in both the network control layer and the network communication layer seriously restrict the extension of the ICVPS, especially when the scale of the platoon expands. Different from the traditional periodical data sampling method that may cause the wastes of communication resources, the event-triggered method provides a more efficient solution to increase communication efficiency and sustain the control performance of the platoon system. By the event-triggered scheme, the trade-off of sampling or data transmission instants is conducted only when the predefined event occurs [16]. In [4], the event-triggered cooperative control method was developed to deal with the actuator delays issues of the ICVPS. In [17], for the problems of the nonzero input of the leader and actuator uncertainties in the ICVPS, the distributed event-triggered control scheme was presented. In [18], the switched event-triggered control scheme was proposed for the vehicular platoon with false data injection attacks in the sense of asymptotic stability. The abovementioned works proposed many method for the stability control of the ICVPS. However, how to achieve better platoon control performance while save more communication resources and extend inter-event execution intervals are still challenging issues and need to be further explored.

## 6.2 Problem Statement

As shown in Fig. 6.1, In this Chapter, the HICVPS composed of connected and automated vehicles driving on a straight road is considered. In this ICVPS, the leader vehicle is represented by the vehicle 0, and the follower vehicle is represented by the vehicle $i$, $i = 1, 2, \ldots, n$. In the communication layer of the HICVPS, based on the PLF information topology flow, the V2V communication interaction is realized. Meanwhile, in this Chapter, the actuator communication faults induced by deception attacks and external disturbances are considered.

First, the longitudinal third-order nonlinear dynamics model of the follower vehicle in the HICVPS is considered as [9, 19],

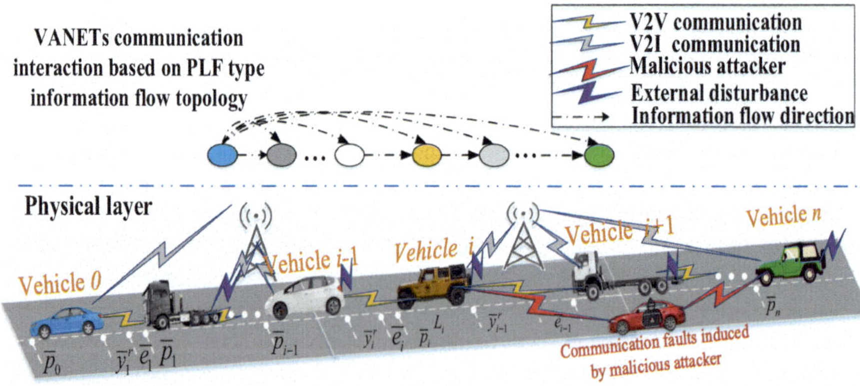


**Fig. 6.1** The HICVPS under the communication fault and disturbances

$$\begin{aligned}\dot{\bar{p}}_i(t) &= \bar{v}_i(t)\\ \dot{\bar{v}}_i(t) &= \bar{a}_i(t)\\ \dot{\bar{a}}_i(t) &= \bar{F}_i(\bar{X}_i) + \bar{\tau}_i^{-1}u_i^A(t) + \bar{\varpi}_i(t)\end{aligned} \tag{6.1}$$

where $\bar{p}_i(t)$, $\bar{v}_i(t)$ and $\bar{a}_i(t)$ are respectively the position, velocity and acceleration of the vehicle. $\bar{\tau}_i$ is the motor time constant; $u_i^A(t)$ represents the control input in case of communication fault, $\bar{F}_i(\bar{X}_i)$ is an unknown nonlinear function related to the system state, where $\bar{X}_i = [\bar{v}_i, \bar{a}_i]^T$. $\bar{\varpi}_i(t)$ indicates the unknown external disturbances in the environment, such as strong winds or uneven roads, which will seriously threaten the stability and safety of the system, assuming it is satisfied that $\left|\ddot{\bar{\varpi}}_i(t)\right| < \bar{\bar{\varpi}}_{i1}$, $\bar{\varpi}_{i1} \in \mathrm{R}^+$.

Due to the openness of the wireless communication channel, malicious attackers are easy to launch the deception attacks on the HICVPS control layer. Thereby, the control signal is maliciously tampered with and the false signals are generated in the feedback loop channel. Finally, the communication fault of the control layer occurs which leads to the serious damage to the control performance of the HICVPS [10, 18]. Therefore, the controller-actuator communication fault model induced by the deception attacks in the control layer of the vehicle $i$ is considered as

$$u_i^A(t) = u_i(t) + \bar{\Phi}_i(\bar{X}_i) \tag{6.2}$$

where $u_i(t) = c_i/M_i$ represents the control input of the engine. The control input is considered independent of the mass of the vehicle. $c_i$ is the input of the engine, $M_i$ is the quality of the vehicle. $\bar{\Phi}_i(\bar{X}_i)$ is an unknown continuous nonlinear communication fault function on the set $\Gamma \in \mathrm{R}^N$, which depends on the system state.

**Remark 6.1** Malicious attackers usually can avoid the abnormal signal detection system of the vehicle and monitor the communication interaction of the vehicle. By this way, the control signal can be tampered with, the communication fault can achieved. Thus, the control performance of the ICVPS can be destroyed. The above technologies mean that malicious attackers need to use the system states information when generating the deception attack signal. Different from the uniformly bounded attack in [20, 21] and the actuator physical bias fault in [22, 23], communication fault induced by the state-dependent attack is more intelligent and targeted.

Considering the dynamics model of the leader vehicle 0 as

$$\begin{aligned}\dot{\bar{p}}_0(t) &= \bar{v}_0(t)\\ \dot{\bar{v}}_0(t) &= \bar{a}_0(t)\\ \dot{\bar{a}}_0(t) &= u_0(t)\end{aligned} \tag{6.3}$$

where $u_0(t)$ is the ideal control input of the vehicle 0.

As discussed in Chaps. 4 and 5, the exponential spacing policy can effectively guarantee the control performance of the platoon system, improve the traffic capacity at both low and high speed, and promote the scheduling and control of the ICVPS. Based on the improved exponential spacing policy, on the one hand, the introduction of exponential function term $\Lambda_i(t)$ can compensate the string instability problems caused by non-zero spacing error and speed. On the other hand, the exponential spacing policy can make the control independent of the constant time headway constant, which reduces the limitation of the spacing policy. The improved exponential spacing policy is

$$\bar{e}_i(t) = \bar{p}_{i-1}(t) - \bar{p}_i(t) - L_{i-1} - \bar{y}_i^r(t) - \bar{\Lambda}_i(t) \tag{6.4}$$

where $\bar{y}_i^r = \bar{d}_{i-1,i} + (2\bar{\Delta}_i)^{-1}\varepsilon(\bar{v}_i^2 - \bar{v}_0^2) + \bar{\varphi}(1 - \bar{E}_i)$, $\bar{\Lambda}_i(t) = \{\bar{e}_i^*(0) + \bar{\Lambda}_i^a t + 0.5\bar{\Lambda}_i^b t^2\} \times e^{-\varsigma_i t}$, $\bar{e}_i^*(t) = \bar{p}_{i-1}(t) - \bar{p}_i(t) - L_{i-1} - \bar{y}_i^r(t)$, $\bar{\Lambda}_i^a = \varsigma_i \bar{e}_i^*(0) + \dot{\bar{e}}_i^*(0)$, $\bar{\Lambda}_i^b = \varsigma_i^2 \bar{e}_i^*(0) + 2\varsigma_i \dot{\bar{e}}_i^*(0) + \ddot{\bar{e}}_i^*(0)$, $\bar{E}_i = \exp(-(\bar{v}_i - \bar{v}_0)/\mu)$. $L_{i-1}$ is the length of the vehicle, $\bar{d}_{i-1,i}$ represents the desired inter-vehicle distance, $\varepsilon > 0$ is the safety factor of the environment, $\bar{\Delta}_i > 0$ is the maximum deceleration, $\bar{\varphi}$, $\mu$ and $\varsigma_i$ are the positive definite constants. Using the formula (6.4), $\bar{e}_i(0) = 0$, $\dot{\bar{e}}_i(0) = 0$ and $\ddot{\bar{e}}_i(0) = 0$ can be derived, which means that the initial spacing error under the exponential spacing policy is always zero.

**Remark 6.2** By the improved the exponential spacing policy (6.4), higher dynamic performance and resilience can be achieved under the conditions of communication faults and disturbances. Combined with the distributed adaptive fixed-time resilient control architecture, the closed-loop control stability and string stability of the HICVPS can be guaranteed in the sense of fixed-time stability no matter at low velocity or high velocity under communication faults and disturbances. However, since the leader vehicle information is introduced into the formula (6.4), the communication load is increased while realizing the above control advantages. Therefore, it is of great significance to improve the event triggered mechanism, extend the triggering time interval and save control energy.

**Lemma 6.1** (Globally fixed-time stability) [15] *For a continuous nonlinear system* $\dot{\bar{x}}(t) = \bar{F}(t, \bar{x})$, $\bar{x}(0) = \bar{x}_0$, *where* $\bar{x} \in \mathrm{R}^n$, $F : \mathrm{R}_+ \times \mathrm{R}^n$ *is a nonlinear function and the origin is the equilibrium point. It is assumed that there exists a continuous positive definite function* $V(\bar{x}) > 0$ *such that* $\dot{V}(\bar{x}) \leq -\phi V^{\alpha}(\bar{x}) - \varphi V^{\beta}(\bar{x})$ *is true, where* $\phi$, $\varphi > 0$, $0 < \alpha < 1$ *and* $\beta > 1$. *Then the nonlinear system is globally fixed time stable and its fixed time upper bound satisfy that* $T \leq (1/\phi(1-\alpha)) + (1/\varphi(\beta - 1))$.

**Lemma 6.2** (Practical fixed-time stability) [19] *For a continuous nonlinear system* $\dot{\bar{x}}(t) = F(t, \bar{x})$, $\bar{x}(0) = \bar{x}_0$, *where* $\bar{x} \in \mathrm{R}^n$, $F : \mathrm{R}_+ \times \mathrm{R}^n$ *is a nonlinear function and the origin is the equilibrium point. It is supposed that there exists a continuous positive definite function* $V(\bar{x}) > 0$ *such that* $\dot{V}(\bar{x}) \leq -\phi V^{\alpha}(\bar{x}) - \varphi V^{\beta}(\bar{x}) + \bar{\lambda}$ *is true, where* $\phi$, $\varphi > 0$, $0 < \alpha < 1$, $\beta > 1$ *and* $\bar{\lambda} > 0$. *Then the nonlinear system is practical fixed time stable. Furthermore, the trajectory of the system will converge to the residual set* $\mho = \{\bar{x} | V(\bar{x}) \leq \min\{(\frac{\bar{\lambda}}{\phi(1-\gamma)})^{1/\alpha}, (\frac{\bar{\lambda}}{\varphi(1-\gamma)})^{1/\beta}\}\}$ *in a fixed time and its fixed time upper bound satisfy that* $T \leq T_{\max} := (1/\phi\gamma(1-\alpha)) + (1/\varphi\gamma(\beta - 1))$, $0 < \gamma < 1$.

**Lemma 6.3** ([15]) $\forall \alpha_1, \alpha_2 \in \mathrm{R}$, *there exists the inequality* $|\bar{\alpha}_1|^{\lambda_1} |\bar{\alpha}_2|^{\lambda_2} \leq (\lambda_1 + \lambda_2)^{-1}(\lambda_1 \pi |\bar{\alpha}_1|^{\lambda_1+\lambda_2} + \lambda_2 \pi^{-(\lambda_1/\lambda_2)} |\bar{\alpha}_2|^{\lambda_1+\lambda_2})$ *is true, where* $\lambda_1, \lambda_2, \pi > 0$ *are positive definite constants.*

**Lemma 6.4** ([24]) $\forall \bar{\lambda}_l \in \mathrm{R}(l = 1, 2, \ldots, n)$, $0 < \iota_1 < 1$ *and* $\iota_2 > 1$, *then, there exists the inequalities* $\left(\sum_{l=1}^{n} |\bar{\lambda}_l|\right)^{\iota_1} \leq \sum_{l=1}^{n} |\bar{\lambda}_l|^{\iota_1}$, *and* $\left(\sum_{l=1}^{n} |\bar{\lambda}_l|\right)^{\iota_2} \leq n^{\iota_2 - 1} \sum_{l=1}^{n} |\bar{\lambda}_l|^{\iota_2}$ *are true.*

**Control Objective** For the unknown communication faults induced by deception attacks and external disturbances, the distributed adaptive fixed-time resilient control scheme based on the HICVPS with improved event triggered is studied to satisfy the follower's velocity and acceleration consistent with that of the leader within a fixed time. The closed-loop control stability in the sense of fixed time can be guaranteed. At the same time, the string stability is ensured, the collision avoidance and the expected safety inter-vehicle spacing under communication faults and external disturbances can also be guaranteed, that is $\lim_{t \to \infty} \bar{e}_i(t) = 0$, $\lim_{t \to \infty} (\bar{p}_{i-1}(t) - \bar{p}_i(t) - L_{i-1}) = \bar{d}_{i-1,i}$.

## 6.3 Distributed Adaptive Resilient Control Design

As shown in Fig. 6.2. In this section, a sliding mode fixed time disturbance observer is constructed to quickly estimate the unknown external disturbances of the ICVPS in a fixed time under any initial conditions. Secondly, based on exponential spacing policy and PLF-type information flow topology, a distributed adaptive fixed-time resilient control scheme based on improved event triggered is designed to ensure the closed-loop control stability and string stability of the HICVPS with the communication faults and disturbances in the sense of Lyapunov fixed-time stability. Finally, the system stability theory is analyzed and verified.

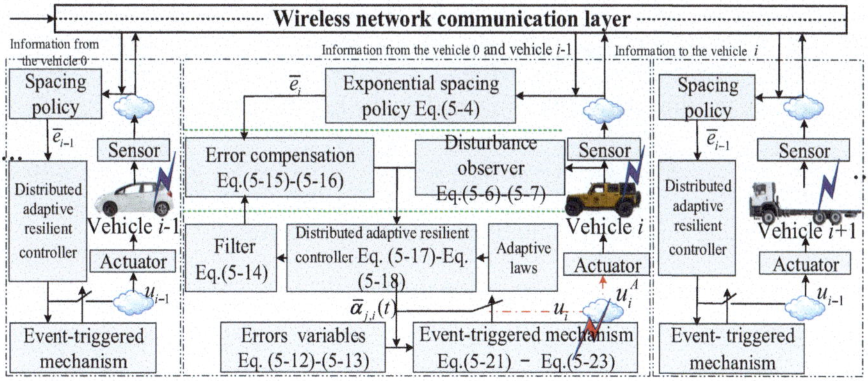


**Fig. 6.2** The control structure of proposed scheme based on the HICVPS

### 6.3.1 Design of Fixed Time Disturbance Observer

In this section, a simple and effective sliding mode fixed time disturbance observer is designed to achieve fast estimation of unknown external disturbance in a fixed time under arbitrary initial conditions.

In order to design a fixed time disturbance observer for the vehicle $i$, the auxiliary variable $\bar{\wp}_i = \bar{a}_i - \bar{\bar{E}}_i$ is defined, we have

$$\dot{\bar{\bar{E}}}_i = \bar{F}_i(\bar{X}_i) + u_i^A(t)\bar{\tau}_i^{-1} - k_1\bar{\wp}_i - k_2 sgn^{(p/q)}(\bar{\wp}_i) + \bar{\eth}_{\varpi i} \tag{6.5}$$

where $k_1, k_2$, $p$ and $q$are positive definite constants, $\forall \bar{x} \in \mathrm{R}, sgn^r(\bar{x}) = sign(\bar{x})\,|\bar{x}|^r$, $r > 0$. In Eq. (6.5), $r = p/q,\ p < q.\ \bar{\eth}_{\varpi i}$ is the auxiliary variable, which will be designed in Theorem 6.1 so that the estimation error of external disturbance $\bar{\varpi}_i(t)$ can converge to zero in a fixed time.

**Theorem 6.1** *For the external disturbances $\bar{\varpi}_i(t)$ existing in the system (6.1), the sliding mode surface is defined as*

$$\bar{s}_{\varpi i}(t) = \dot{\bar{\wp}}_i - k_1\bar{\wp}_i - k_2 sgn^{(p/q)}(\bar{\wp}_i) \tag{6.6}$$

*By using the super-twisting algorithm, the fixed time disturbance observer is designed as*

$$\begin{aligned} \dot{\bar{\eth}}_{\varpi i} &= \bar{\mu}_{\varpi i1} sgn^{\eth_{i,1}}(\bar{s}_{\varpi i}(t)) + \bar{\mu}_{\varpi i2} sgn^{\eth_{i,2}}(\bar{s}_{\varpi i}(t)) + \bar{h}_{\varpi i} \\ \dot{\bar{h}}_{\varpi i} &= \lambda_{\varpi i} sign(\bar{s}_{\varpi i}(t)) \end{aligned} \tag{6.7}$$

*where $\hat{\bar{\varpi}}_i(t) = \bar{\eth}_{\varpi i}$, $\bar{\mu}_{\varpi i1}$ and $\bar{\mu}_{\varpi i2}$ are positive definite constants, $\lambda_{\varpi i}$ satisfy that $\bar{\bar{\varpi}}_{i1} < \lambda_{\varpi i}$, $\eth_{i,1} > 1$ and $0 < \eth_{i,2} < 1$. $\bar{s}_{\varpi i}(t)$ and $\bar{\wp}_i$ can converge to the origin in a fixed time, where the fixed time meets the conditions that $T_{\varpi i} \le \sqrt{2}(\left(\sqrt{2}\right)^{\eth_{i,1}} \bar{\mu}_{\varpi i1}(\eth_{i,1} - 1))^{-1} + \sqrt{2}(\left(\sqrt{2}\right)^{\eth_{i,2}} \bar{\mu}_{\varpi i2}(1 - \eth_{i,2}))^{-1}$, then the fixed time estimation of external disturbances $\bar{\varpi}_i(t)$ can be guaranteed.*

***Proof*** Taking the time derivative of $\bar{\wp}_i$ and by the Eq. (6.5), we get

$$\dot{\bar{\wp}}_i = \bar{\varpi}_i(t) + k_1\bar{\wp}_i + k_2 sgn^{(p/q)}(\bar{\wp}_i) - \bar{\eth}_{\varpi i} \tag{6.8}$$

By substituting formula (6.8) into formula (6.6), we have

$$\bar{s}_{\varpi i}(t) = \bar{\varpi}_i(t) - \bar{\eth}_{\varpi i} \tag{6.9}$$

Secondly, taking the time derivative of Eq. (6.9), and combine it with Eq. (6.7) to obtain,

$$\dot{\bar{s}}_{\varpi i}(t) = \dot{\bar{\varpi}}_i - \bar{\mu}_{\varpi i1} sgn^{\partial_{i,1}}(\bar{s}_{\varpi i}(t)) - \bar{\mu}_{\varpi i2} sgn^{\partial_{i,2}}(\bar{s}_{\varpi i}(t)) - \int \lambda_{\varpi i} sign(\bar{s}_{\varpi i}(t))dt \tag{6.10}$$

Considering the Lyapunov candidate function $\bar{V}_{\varpi i} = 0.5\bar{s}^2_{\varpi i}(t)$ and differentiating it with the Eq. (6.10), it is obtained

$$\begin{aligned}\dot{\bar{V}}_{\varpi i} &\leqslant \left|\dot{\bar{\varpi}}_i\right| |\bar{s}_{\varpi i}| - |\bar{s}_{\varpi i}| \int \lambda_{\varpi i} dt - \bar{\mu}_{\varpi i1} |\bar{s}_{\varpi i}(t)|^{\partial_{i,1}+1} - \bar{\mu}_{\varpi i2} |\bar{s}_{\varpi i}(t)|^{\partial_{i,2}+1} \\ &\leqslant -\bar{\mu}_{\varpi i1} |\bar{s}_{\varpi i}(t)|^{\partial_{i,1}+1} - \bar{\mu}_{\varpi i2} |\bar{s}_{\varpi i}(t)|^{\partial_{i,2}+1} - |\bar{s}_{\varpi i}| \left( \int \lambda_{\varpi i} dt - \int \left|\ddot{\bar{\varpi}}_i\right| dt \right) \\ &\leqslant -2^{\frac{\partial_{i,1}+1}{2}} \bar{\mu}_{\varpi i1} \bar{V}_{\varpi i}^{\frac{\partial_{i,1}+1}{2}} - 2^{\frac{\partial_{i,2}+1}{2}} \bar{\mu}_{\varpi i2} \bar{V}_{\varpi i}^{\frac{\partial_{i,2}+1}{2}} \end{aligned} \tag{6.11}$$

where $\left|\ddot{\bar{\varpi}}_i\right| < \bar{\bar{\varpi}}_{i1}$, $\bar{\bar{\varpi}}_{i1} < \lambda_{\varpi i}$.

According to Lemma 6.1 global fixed time stability principle, the above proof shows that the sliding mode surface $\bar{s}_{\varpi i}(t)$ and $\bar{\wp}_i$ can converge to the origin in a fixed time $T_{\varpi i}$. When $\bar{s}_{\varpi i}(t) = 0$, the estimation error of external disturbances $\bar{\varpi}_i(t)$ is considered. Based on Eq. (6.9) and $\hat{\bar{\varpi}}_i(t) = \bar{\partial}_{\varpi i}$, it can be seen that the estimation error of external disturbances $\tilde{\bar{\varpi}}_i(t) = \bar{\varpi}_i(t) - \hat{\bar{\varpi}}_i(t)$ can also be guaranteed to converge to the origin within a fixed time. Proof completed. □

### 6.3.2 Design of Distributed Adaptive Fixed-Time Resilient Controller

In this section, based on command filtering control and improved event triggered control method, an event triggered based distributed adaptive fixed-time resilient controller is designed. First, we define the following coordinate transformation as

$$\bar{z}_{i,1} = \bar{e}_i(t) \tag{6.12}$$

$$\bar{z}_{i,j} = \bar{v}_{i-1}(t) - \bar{v}_i(t) - \bar{H}_{1i}\bar{a}_i + \bar{H}_{2i}\bar{a}_0 - \dot{\bar{\Lambda}}_i(t) - \bar{w}_{j,i} \tag{6.13}$$

where $i = 1, 2, \ldots, n$, $\bar{\Omega}_{1i} = \varepsilon/\bar{\Delta}_i$, $\bar{\Omega}_{2i} = \bar{\varphi}/\mu$, $\bar{H}_{1i} = \bar{\Omega}_{1i}\bar{v}_i + \bar{\Omega}_{2i}\bar{E}_i$ and $\bar{H}_{2i} = \bar{\Omega}_{1i}\bar{v}_0 + \bar{\Omega}_{2i}\bar{E}_i$. $\bar{w}_{j,i}$ is not only the output of the filter, but also the control signal of the filter, $j = 2$. For avoiding the problem of "dimensional explosion" caused by repeated differentiation of virtual control signals in the back-stepping control method, Uniform Exact Differentiator (UED) [25] is introduced to filter virtual control signals $\bar{\alpha}_{j-1,i}(t)$ within a fixed time. The UED-based filter is defined as

$$\dot{\bar{\ell}}_{j,i,1} = -\lambda_{j,i,1}\beta_{j,i,1}(\bar{\Im}_{j-1,i}) + \bar{\ell}_{j,i,2}$$
$$\dot{\bar{\ell}}_{j,i,2} = -\lambda_{j,i,2}\beta_{j,i,2}(\bar{\Im}_{j-1,i}) \tag{6.14}$$

where $j=2$, $\bar{\Im}_{j-1,i} = \bar{\ell}_{j,i,1} - \bar{\alpha}_{j-1,i}$, $\bar{\alpha}_{j-1,i}(t)$ is the input of the filter, $\bar{w}_{j,i}=\bar{\ell}_{j,i,1}$, $\dot{\bar{w}}_{j,i}=\bar{\ell}_{j,i,2}$ are the output of the filter. $\lambda_{j,i,1}$, $\lambda_{j,i,2}$, and $\iota_{j,i}$ are positive definite constants. In addition, for bounded independent variables $\bar{\alpha}_{j-1,i}(t)$, the filter (6.14) has a bounded output $\bar{\Im}_{j-1,i}$, which satisfies the condition that $\left|\bar{\Im}_{j-1,i}\right| \leq \bar{\bar{\ell}}_{i,1}\bar{\bar{\alpha}}_i$. $\left|\bar{\ell}_{j,i,2} - \dot{\bar{\alpha}}_{j-1,i}\right| \leq \bar{\bar{\ell}}_{i,2}\bar{\bar{\alpha}}_i^{0.5}$ can be satisfied within a fixed time $T_\ell$, where $\bar{\bar{\ell}}_{i,1}$, $\bar{\bar{\ell}}_{i,2}$ and $\bar{\bar{\alpha}}_{i,1}$ are positive definite constants. $\beta_{j,i,1}(\bar{\Im}_{j-1,i})$, $\beta_{j,i,2}(\bar{\Im}_{j-1,i})$ are defined as $\beta_{j,i,1}(\bar{\Im}_{j-1,i}) = sgn^{0.5}(\bar{\Im}_{j-1,i}) + \iota_{j,i}sgn^{1.5}(\bar{\Im}_{j-1,i})$, and $\beta_{j,i,2}(\bar{\Im}_{j-1,i}) = 0.5sgn(\bar{\Im}_{j-1,i}) + 2\iota_{j,i}\bar{\Im}_{j-1,i} + 1.5\iota_{j,i}^2 sgn^2(\bar{\Im}_{j-1,i})$, respectively.

Further, for reducing the errors caused by the introduced filters (6.14), the error compensation scheme is designed. The error surfaces to be compensated are defined as $\bar{\mathfrak{V}}_{i,1} = \bar{z}_{i,1} - \bar{\Upsilon}_{i,1}$, $\bar{\mathfrak{V}}_{i,j} = \bar{z}_{i,j} - \bar{\Upsilon}_{i,j}$, $\bar{\Upsilon}_{i,1}$ and $\bar{\Upsilon}_{i,j}$ are compensated signals. The error compensation scheme is constructed as follows,

$$\dot{\bar{\Upsilon}}_{i,1} = -\kappa_{i,1}\bar{\Upsilon}_{i,1} + \bar{\Upsilon}_{i,j} + \bar{w}_{j,i} - \bar{\alpha}_{j-1,i} - \kappa_{i,5}sgn^{\upsilon_1}(\bar{\Upsilon}_{i,1}) - \kappa_{i,6}sgn^{\upsilon_2}(\bar{\Upsilon}_{i,1}) \tag{6.15}$$
$$\dot{\bar{\Upsilon}}_{i,j} = -\kappa_{i,j}\bar{\Upsilon}_{i,j} - \bar{\Upsilon}_{i,1} - 1.5\bar{\Upsilon}_{i,j} - \kappa_{i,j+1}sgn^{\upsilon_1}(\bar{\Upsilon}_{i,j}) - \kappa_{i,j+2}sgn^{\upsilon_2}(\bar{\Upsilon}_{i,j}) \tag{6.16}$$

where $\kappa_{i,1} > 0.5$, $\kappa_{i,5}$, $\kappa_{i,6}$, $\kappa_{i,j}$, $\kappa_{i,j+1}$ and $\kappa_{i,j+2} > 0$ are positive definite constants. $0 < \upsilon_1 < 1$ and $\upsilon_2 > 1$, $\bar{\Upsilon}_{i,1}(0) = \bar{\Upsilon}_{i,j}(0) = 0$.

The constructed virtual distributed adaptive fixed time resilient controllers with event triggered strategy are respectively,

$$\bar{\alpha}_{j-1,i} = -\kappa_{i,1}\bar{z}_{i,1} - \kappa_{i,5}\text{sgn}^{\upsilon_1}(\bar{\Upsilon}_{i,1}) - \kappa_{i,6}\text{sgn}^{\upsilon_2}(\bar{\Upsilon}_{i,1}) - \mathfrak{v}_{i,1}\text{sgn}^{\gamma_1}(\bar{\mathfrak{V}}_{i,1}) - \mathfrak{v}_{i,2}\text{sgn}^{\gamma_2}(\bar{\mathfrak{V}}_{i,1}) \tag{6.17}$$

$$\begin{aligned}\bar{\alpha}_{j,i} = & -\kappa_{i,j}\bar{z}_{i,j} - \bar{z}_{i,1} - \bar{C}_i + \bar{H}_{1i}\hat{\bar{\varpi}}_i + \dot{\bar{w}}_{j,i} - \kappa_{i,j+2}\text{sgn}^{\upsilon_2}(\bar{\Upsilon}_{i,j}) - (2m_{2i}^2)^{-1}\bar{\mathfrak{V}}_{i,j}\hat{\bar{\vartheta}}_i\bar{\delta}_i^T\bar{\delta}_i \\ & - (2m_{1i}^2)^{-1}\bar{\mathfrak{V}}_{i,j}\hat{\bar{\bar{\theta}}}_{\Phi,i}\bar{\delta}_{\Phi,i}^T\bar{\delta}_{\Phi,i} - 1.5\bar{z}_{i,j} - \mathfrak{v}_{i,j+2}\text{sgn}^{\gamma_2}(\bar{\mathfrak{V}}_{i,j}) - \mathfrak{v}_{i,j+1}\text{sgn}^{\gamma_1}(\bar{\mathfrak{V}}_{i,j}) \\ & - \kappa_{i,j+1}\text{sgn}^{\upsilon_1}(\bar{\Upsilon}_{i,j})\end{aligned} \tag{6.18}$$

where $\mathfrak{v}_{i,1}$, $\mathfrak{v}_{i,2}$, $\mathfrak{v}_{i,j+1}$, $\mathfrak{v}_{i,j+2}$, $m_{1i}$ and $m_{2i}$ are positive definite constants. $0 < \gamma_1 < 1$, $\gamma_2 > 1$. $\hat{\bar{\vartheta}}_i$ and $\hat{\bar{\bar{\theta}}}_{\Phi i}$ are the estimation of $\bar{\vartheta}_i$ and $\bar{\bar{\theta}}_{\Phi i}$, respectively. $\bar{\vartheta}_i = \left\|\bar{\vartheta}_i^*\right\|^2$, $\bar{\bar{\theta}}_{\Phi i} = \left\|\bar{\theta}_{\Phi i}^*\right\|^2$, where $\bar{\theta}_{\Phi i}^*$ is related to the communication faults induced by cyberattacks function term, which will be defined later.

The constructed adaptive compensated and updated laws $\hat{\hat{\theta}}_{\Phi i}$ and $\hat{\vartheta}_i$ are

$$\dot{\hat{\hat{\theta}}}_{\Phi i}= (2m_{1i}^2)^{-1}\varsigma_{\theta i}\bar{\mathfrak{V}}_{i,j}^2\bar{\delta}_{\Phi,i}^T\bar{\delta}_{\Phi,i}- \widehat{\varsigma}_{\theta i}\,\hat{\hat{\theta}}_{\Phi i} \tag{6.19}$$

$$\dot{\hat{\vartheta}}_i= (2m_{2i}^2)^{-1}\varsigma_{\vartheta i}\bar{\mathfrak{V}}_{i,j}^2\bar{\delta}_i^T\bar{\delta}_i- \widehat{\varsigma}_{\vartheta i}\,\hat{\vartheta}_i \tag{6.20}$$

where $\varsigma_{\vartheta i}$, $\varsigma_{\theta i}$, $\widehat{\varsigma}_{\vartheta i}$ and $\widehat{\varsigma}_{\theta i}$ are designed positive definite constants. In the subsequent design process, for simplify, we define $X_i = [\bar{X}_i, \bar{v}_0]^T$, $\bar{\delta}_i$ and $\bar{\delta}_{\Phi,i}$ are used to represent $\bar{\delta}_i(X_i)$ and $\bar{\delta}_{\Phi,i}(X_i)$, respectively.

Meanwhile, for reducing the computational burden, communication resources and control energy use of the HICVPS, and maintaining the control performance, the event triggered controller is designed as

$$\begin{aligned}\bar{\mathfrak{W}}_i(t) = b_i^{-1}(1+\mathfrak{d}_{i,0})\{\bar{\alpha}_{j,i}\tanh(\mathfrak{d}_{i,1}^{-1}\bar{\alpha}_{j,i}\bar{\mathfrak{V}}_{i,j})\\ +\bar{\mathfrak{O}}_i^{-1}(t)\tanh((\bar{\mathfrak{O}}_i\mathfrak{d}_{i,1})^{-1}\bar{\mathfrak{V}}_{i,j})+\mathfrak{d}_{i,2}\tanh(\mathfrak{d}_{i,1}^{-1}\mathfrak{d}_{i,2}\bar{\mathfrak{V}}_{i,j})\}\end{aligned} \tag{6.21}$$

The event triggered conditions are designed as

$$u_i(t) = \bar{\mathfrak{W}}_i(t_{\mathfrak{k}}) \quad \forall t \in [t_{\mathfrak{k}}, t_{\mathfrak{k}+1}) \tag{6.22}$$

$$t_{\mathfrak{k}+1} = \inf\{t \in \mathbb{R}^+ \,||\bar{e}_{ui}(t)| \geqslant \mathfrak{d}_{i,0}\,|u_i(t)| + \bar{\mathfrak{O}}_i^{-1}(t)h_i + \widehat{\bar{\Delta}}_i\} \tag{6.23}$$

where $b_i = \bar{H}_{1i}/\bar{\tau}_i > 0$, $\bar{e}_{ui}(t)=\bar{\bar{\mathfrak{W}}}_i(t) - u_i(t)$ is triggered sampling error, $\bar{\mathfrak{O}}_i(t) = \left|\bar{z}_{i,1}\right| + \left|\bar{z}_{i,j}\right| + \widehat{\bar{d}}_{i,0}$, $t_{\mathfrak{k}}$, $\mathfrak{k} \in \mathbb{Z}^+$indicates the signal update interval. During the time range of $t \in [t_{\mathfrak{k}}, t_{\mathfrak{k}+1})$, the control input of the vehicle is not updated through the communication network, and will be kept at a constant value. For example; When $u_i(t) = \bar{\mathfrak{W}}_i(t_{\mathfrak{k}})$, and the event triggered condition (6.23) is met, $u_i(t_{\mathfrak{k}+1})$ will be applied in the control system, and the triggered moment at this time instant is defined as $t_{\mathfrak{k}+1}$. $\mathfrak{d}_{i,0}$, $\mathfrak{d}_{i,1}$, $\mathfrak{d}_{i,2}$, $\widehat{\bar{d}}_{i,0}$, $h_i$ and $\widehat{\bar{\Delta}}_i$ are positive defined design parameters, where $0 < \mathfrak{d}_{i,0} < 1$, $\mathfrak{d}_{i,0} < \mathfrak{d}_{i,2}$, and satisfy the conditions that $b_i h_i < 1 - \mathfrak{d}_{i,0}$ and $b_i\ \widehat{\bar{\Delta}}_i< \mathfrak{d}_{i,2}(1 - \mathfrak{d}_{i,0})$.

**Remark 6.3** By selecting appropriate control parameters, such as$\mathfrak{v}_{i,1}$, $\mathfrak{v}_{i,2}$, $\mathfrak{v}_{i,j+1}$, $\mathfrak{v}_{i,j+2}$, $\varsigma_{\vartheta i}$, $\varsigma_{\theta i}$, $\widehat{\varsigma}_{\vartheta i}$ and $\widehat{\varsigma}_{\theta i}$, etc. in Eqs. (6.17) and (6.18), the HICVPS can obtain closed-loop system control stability and string stability. A suitable convergence time $T$can be obtained by using Theorem 6.2 and increasing the control parameters $\mathfrak{v}_{i,1}$, $\mathfrak{v}_{i,2}$, $\mathfrak{v}_{i,j+1}$, $\mathfrak{v}_{i,j+2}$, $\varsigma_{\vartheta i}$ and $\varsigma_{\theta i}$ that affect the rate of convergence under the given constraints while decreasing $\widehat{\varsigma}_{\vartheta i}$ and $\widehat{\varsigma}_{\theta i}$. Different from the event triggered scheme in [4, 5], by introducing the error decaying function $\bar{\mathfrak{O}}_i^{-1}(t)h_i$, on the one hand, when the errors $\bar{z}_{i,1}$ and $\bar{z}_{i,j}$ converge to zero, the "Zeno" phenomenon that causes the event to be triggered incorrectly can be effectively avoided. On the other hand, by adjusting the parameters $\widehat{\bar{d}}_{i,0}$, $h_i$ and $\widehat{\bar{\Delta}}_i$, a larger triggered threshold can be obtained. Thus, the event triggered time interval can be extended and the control energy transmission can be saved.

In order to clarify the control scheme design, taking the $i$th vehicle as an example, its specific distributed adaptive event-triggered fixed-time resilient control scheme design is described below.

First, for ensuring the convergence of $\bar{z}_{i,1}$ and $\bar{e}_i(t)$ in Eq. (6.12), the Lyapunov function is selected as $\bar{V}_{i,1} = 0.5\bar{\mathfrak{V}}_{i,1}^2$, where the compensation error is $\bar{\mathfrak{V}}_{i,1} = \bar{z}_{i,1} - \bar{\Upsilon}_{i,1}$. And by taking the time derivative of $\bar{V}_{i,1}$, we get

$$\begin{aligned}\dot{\bar{V}}_{i,1} &= \bar{\mathfrak{V}}_{i,1}(\bar{v}_{i-1} - \bar{v}_i - \bar{H}_{1i}\bar{a}_i + \bar{H}_{2i}\bar{a}_0 - \dot{\bar{\Lambda}}_i - \dot{\bar{\Upsilon}}_{i,1}) \\ &= \bar{\mathfrak{V}}_{i,1}(\bar{z}_{i,j} + \bar{w}_{j,i} - \dot{\bar{\Upsilon}}_{i,1})\end{aligned} \tag{6.24}$$

By substituting the error compensation $\dot{\bar{\Upsilon}}_{i,1}$ in Eq. (6.15) into the above equation, we can get

$$\dot{\bar{V}}_{i,1} = \bar{\mathfrak{V}}_{i,1}(\bar{z}_{i,j} - \bar{\Upsilon}_{i,j} + \kappa_{i,1}\bar{\Upsilon}_{i,1} + \kappa_{i,5}\mathrm{sgn}^{\upsilon_1}(\bar{\Upsilon}_{i,1}) + \kappa_{i,6}\mathrm{sgn}^{\upsilon_2}(\bar{\Upsilon}_{i,1}) + \bar{\alpha}_{j-1,i}) \tag{6.25}$$

Combined with the virtual control law $\bar{\alpha}_{j-1,i}$ in Eqs. (6.17), (6.25) can be further written as

$$\begin{aligned}\dot{\bar{V}}_{i,1} &= \bar{\mathfrak{V}}_{i,1}(\bar{\mathfrak{V}}_{i,j} - \kappa_{i,1}\bar{\mathfrak{V}}_{i,1} - \mathfrak{v}_{i,1}\mathrm{sgn}^{\gamma_1}(\bar{\mathfrak{V}}_{i,1}) - \mathfrak{v}_{i,2}\mathrm{sgn}^{\gamma_2}(\bar{\mathfrak{V}}_{i,1})) \\ &= -\kappa_{i,1}\bar{\mathfrak{V}}_{i,1}^2 + \bar{\mathfrak{V}}_{i,1}\bar{\mathfrak{V}}_{i,j} - \mathfrak{v}_{i,1}\left|\bar{\mathfrak{V}}_{i,1}\right|^{\gamma_1+1} - \mathfrak{v}_{i,2}\left|\bar{\mathfrak{V}}_{i,1}\right|^{\gamma_2+1}\end{aligned} \tag{6.26}$$

Similar to the calculation of formula (6.24), based on the formula (6.13), and obtaining the time derivative of $\bar{\mathfrak{V}}_{i,j} = \bar{z}_{i,j} - \bar{\Upsilon}_{i,j}$, we have

$$\begin{aligned}\dot{\bar{\mathfrak{V}}}_{i,j} &= \bar{a}_{i-1}(t) - \bar{a}_i(t) - \bar{\Omega}_{1i}(\bar{a}_i^2 - \bar{a}_0^2) - \bar{H}_{1i}\dot{\bar{a}}_i + \bar{H}_{2i}\dot{\bar{a}}_0 \\ &+ \bar{\Omega}_{2i}\bar{E}_i(\bar{a}_i - \bar{a}_0)^2\mu^{-1} - \ddot{\bar{\Lambda}}_i(t) - \dot{\bar{w}}_{j,i} - \dot{\bar{\Upsilon}}_{i,j}\end{aligned} \tag{6.27}$$

By substituting Eqs. (6.1) to (6.3) into the above Eq. (6.27), we get

$$\dot{\bar{\mathfrak{V}}}_{i,j} = \bar{C}_i + \bar{\bar{\Phi}}_i(X_i) - b_i u_i(t) - \bar{\bar{F}}_i(X_i) - \bar{H}_{1i}\bar{\varpi}_i(t) - \dot{\bar{w}}_{j,i} - \dot{\bar{\Upsilon}}_{i,j} \tag{6.28}$$

where $\bar{C}_i = \bar{a}_{i-1}(t) - \bar{a}_i(t) - \bar{\Omega}_{1i}(\bar{a}_i^2 - \bar{a}_0^2) + \bar{H}_{2i}\bar{u}_0(t) + \bar{\Omega}_{2i}\bar{E}_i(\bar{a}_i - \bar{a}_0)^2\mu^{-1} - \ddot{\bar{\Lambda}}_i(t)$, $\bar{\bar{F}}_i(X_i) = \bar{H}_{1i}\bar{F}_i(\bar{X}_i)$, $\bar{\bar{\Phi}}_i(X_i) = -\bar{H}_{1i}\bar{\Phi}_i(\bar{X}_i)\bar{\tau}_i^{-1}$. For the unknown communication fault compound term $\bar{\bar{\Phi}}_i(X_i)$, according to the fuzzy logic system approximation Lemma 4.1, there exists the fuzzy logic systems $\bar{\theta}_{\Phi,i}^{*T}\bar{\delta}_{\Phi,i}$ and $\bar{\vartheta}_i^{*T}\bar{\delta}_i$ that satisfy the following conditions

$$\bar{\bar{F}}_i(X_i) = \bar{\vartheta}_i^{*T}\bar{\delta}_i + \chi_i(X_i) \tag{6.29}$$

$$\bar{\bar{\Phi}}_i(X_i) = \bar{\theta}_{\Phi,i}^{*T}\bar{\delta}_{\Phi,i} + \chi_{\Phi,i}(X_i) \tag{6.30}$$

where $\chi_i(X_i)$, $\chi_{\Phi,i}(X_i)$ are the approximation errors of fuzzy logic system, $\chi_i(X_i)$, $\chi_{\Phi,i}(X_i)$ satisfy the conditions $|\chi_i(X_i)| \leqslant \widehat{\delta}_i$ and $\left|\chi_{\Phi,i}(X_i)\right| \leqslant \widehat{\delta}_{\Phi,i}$, respectively. $\widehat{\delta}_i > 0$, $\widehat{\delta}_{\Phi,i} > 0$ are the upper bound constant of the approximation error. Furthermore, Eq. (6.28) can become

$$\begin{aligned}\bar{\mathfrak{V}}_{i,j}\dot{\bar{\mathfrak{V}}}_{i,j} = \bar{\mathfrak{V}}_{i,j}(\bar{C}_i + \bar{\theta}_{\Phi,i}^{*T}\bar{\delta}_{\Phi,i}(X_i) + \chi_{\Phi,i}(X_i) - \bar{\vartheta}_i^{*T}\bar{\delta}_i(X_i) \\ - b_i u_i(t) - \bar{H}_{1i}\hat{\bar{\varpi}}_i - \bar{H}_{1i}\bar{s}_{\varpi i} - \chi_i(X_i) - \dot{\bar{w}}_{j,i} - \dot{\bar{\Upsilon}}_{i,j})\end{aligned} \tag{6.31}$$

It is worth noting that according to Young's inequality, the following conclusion can be obtained,

$$\begin{aligned}&\bar{\mathfrak{V}}_{i,j}(-\bar{\vartheta}_i^{*T}\bar{\delta}_i(X_i) + \bar{\theta}_{\Phi,i}^{*T}\bar{\delta}_{\Phi,i}(X_i)) \\ &\leqslant (2m_{1i}^2)^{-1}\bar{\mathfrak{V}}_{i,j}^2\bar{\bar{\theta}}_{\Phi,i}\bar{\delta}_{\Phi,i}^T\bar{\delta}_{\Phi,i} + (2m_{2i}^2)^{-1}\bar{\mathfrak{V}}_{i,j}^2\bar{\vartheta}_i\bar{\delta}_i^T\bar{\delta}_i + 2^{-1}(m_{1i}^2 + m_{2i}^2)\end{aligned} \tag{6.32}$$

$$\bar{\mathfrak{V}}_{i,j}(\chi_{\Phi,i}(X_i) - \chi_i(X_i)) \leqslant \bar{\mathfrak{V}}_{i,j}^2 + 2^{-1}(\widehat{\delta}_{\Phi,i}^2 + \widehat{\delta}_i^2) \tag{6.33}$$

$$-\bar{\mathfrak{V}}_{i,j}\bar{H}_{1i}\bar{s}_{\varpi i} \leqslant 2^{-1}\bar{\mathfrak{V}}_{i,j}^2 + 2^{-1}\bar{H}_{1i}^2\bar{s}_{\varpi i}^2 \tag{6.34}$$

By selecting the Lyapunov function $\bar{V}_{i,2} = \bar{V}_{i,1} + 0.5\bar{\mathfrak{V}}_{i,j}^2$ and obtaining its time derivative, and based on the Eqs. (6.32) to (6.34), we can get

$$\begin{aligned}\dot{\bar{V}}_{i,2} \leqslant \dot{\bar{V}}_{i,1} + \bar{\mathfrak{V}}_{i,j}(\bar{C}_i + (2m_{1i}^2)^{-1}\bar{\mathfrak{V}}_{i,j}\bar{\bar{\theta}}_{\Phi,i}\bar{\delta}_{\Phi,i}^T\bar{\delta}_{\Phi,i} + (2m_{2i}^2)^{-1}\bar{\mathfrak{V}}_{i,j}\bar{\vartheta}_i\bar{\delta}_i^T\bar{\delta}_i \\ - b_i u_i(t) - \bar{H}_{1i}\hat{\bar{\varpi}}_i + 1.5\bar{\mathfrak{V}}_{i,j} - \dot{\bar{w}}_{j,i} - \dot{\bar{\Upsilon}}_{i,j}) + \widehat{\bar{M}}_i + 2^{-1}\bar{H}_{1i}^2\bar{s}_{\varpi i}^2\end{aligned} \tag{6.35}$$

where $\widehat{\bar{M}}_i = 2^{-1}(m_{1i}^2 + m_{2i}^2 + \widehat{\delta}_{\Phi,i}^2 + \widehat{\delta}_i^2)$.

In combination with the error compensation scheme $\dot{\bar{\Upsilon}}_{i,j}$ in Eqs. (6.16) and (6.35) can be derived as

$$\begin{aligned}\dot{\bar{V}}_{i,2} \leqslant &-\kappa_{i,1}\bar{\mathfrak{V}}_{i,1}^2 - \mathfrak{v}_{i,1}\left|\bar{\mathfrak{V}}_{i,1}\right|^{\gamma_1+1} - \mathfrak{v}_{i,2}\left|\bar{\mathfrak{V}}_{i,1}\right|^{\gamma_2+1} + \bar{\mathfrak{V}}_{i,j}(\bar{z}_{i,1} \\ &+ \bar{C}_i + (2m_{1i}^2)^{-1}\bar{\mathfrak{V}}_{i,j}\bar{\bar{\theta}}_{\Phi,i}\bar{\delta}_{\Phi,i}^T\bar{\delta}_{\Phi,i} + (2m_{2i}^2)^{-1}\bar{\mathfrak{V}}_{i,j}\bar{\vartheta}_i\bar{\delta}_i^T\bar{\delta}_i \\ &- H_{1i}\hat{\bar{\varpi}}_i - b_i u_i(t) + 1.5\bar{z}_{i,j} - \dot{\bar{w}}_{j,i} + \kappa_{i,j}\bar{\Upsilon}_{i,j} + \widehat{\bar{M}}_i \\ &+ \kappa_{i,j+1}\mathrm{sgn}^{\upsilon_1}(\bar{\Upsilon}_{i,j}) + \kappa_{i,j+2}\mathrm{sgn}^{\upsilon_2}(\bar{\Upsilon}_{i,j})) + 2^{-1}\bar{H}_{1i}^2\bar{s}_{\varpi i}^2\end{aligned} \tag{6.36}$$

According to the distributed adaptive virtual control law $\bar{\alpha}_{j,i}$ in formula (6.18), and based on Theorem 6.1, it can be seen that when $t \geqslant T_{\varpi i}$, $\bar{s}_{\varpi i}(t) = \bar{\varpi}_i(t) - \bar{\mathfrak{d}}_{\varpi i}$ will converge to the origin in a fixed time $T_{\varpi i}$, formula (6.36) can be derived as,

$$
\begin{aligned}
\dot{\bar{V}}_{i,2} \leqslant & -\kappa_{i,1}\bar{\mathfrak{V}}_{i,1}^2 - \kappa_{i,j}\bar{\mathfrak{V}}_{i,j}^2 - \mathfrak{v}_{i,1}\left|\bar{\mathfrak{V}}_{i,1}\right|^{\gamma_1+1} - \mathfrak{v}_{i,2}\left|\bar{\mathfrak{V}}_{i,1}\right|^{\gamma_2+1} \\
& - \mathfrak{v}_{i,j+1}\left|\bar{\mathfrak{V}}_{i,j}\right|^{\gamma_1+1} - \mathfrak{v}_{i,j+2}\left|\bar{\mathfrak{V}}_{i,j}\right|^{\gamma_2+1} + (2m_{2i}^2)^{-1}\bar{\mathfrak{V}}_{i,j}^2\bar{\bar{\vartheta}}_i\bar{\delta}_i^T\bar{\delta}_i \\
& + (2m_{1i}^2)^{-1}\bar{\mathfrak{V}}_{i,j}^2\bar{\bar{\bar{\theta}}}_{\Phi,i}\bar{\delta}_{\Phi,i}^T\bar{\delta}_{\Phi,i} + \widehat{\bar{M}}_i - \bar{\mathfrak{V}}_{i,j}b_iu_i(t) - \bar{\mathfrak{V}}_{i,j}\bar{\alpha}_{j,i}
\end{aligned} \tag{6.37}
$$

Selecting the Lyapunov function $\bar{V}_i = \bar{V}_{i,2} + (2\varsigma_{\vartheta i})^{-1}\tilde{\bar{\vartheta}}_i^2 + (2\varsigma_{\theta i})^{-1}\tilde{\bar{\bar{\theta}}}_{\Phi i}^2$, $\tilde{\bar{\vartheta}}_i = \bar{\vartheta}_i - \hat{\bar{\vartheta}}_i$ and $\tilde{\bar{\bar{\theta}}}_{\Phi i} = \bar{\bar{\theta}}_{\Phi i} - \hat{\bar{\bar{\theta}}}_{\Phi i}$. Combining with formula (6.37), (6.19) to (6.20), the time derivative of $\bar{V}_i$ can be derived as

$$
\begin{aligned}
\dot{\bar{V}}_i \leqslant & -\kappa_{i,1}\bar{\mathfrak{V}}_{i,1}^2 - \kappa_{i,j}\bar{\mathfrak{V}}_{i,j}^2 - \mathfrak{v}_{i,1}\left|\bar{\mathfrak{V}}_{i,1}\right|^{\gamma_1+1} - \mathfrak{v}_{i,2}\left|\bar{\mathfrak{V}}_{i,1}\right|^{\gamma_2+1} \\
& - \bar{\mathfrak{V}}_{i,j}\bar{\alpha}_{j,i} - \bar{\mathfrak{V}}_{i,j}b_iu_i(t) - \mathfrak{v}_{i,j+1}\left|\bar{\mathfrak{V}}_{i,j}\right|^{\gamma_1+1} - \mathfrak{v}_{i,j+2}\left|\bar{\mathfrak{V}}_{i,j}\right|^{\gamma_2+1} \\
& + \widehat{\bar{M}}_i + \varsigma_{\theta i}^{-1}\widehat{\varsigma}_{\theta i}\tilde{\bar{\bar{\theta}}}_{\Phi i}\hat{\bar{\bar{\theta}}}_{\Phi i} + \varsigma_{\vartheta i}^{-1}\widehat{\varsigma}_{\vartheta i}\tilde{\bar{\vartheta}}_i\hat{\bar{\vartheta}}_i
\end{aligned} \tag{6.38}
$$

According to the event triggered conditions in Eq. (6.23), the inequality $\left|\bar{\mathfrak{W}}_i(t) - u_i(t)\right| \leqslant \mathfrak{d}_{i,0}\left|u_i(t)\right| + h_i\bar{\mathfrak{O}}_i^{-1}(t) + \widehat{\bar{\Delta}}_i$ is established within the time interval $[t_{\mathfrak{k}}, t_{\mathfrak{k}+1})$ . So we know that there is a continuous function $\bar{\mathfrak{g}}_{i,0}(t)$ that holds the following conditions,

$$
u_i(t) = (\bar{\mathfrak{W}}_i(t) - \bar{\mathfrak{g}}_{i,0}(t)h_i\bar{\mathfrak{O}}_i^{-1}(t) - \bar{\mathfrak{g}}_{i,0}(t)\widehat{\bar{\Delta}}_i)(1 + \mathfrak{g}_{i,1}(t)\mathfrak{d}_{i,0})^{-1} \tag{6.39}
$$

where $\bar{\mathfrak{g}}_{i,0}(t_{\mathfrak{k}}) = 0$, $\bar{\mathfrak{g}}_{i,0}(t_{\mathfrak{k}+1}) = \pm 1$, $\left|\bar{\mathfrak{g}}_{i,0}(t)\right| \leqslant 1$. $\mathfrak{g}_{i,1}(t) = \pm\bar{\mathfrak{g}}_{i,0}(t)$ meets the condition $\left|\mathfrak{g}_{i,1}(t)\right| \leqslant 1$. According to Lemma 2.1?$\forall x \in \mathbb{R}$ and $\mu > 0$?$b_i > 0$? there exists inequality $x\tanh(x/\mu) \geqslant 0$ is true. Based on Eq. (6.21), we have $\bar{\mathfrak{V}}_{i,j}\bar{\mathfrak{W}}_i(t) \geqslant 0$. Based on the above, it can be concluded that $-b_i\bar{\mathfrak{V}}_{i,j}\bar{\mathfrak{W}}_i(t)(1 + \mathfrak{g}_{i,1}\mathfrak{d}_{i,0})^{-1} \leqslant -b_i\bar{\mathfrak{V}}_{i,j}\bar{\mathfrak{W}}_i(t)(1 + \mathfrak{d}_{i,0})^{-1}$. The formula (6.38) can be transformed into

$$
\begin{aligned}
\dot{\bar{V}}_i \leqslant & -\kappa_{i,1}\bar{\mathfrak{V}}_{i,1}^2 - \kappa_{i,j}\bar{\mathfrak{V}}_{i,j}^2 - \mathfrak{v}_{i,1}\left|\bar{\mathfrak{V}}_{i,1}\right|^{\gamma_1+1} - \mathfrak{v}_{i,2}\left|\bar{\mathfrak{V}}_{i,1}\right|^{\gamma_2+1} \\
& - \mathfrak{v}_{i,j+1}\left|\bar{\mathfrak{V}}_{i,j}\right|^{\gamma_1+1} - \mathfrak{v}_{i,j+2}\left|\bar{\mathfrak{V}}_{i,j}\right|^{\gamma_2+1} + \widehat{\bar{M}}_i - b_i\bar{\mathfrak{V}}_{i,j}\bar{\mathfrak{W}}_i(t)(1 + \mathfrak{d}_{i,0})^{-1} \\
& + \left|\bar{\mathfrak{V}}_{i,j}\right| b_ih_i((1 - \mathfrak{d}_{i,0})\bar{\mathfrak{O}}_i(t))^{-1} - \bar{\mathfrak{V}}_{i,j}\bar{\alpha}_{j,i} + \left|\bar{\mathfrak{V}}_{i,j}\right| b_i\widehat{\bar{\Delta}}_i(1 - \mathfrak{d}_{i,0})^{-1} \\
& + \varsigma_{\vartheta i}^{-1}\widehat{\varsigma}_{\vartheta i}\tilde{\bar{\vartheta}}_i\hat{\bar{\vartheta}}_i + \varsigma_{\theta i}^{-1}\widehat{\varsigma}_{\theta i}\tilde{\bar{\bar{\theta}}}_{\Phi i}\hat{\bar{\bar{\theta}}}_{\Phi i}
\end{aligned} \tag{6.40}
$$

Substitutting the formula (6.21) into (6.40). Combined with Lemma 2.1, and conditions $b_i h_i < 1 - \mathfrak{d}_{i,0}$, $b_i \widehat{\bar{\Delta}}_i < \mathfrak{d}_{i,2}(1 - \mathfrak{d}_{i,0})$, $\bar{\mathfrak{V}}_{i,j}(-b_i u_i(t) - \bar{\alpha}_{j,i}) \leqslant 0.8355\mathfrak{d}_{i,1}$ can be obtained. Thus, the Eq. (6.40) can be further derived as

$$
\begin{aligned}
\dot{\bar{V}}_i \leqslant & -\kappa_{i,1}\bar{\mathfrak{V}}_{i,1}^2 - \kappa_{i,j}\bar{\mathfrak{V}}_{i,j}^2 - \mathfrak{v}_{i,1}\left|\bar{\mathfrak{V}}_{i,1}\right|^{\gamma_1+1} - \mathfrak{v}_{i,2}\left|\bar{\mathfrak{V}}_{i,1}\right|^{\gamma_2+1} \\
& - \mathfrak{v}_{i,j+1}\left|\bar{\mathfrak{V}}_{i,j}\right|^{\gamma_1+1} - \mathfrak{v}_{i,j+2}\left|\bar{\mathfrak{V}}_{i,j}\right|^{\gamma_2+1} + \widehat{\bar{M}}_i + 0.8355\mathfrak{d}_{i,1} \\
& + \varsigma_{\vartheta i}^{-1}\widehat{\varsigma}_{\vartheta i}\tilde{\bar{\vartheta}}_i\hat{\bar{\vartheta}}_i + \varsigma_{\theta i}^{-1}\widehat{\varsigma}_{\theta i}\tilde{\bar{\bar{\theta}}}_{\Phi i}\hat{\bar{\bar{\theta}}}_{\Phi i}
\end{aligned}
\tag{6.41}
$$

By using Young's inequality, we can obtain the conclusions $\varsigma_{\vartheta i}^{-1}\widehat{\varsigma}_{\vartheta i}\tilde{\bar{\vartheta}}_i\hat{\bar{\vartheta}}_i \leqslant (2\varsigma_{\vartheta i})^{-1}\widehat{\varsigma}_{\vartheta i}(\bar{\vartheta}_i^2 - \tilde{\bar{\vartheta}}_i^2)$, $\varsigma_{\theta i}^{-1}\widehat{\varsigma}_{\theta i}\tilde{\bar{\bar{\theta}}}_{\Phi i}\hat{\bar{\bar{\theta}}}_{\Phi i} \leqslant (2\varsigma_{\theta i})^{-1}\widehat{\varsigma}_{\theta i}(\bar{\bar{\theta}}_{\Phi i}^2 - \tilde{\bar{\bar{\theta}}}_{\Phi i}^2)$. Equation (6.41) can be derived as

$$
\begin{aligned}
\dot{\bar{V}}_i \leqslant & -\kappa_{i,1}\bar{\mathfrak{V}}_{i,1}^2 - \kappa_{i,j}\bar{\mathfrak{V}}_{i,j}^2 - \mathfrak{v}_{i,1}\left|\bar{\mathfrak{V}}_{i,1}\right|^{\gamma_1+1} - \mathfrak{v}_{i,2}\left|\bar{\mathfrak{V}}_{i,1}\right|^{\gamma_2+1} \\
& + \bar{N}_i^* - \mathfrak{v}_{i,j+1}\left|\bar{\mathfrak{V}}_{i,j}\right|^{\gamma_1+1} - \mathfrak{v}_{i,j+2}\left|\bar{\mathfrak{V}}_{i,j}\right|^{\gamma_2+1} \\
& - (2\varsigma_{\vartheta i})^{-1}\widehat{\varsigma}_{\vartheta i}\tilde{\bar{\vartheta}}_i^2 - (2\varsigma_{\theta i})^{-1}\widehat{\varsigma}_{\theta i}\tilde{\bar{\bar{\theta}}}_{\Phi i}^2
\end{aligned}
\tag{6.42}
$$

where $\bar{N}_i^* = \widehat{\bar{M}}_i + 0.8355\mathfrak{d}_{i,1} + (2\varsigma_{\vartheta i})^{-1}\widehat{\varsigma}_{\vartheta i}\bar{\vartheta}_i^2 + (2\varsigma_{\theta i})^{-1}\widehat{\varsigma}_{\theta i}\bar{\bar{\theta}}_{\Phi i}^2$.

### *6.3.3 Analysis of System Stability*

The distributed control design is completed in the above section, which leads to Theorem 6.2.

**Theorem 6.2** *Considering the HICVPS consists of the vehicle longitudinal dynamics model with external disturbances (6.1), and the unknown communication faults induced by the deception attacks (6.2). Based on the improved exponential spacing policy (6.4) and the PLF-type information flow topology, the designed fixed-time disturbance observer (6.6) and (6.7), filter error compensation scheme (6.15) and (6.16), distributed adaptive fixed-time resilient controller (6.17) and (6.18), improved event triggered mechanism (6.21) to (6.23), and adaptive updated and compensation laws (6.19) and (6.20) can all work together to guarantee the closed-loop control stability of the system in the sense of practical fixed time stability and the string stability of the compensated spacing error. The upper bound of fixed time satisfy that* $T \leqslant T_{\max} := 2(\bar{\aleph}_a\bar{\varepsilon}(1-\gamma_1))^{-1} + 2(\bar{\aleph}_b\bar{\varepsilon}(\gamma_2-1))^{-1}$, *where* $\bar{\aleph}_a$, $\bar{\aleph}_b$ *and* $\bar{\varepsilon}$ *will be defined later. The time interval* $\{t_{\mathfrak{k}+1} - t_{\mathfrak{k}}\}$ *between the triggered events is the lower bound of* $T_e > 0$, $\forall \mathfrak{k} \in \mathbb{Z}^+$.

***Proof*** For verifying the fixed-time closed-loop control stability and string stability of the entire HICVPS, the Lyapunov function is selected as $\bar{V}(t) = \sum_{i=1}^{n}(\bar{V}_i(t))$, based on Eq. (6.42), $\dot{\bar{V}}(t)$ can be,

$$\begin{aligned}\dot{\bar{V}}(t) \leqslant\ & 2^{\frac{\gamma_1+1}{2}} \sum_{i=1}^{n} \left\{ -\mathfrak{v}_{i,1}(2^{-1}|\bar{\mathfrak{V}}_{i,1}|^2)^{\frac{\gamma_1+1}{2}} - \mathfrak{v}_{i,j+1}(2^{-1}|\bar{\mathfrak{V}}_{i,j}|^2)^{\frac{\gamma_1+1}{2}} \right\} \\ & + 2^{\frac{\gamma_2+1}{2}} \sum_{i=1}^{n} \left\{ -\mathfrak{v}_{i,2}(2^{-1}|\bar{\mathfrak{V}}_{i,1}|^2)^{\frac{\gamma_2+1}{2}} - \mathfrak{v}_{i,j+2}(2^{-1}|\bar{\mathfrak{V}}_{i,j}|^2)^{\frac{\gamma_2+1}{2}} \right\} \\ & - 2\widehat{\varsigma}_{\vartheta} \sum_{i=1}^{n} \left\{ (2\varsigma_{\vartheta i})^{-1}\tilde{\bar{\vartheta}}_i^2 \right\} - 2\widehat{\varsigma}_{\theta} \sum_{i=1}^{n} \left\{ (2\varsigma_{\theta i})^{-1}\tilde{\bar{\bar{\theta}}}_{\Phi i}^2 \right\} + \sum_{i=1}^{n} \bar{N}_i^* \end{aligned} \tag{6.43}$$

where $\widehat{\varsigma}_{\vartheta} = \min\limits_{1\leqslant i\leqslant n} \{2^{-1}\widehat{\varsigma}_{\vartheta i}\}$, $\widehat{\varsigma}_{\theta} = \min\limits_{1\leqslant i\leqslant n} \{2^{-1}\widehat{\varsigma}_{\theta i}\}$.

Using Lemma 6.3, the following inequalities can be established

$$-(2\varsigma_{\vartheta i})^{-1}\tilde{\bar{\vartheta}}_i^2 \leqslant -((2\varsigma_{\vartheta i})^{-1}\tilde{\bar{\vartheta}}_i^2)^{\frac{\gamma_2+1}{2}} + (2^{-1}(1-\gamma_2))(2^{-1}(\gamma_2+1))^{-\frac{1+\gamma_2}{1-\gamma_2}} < 0 \tag{6.44}$$

$$0 < ((2\varsigma_{\vartheta i})^{-1}\tilde{\bar{\vartheta}}_i^2)^{\frac{\gamma_1+1}{2}} - (2\varsigma_{\vartheta i})^{-1}\tilde{\bar{\vartheta}}_i^2 \leqslant (2^{-1}(1-\gamma_1))(2^{-1}(\gamma_1+1))^{-\frac{1+\gamma_1}{1-\gamma_1}} \tag{6.45}$$

where $2^{-1}(\gamma_2+1) \in (1, +\infty)$, $2^{-1}(\gamma_1+1) \in (0.5, 1)$. As for $(2\varsigma_{\theta i})^{-1}\tilde{\bar{\bar{\theta}}}_{\Phi i}^2$, we can obtain the same results. Based on the above results, Eq. (6.43) can be derived as

$$\begin{aligned}\dot{\bar{V}}(t) \leqslant\ & 2^{\frac{\gamma_1+1}{2}} \sum_{i=1}^{n} \left\{ -\mathfrak{v}_{i,1}(2^{-1}|\bar{\mathfrak{V}}_{i,1}|^2)^{\frac{\gamma_1+1}{2}} - \mathfrak{v}_{i,j+1}(2^{-1}|\bar{\mathfrak{V}}_{i,j}|^2)^{\frac{\gamma_1+1}{2}} \right\} \\ & + 2^{\frac{\gamma_2+1}{2}} \sum_{i=1}^{n} \left\{ -\mathfrak{v}_{i,2}(2^{-1}|\bar{\mathfrak{V}}_{i,1}|^2)^{\frac{\gamma_2+1}{2}} - \mathfrak{v}_{i,j+2}(2^{-1}|\bar{\mathfrak{V}}_{i,j}|^2)^{\frac{\gamma_2+1}{2}} \right\} \\ & - \widehat{\varsigma}_{\vartheta} \sum_{i=1}^{n} \left\{ ((2\varsigma_{\vartheta i})^{-1}\tilde{\bar{\vartheta}}_i^2)^{\frac{\gamma_1+1}{2}} \right\} - \widehat{\varsigma}_{\theta} \sum_{i=1}^{n} \left\{ ((2\varsigma_{\theta i})^{-1}\tilde{\bar{\bar{\theta}}}_{\Phi i}^2)^{\frac{\gamma_1+1}{2}} \right\} \\ & - \widehat{\varsigma}_{\vartheta} \sum_{i=1}^{n} \left\{ ((2\varsigma_{\vartheta i})^{-1}\tilde{\bar{\vartheta}}_i^2)^{\frac{\gamma_2+1}{2}} \right\} - \widehat{\varsigma}_{\theta} \sum_{i=1}^{n} \left\{ ((2\varsigma_{\theta i})^{-1}\tilde{\bar{\bar{\theta}}}_{\Phi i}^2)^{\frac{\gamma_2+1}{2}} \right\} + \bar{\bar{N}}^* \end{aligned} \tag{6.46}$$

where $\bar{\bar{N}}^* = \sum_{i=1}^{n}\{\bar{N}_i^*\} + (\widehat{\varsigma}_{\vartheta} + \widehat{\varsigma}_{\theta})n(2^{-1}(1-\gamma_1))(2^{-1}(\gamma_1+1))^{-\frac{1+\gamma_1}{1-\gamma_1}}$.

Using Lemma 6.4, Eq. (6.46) can be further derived as

$$\begin{aligned}\dot{\bar{V}}(t) \leqslant & -\widehat{\varsigma}_a\left\{\sum_{i=1}^{n}((2\varsigma_{\vartheta i})^{-1}\bar{\bar{\vartheta}}_i^2 + (2\varsigma_{\theta i})^{-1}\bar{\bar{\bar{\theta}}}_{\Phi i}^2)\right\}^{\frac{\gamma_1+1}{2}} \\ & - n^{\frac{1-\gamma_2}{2}}2^{\frac{1-\gamma_2}{2}}\widehat{\varsigma}_a\left\{\sum_{i=1}^{n}((2\varsigma_{\vartheta i})^{-1}\bar{\bar{\vartheta}}_i^2 + (2\varsigma_{\theta i})^{-1}\bar{\bar{\bar{\theta}}}_{\Phi i}^2)\right\}^{\frac{\gamma_2+1}{2}} \\ & - \mathfrak{v}_1^*\left\{\sum_{i=1}^{n}(2^{-1}\left|\bar{\mathfrak{V}}_{i,1}\right|^2 + 2^{-1}\left|\bar{\mathfrak{V}}_{i,j}\right|^2)\right\}^{\frac{\gamma_1+1}{2}} \\ & - 2^{\frac{1-\gamma_2}{2}}\mathfrak{v}_2^*\left\{\sum_{i=1}^{n}(2^{-1}\left|\bar{\mathfrak{V}}_{i,1}\right|^2 + 2^{-1}\left|\bar{\mathfrak{V}}_{i,j}\right|^2)\right\}^{\frac{\gamma_2+1}{2}} + \bar{\bar{N}}^*\end{aligned} \tag{6.47}$$

where $\widehat{\varsigma}_a = \min\{\widehat{\varsigma}_\vartheta, \widehat{\varsigma}_\theta\}$, $\mathfrak{v}_1^* = 2^{\frac{\gamma_1+1}{2}}\min\{\min\limits_{1\leqslant i\leqslant n}\mathfrak{v}_{i,1}, \min\limits_{1\leqslant i\leqslant n}\mathfrak{v}_{i,j+1}\}$, $\mathfrak{v}_2^* = 2^{\frac{\gamma_2+1}{2}}n^{\frac{1-\gamma_2}{2}}$ $\min\{\min\limits_{1\leqslant i\leqslant n}\mathfrak{v}_{i,2}, \min\limits_{1\leqslant i\leqslant n}\mathfrak{v}_{i,j+2}\}$.

Based on the above analysis, we have

$$\dot{\bar{V}}(t) \leqslant -\bar{\aleph}_a\bar{V}^{\frac{\gamma_1+1}{2}}(t) - \bar{\aleph}_b\bar{V}^{\frac{\gamma_2+1}{2}}(t) + \bar{\bar{N}}^* \tag{6.48}$$

where $\bar{\aleph}_a = \min\{\mathfrak{v}_1^*, \widehat{\varsigma}_a\}$, $\bar{\aleph}_b = 2^{\frac{1-\gamma_2}{2}}\min\{2^{\frac{1-\gamma_2}{2}}\mathfrak{v}_2^*, n^{\frac{1-\gamma_2}{2}}2^{\frac{1-\gamma_2}{2}}\widehat{\varsigma}_a\}$. According to the practical fixed time stability Lemma 6.2, $\bar{\mathfrak{V}}_{i,1}$, $\bar{\mathfrak{V}}_{i,j}$, $\bar{\bar{\vartheta}}_i$, and $\bar{\bar{\bar{\theta}}}_{\Phi i}$ of the HICVPS are the practical fixed time stability, and the upper bound of its fixed convergence time is $T \leq T_{\max} := 2(\bar{\aleph}_a\bar{\varepsilon}(1-\gamma_1))^{-1} + 2(\bar{\aleph}_b\bar{\varepsilon}(\gamma_2-1))^{-1}$, where $0 < \bar{\varepsilon} < 1$. Thus, $\bar{\mathfrak{V}}_{i,1}$, $\bar{\mathfrak{V}}_{i,j}$, $\bar{\bar{\vartheta}}_i$, and $\bar{\bar{\bar{\theta}}}_{\Phi i}$ are uniformly ultimately bounded stability. Based on the above conclusion, and according to the bounded constant $\bar{\vartheta}_i$ and $\bar{\bar{\theta}}_{\Phi i}$, we can see that the bounded property of $\hat{\bar{\vartheta}}_i$ and $\hat{\bar{\bar{\theta}}}_{\Phi i}$ is guaranteed.

Secondly, for ensuring the fixed time stability of the designed filter compensation signal $\bar{\Upsilon}_{i,1}$ and $\bar{\Upsilon}_{i,j}$ in the whole closed loop system, the Lyapunov function is selected as $\bar{V}_\Upsilon = \sum_{i=1}^{n}\bar{V}_{\Upsilon i}$, where $\bar{V}_{\Upsilon i} = 0.5\left\|\bar{\Upsilon}_i\right\|^2$, $\bar{\Upsilon}_i = [\bar{\Upsilon}_{i,1}, \bar{\Upsilon}_{i,j}]^T$. By using Eqs. (6.14) to (6.16), it can be seen that $\left|\bar{w}_{j,i} - \bar{\alpha}_{j-1,i}\right| \leq \bar{\bar{\ell}}_{i,1}\bar{\bar{\alpha}}_i$ can be satisfied in a fixed time $T_\ell$. Taking the time derivative of $\bar{V}_\Upsilon$, when $t \geq T_\ell$,

$$\begin{aligned}\dot{\bar{V}}_\Upsilon \leqslant & \sum_{i=1}^{n}(-(\kappa_{i,1}-0.5)\bar{\Upsilon}_{i,1}^2 - (\kappa_{i,j}+1.5)\bar{\Upsilon}_{i,j}^2 - \kappa_{i,5}\left|\bar{\Upsilon}_{i,1}\right|^{\upsilon_1+1}) \\ & + \sum_{i=1}^{n}(-\kappa_{i,6}\left|\bar{\Upsilon}_{i,1}\right|^{\upsilon_2+1} - \kappa_{i,j+1}\left|\bar{\Upsilon}_{i,j}\right|^{\upsilon_1+1} - \kappa_{i,j+2}\left|\bar{\Upsilon}_{i,j}\right|^{\upsilon_2+1} + 0.5\bar{\bar{\ell}}_{i,1}^2\bar{\bar{\alpha}}_i^2) \\ \leqslant & -\kappa_\Upsilon\bar{V}_\Upsilon^{\frac{\upsilon_1+1}{2}} - \mu_\Upsilon\bar{V}_\Upsilon^{\frac{\upsilon_2+1}{2}} + n_0\bar{\bar{\ell}}_{i,1}^2\bar{\bar{\alpha}}_i^2\end{aligned} \tag{6.49}$$

where $n_0 = n/2$. According to the initial condition $\bar{\Upsilon}_{i,1}(0) = \bar{\Upsilon}_{i,j}(0) = 0$, we know that $\bar{\Upsilon}_{i,1}$, $\bar{\Upsilon}_{i,j}$are bounded and can converge to the origin in fixed time, the fixed time $T_{\Upsilon} \leq 2(\kappa_{\Upsilon}\Delta(1-\upsilon_1))^{-1} + 2(\mu_{\Upsilon}\Delta(\upsilon_2-1))^{-1}$ is satisfied, where $0 < \Delta < 1$. $\kappa_{\Upsilon} = 2^{(\upsilon_1+1)/2} \min_{1\leq i\leq n}\{\kappa_{i,5}, \kappa_{i,j+1}\}$, and $\mu_{\Upsilon} = 2^{(1+\upsilon_2)/2} \min_{1\leq i\leq n}\{\kappa_{i,6}, \kappa_{i,j+2}\}$.

According to the above analysis, it is known that $\bar{z}_{i,1} = \bar{\mathfrak{V}}_{i,1} + \bar{\Upsilon}_{i,1}$, $\bar{z}_{i,j} = \bar{\mathfrak{V}}_{i,j} + \bar{\Upsilon}_{i,j}$, based on the fixed-time convergence analysis in the closed-loop system of $\bar{\Upsilon}_{i,1}$, $\bar{\Upsilon}_{i,j}$, $\bar{\mathfrak{V}}_{i,1}$ and $\bar{\mathfrak{V}}_{i,j}$, $\bar{z}_{i,1} = \bar{e}_i(t)$ and $\bar{z}_{i,j}$ both are uniformly ultimately bounded and can converge to a small neighborhood of the origin within a fixed time, which means that the closed-loop control stability of the HICVPS is guaranteed. The final inter-vehicle distance can also converge to the desired constant inter-vehicle distance within a fixed time, that is $\lim_{t\to\infty}(\bar{p}_{i-1}(t) - \bar{p}_i(t) - L_{i-1}) = \bar{d}_{i-1,i}$. According to the boundedness of $\dot{\bar{z}}_{i,1} = \bar{z}_{i,j} + \bar{w}_{j,i}$ in formula (6.25) and the boundedness of $\bar{w}_{j,i}$ in formula (6.14), there exists that $\dot{\bar{z}}_{i,1} = \dot{\bar{e}}_i$ is bounded. Therefore, combining the known conditions $\bar{e}_i(0) = 0$, $\dot{\bar{e}}_i(0) = 0$, and according to Definition 3.1 of string stability, it can be seen that there are constants $\phi > 0$ and $\varphi > 0$, and condition $\sup_{i\in(1,2,\cdots,n)}\{|\bar{e}_i(0)|, |\dot{\bar{e}}_i(0)|\} < \phi$, which make $\sup_{i\in(1,2,\cdots,n)}\{|\bar{e}_i(0)|, |\dot{\bar{e}}_i(0)|\} < \varphi$ be true, that is the string stability of the system can be guaranteed within a fixed time.

Finally, the proposed control scheme can effectively avoid the "Zeno" phenomenon caused by the event mistake triggered is verified theoretically, which means that the exists the time $T_e > 0$ makes $T_e \leq t_{\mathfrak{k}+1} - t_{\mathfrak{k}}$ valid, $\forall \mathfrak{k} \in Z^+$. By $\bar{e}_{ui}(t) = \bar{\mathfrak{W}}_i(t) - u_i(t), \forall t \in [t_{\mathfrak{k}}, t_{\mathfrak{k}+1}), d\,|\bar{e}_{ui}(t)|/dt = \dot{\bar{e}}_{ui} sign(\bar{e}_{ui}) \leq \left|\dot{\bar{\mathfrak{W}}}_i(t)\right|$ can be obtained. At the same time, according to Eq. (6.22), it can be seen that $\bar{\mathfrak{W}}_i(t)$ is differentiable and $\dot{\bar{\mathfrak{W}}}_i(t)$ is a continuous compound function, whose variable signals are bounded. Based on the above analysis, there must be a constant $\nabla^* > 0$ that satisfies the condition $\left|\dot{\bar{\mathfrak{W}}}_i(t)\right| \leq \nabla^*$. By $\bar{e}_{ui}(t_{\mathfrak{k}}) = 0$, $\lim_{t\to t_{\mathfrak{k}+1}} \bar{e}_{ui}(t) = \mathfrak{d}_{i,0}|u_i(t_{\mathfrak{k}+1})| + h_i\bar{\mathfrak{H}}_i^{-1}(t_{\mathfrak{k}+1}) + \widehat{\bar{\Delta}}_i$, then, the triggered event interval $T_e$ satisfy $T_e \geq (\mathfrak{d}_{i,0}|u_i(t_{\mathfrak{k}+1})| + h_i\bar{\mathfrak{H}}_i^{-1}(t_{\mathfrak{k}+1}) + \widehat{\bar{\Delta}}_i)(\nabla^*)^{-1} \geq \widehat{\bar{\Delta}}_i\,(\nabla^*)^{-1}$, which means that the proposed control scheme can effectively avoid the "Zeno" phenomenon of the event mistake triggered. Therefore, the theoretical proof of Theorem 6.2 is complete. □

## 6.4 Method Verification and Analysis

In this section, the feasibility of the proposed scheme is verified by the comparative experimental analysis and the road traffic simulator SUMO-MATLAB joint simulation experiment. The HICVPS considered is composed of one leader vehicle and five heterogeneous follower vehicles. Considering the nonlinear dynamic function of the follower vehicle as $\bar{F}_i(\bar{X}_i) = -\bar{v}_i\bar{a}_i\bar{\varphi}_a Ae_i\bar{K}_{ai}M_i^{-1} - \bar{a}_i\bar{\tau}_i^{-1} - \bar{v}_i^2\bar{\varphi}_a Ae_i\bar{K}_{ai}(2\bar{\tau}_i M_i)^{-1} - \bar{d}_{mi}(\bar{\tau}_i M_i)^{-1}$. The longitudinal dynamics parameters of the follower vehicle are $\bar{\varphi}_a = 1.2$, $Ae_i = 2.2$,

$\bar{d}_{mi} = 5$, $\bar{K}_{ai} = 0.35$, respectively. The longitudinal dynamic heterogeneous parameters ($i = 1, 2, 3, 4, 5$) of the five follower vehicles are: the mass $M_i$ of five vehicles are $1000, 1200, 1150, 900, 1100$ kg, the vehicle engine inertia delay $\bar{\tau}_i$ are $0.4, 0.5, 0.45, 0.51, 0.42$ s, and the vehicle's length $L_i$ are $3.5, 4, 5, 4, 5$ m.

For the improved exponential spacing policy, the actual measured inter-vehicle distance is $\bar{D}_i(t) = \bar{p}_{i-1}(t) - \bar{p}_i(t) - L_{i-1}$, the desired inter-vehicle distance is $\bar{d}_{i-1,i} = 5$ m, $\varepsilon = 0.5$ and $\bar{\Delta}_i = 7\text{m/s}^2$. For the fixed time disturbance observer, $\bar{\varpi}_i(t) = \bar{\varpi}^o + A^*$, $A^* = 0.18\sin(2\pi t) + 0.13\cos(0.5\pi t)$, the initial external disturbance values $\bar{\varpi}^o$ of five follower vehicles are $0.5, 0.43, 0.4, 0.3, 0.1$, $i = 1, 2, 3, 4, 5$. The fuzzy membership function is selected as $\kappa_{F_i^{\bar{l}}}(X_i) = \exp[-(X_i + 2\bar{l}_s - 6)^2 2^{-1}]$, $\bar{l}_s = 1, \ldots, 5, \bar{l} = 1, \ldots, 9$. For the improved event triggered mechanism parameters, we select $\eth_{i,0} = 0.1$,$\eth_{i,1} = 0.038$, $\eth_{i,2} = 5.8$, $\widehat{\bar{d}}_{i,0} = 2.1$, $h_i = 0.01$, and $\widehat{\bar{\Delta}}_i = 0.02$. For the leader vehicle 0, its vehicle length is selected as $L_0 = 2.5$ m, and its initial states of position, velocity and acceleration of leader vehile are 145 m, 1.2 m/s, 0 m/s$^2$, respectively. The reference control input of the leader vehicle is

$$u_0(t) = \begin{cases} 0.3 \quad 5s \le t < 9s; \; -0.3 \quad 14s \le t < 18s \\ 0, \quad otherwise \end{cases}.$$

In addition, the case of multiple communication faults suffered by the HICVPS is considered in the comparative experimental case to further verify the string stability performance of the proposed scheme under unknown communication faults and disturbance conditions, as well as the resilience performance and rapid recovery of closed-loop control. When $t = [20, 22)$, the vehicle $i = 3$ suffers from the controller-actuator communication fault induced by the deception attack, the fault function is $\bar{\Phi}_3(\bar{X}_3) = 0.2\sqrt{\bar{v}_3^2 + \bar{a}_3^2}\sin(t)$. When $t = [30, 32)$, the vehicle $i = 5$ suffers from the communication faults, the fault function is $\bar{\Phi}_5(X_5) = 0.3\sqrt{\bar{v}_5^2 + \bar{a}_5^2}\sin(t)$.

The SUMO-MATLAB joint simulation experiment platform is developed based on intelligent traffic simulator SUMO and Matlab for verifying the applicability of the scheme proposed in this Chapter, and the actual road scene of S6 Chengdu Airport Expressway is taken as an example to conduct experimental verification of the proposed scheme. Under the conditions of communication faults and disturbances, the control results of the scheme proposed in this Chapter on the SUMO-MATLAB joint simulation experiment platform are shown in Fig. 6.3.

Figure 6.4 shows the control results of the proposed scheme for the HICVPS that is only with external disturbances. Meanwhile, for further verifying the effectiveness and superiority of the proposed control scheme, under the same information flow topology and based on the HICVPS with multiple communication faults and disturbances, the proposed control scheme Theorem 6.2, the existing fractional fixed-time control method [26], and the existing practical finite time control method [27] are respectively used to analysis. The experimental results were shown in Figs. 6.5, 6.6 and 6.7. The improved event triggered mechanism is also compared with the traditional event triggered method [5] in terms of triggered time interval, and the results are shown in Fig. 6.8.

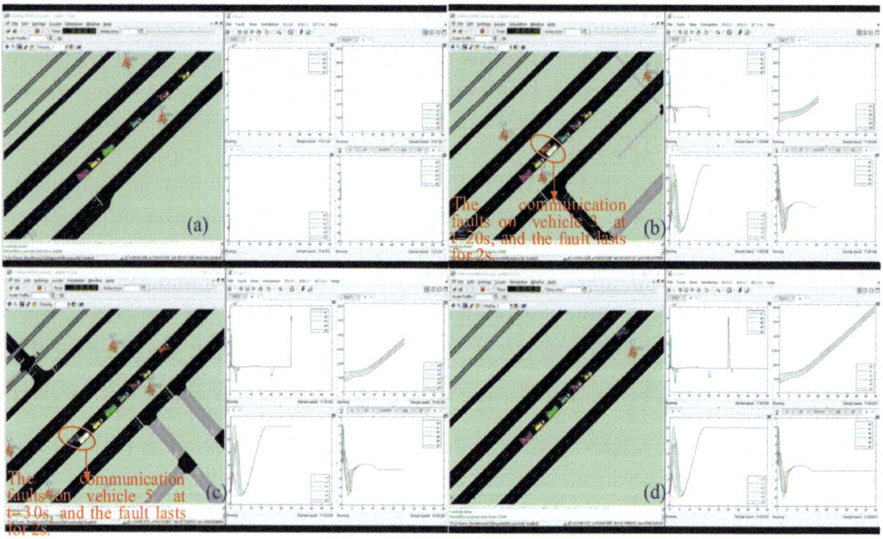


Fig. 6.3 State trajectory of the proposed scheme on the SUMO-MATLAB joint simulation experiment platform

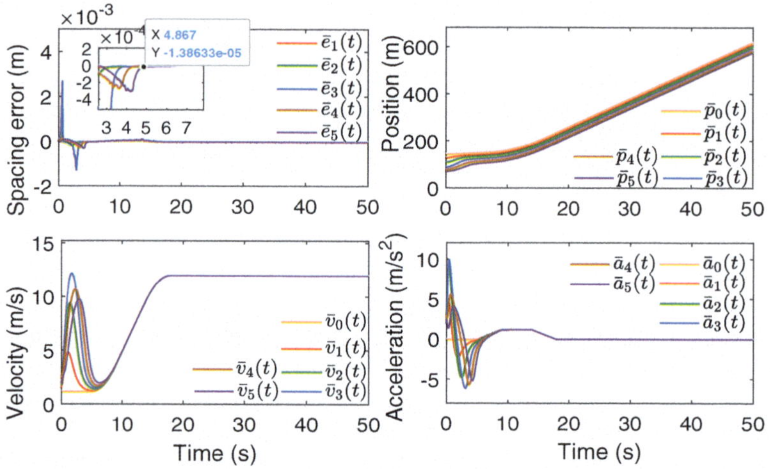


Fig. 6.4 Inter-vehicle spacing error and system state of proposed method in the absence of communication fault

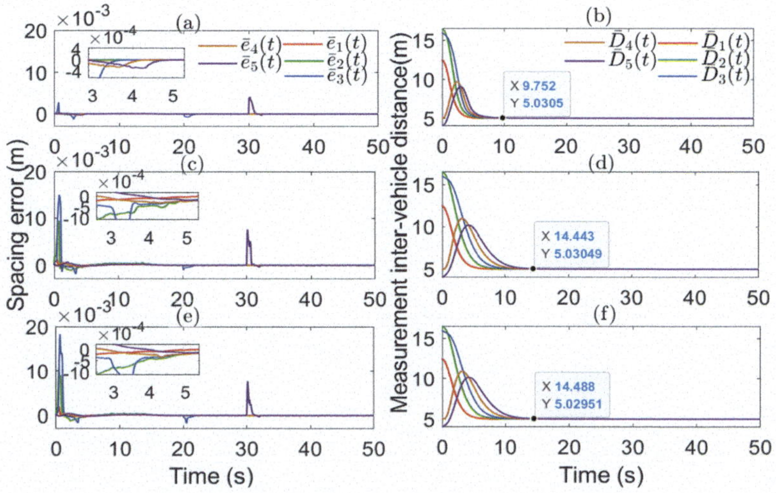


**Fig. 6.5** Comparative experimental results of the HICVPS control. **a** Inter-vehicle spacing error under proposed method, **b** Actual measurement inter-vehicle distance under proposed method, **c** Inter-vehicle spacing error under the existing fractional fixed-time control method, **d** Actual measurement inter-vehicle distance under the existing fractional fixed-time control method, **e** Inter-vehicle spacing error under the existing practical finite-time control method, **f** Actual measurement inter-vehicle distance under the existing practical finite time control method

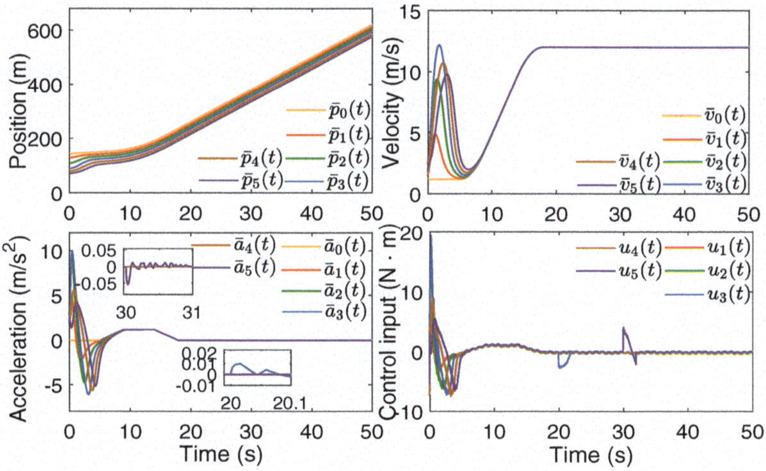


**Fig. 6.6** System state and control input under the proposed method

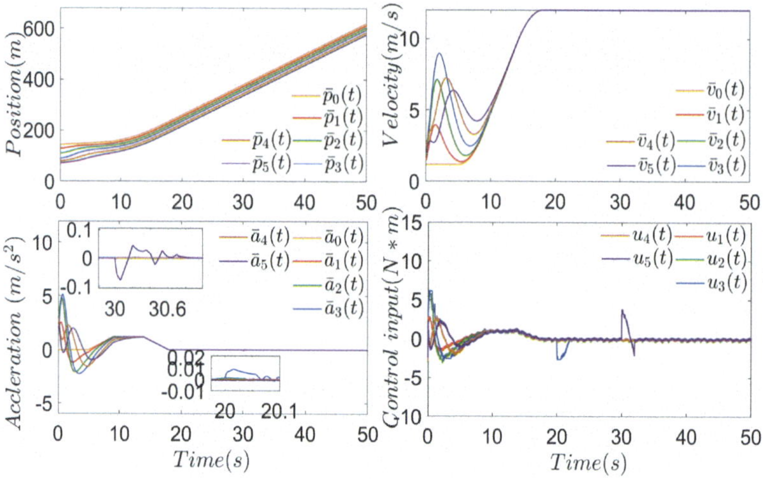

**Fig. 6.7** System state and control input under the existing practical finite time control method

As shown in Fig. 6.5a, the inter-vehicle spacing error of the follower vehicle is practical fixed time stable in a small origin neighborhood, and the accuracy of the bounded error is less than 0.001 m. Even if the disturbances and multiple communication faults occur, the proposed method can still ensure the closed-loop control stability and string stability of the HICVPS. Compared with the results of Fig. 6.5a, the experimental results of Figs. 6.5c and 6.5e show a relatively slow convergence rate ($T \geqslant 14$s) and a relatively larger transient spacing error. Moreover, when the communication faults occur, the proposed method mentioned in Fig. 6.5a obviously has stronger resilience, fault recovery and faster convergence rate.

As shown in Fig. 6.5b, under the conditions of disturbances and multiple communication faults, the proposed method can achieve the desired inter-vehicle safety spacing in 9.7 s, which is faster than 14.44 s under the fractional fixed-time control method in Fig. 6.5d, and 14.48 s under the practical finite time control method in Fig. 6.5f. The convergence time and root mean square error control performance indexes under different control methods are shown in Table 6.1, which shows that the proposed method has advantages in convergence time and control accuracy, and proves the effectiveness of the proposed method.

Figures 6.6 and 6.7 respectively show the system state and control input of the HICVPS under the proposed method and the existing practical finite time control method. Under the conditions of unknown communication faults and unknown external disturbances, as shown in Fig. 6.6, it can be clearly seen that the velocity and acceleration of the five follower vehicles can quickly track the leader vehicles within, and the convergence velocity is significantly higher than that in Fig. 6.7. At the same time, Fig. 6.6 also shows that the proposed method can also ensure that no collision occurs during the entire driving process. Figures 6.6 and 6.7 verify that the convergence rate and tracking accuracy of the proposed method are significantly better than

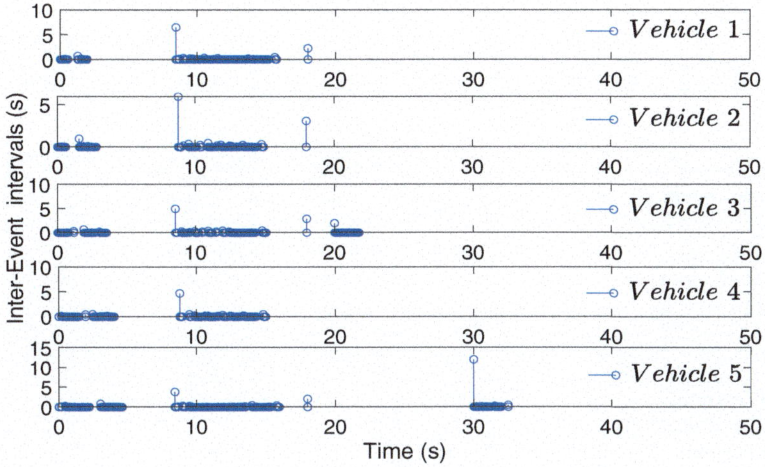


(a)

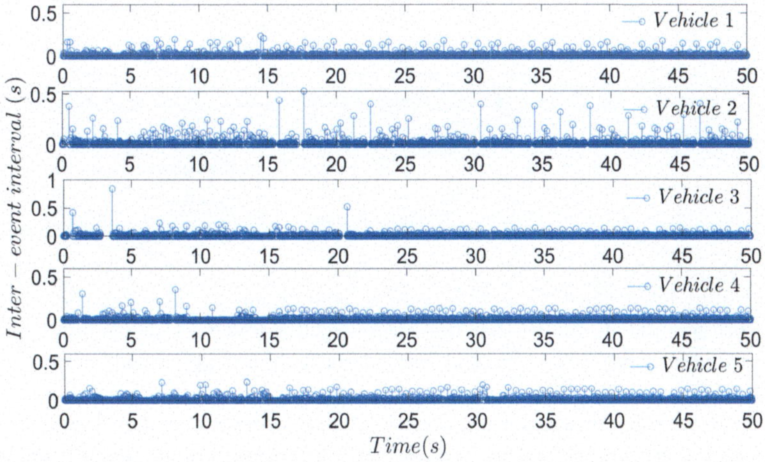


(b)

**Fig. 6.8** Comparative experimental results of event triggered time intervals. **a** Proposed method, **b** Traditional method

the existing practical finite time control method. Figure 6.8a, b respectively show the triggered event time interval curve under the proposed event triggered mechanism and the traditional event triggered mechanism.

As shown in Fig. 6.8a, the triggered frequency of the proposed improved event triggered mechanism is much smaller than that of the traditional event triggered mechanism, and the execution interval of the triggered event is significantly larger than that the traditional mechanism, in which the maximum interval of the triggered event execution interval under the proposed method reaches 14 s. However,

**Table 6.1** The dynamic parameters of the HICVPS

| Vehicle $i = 1, 2, \ldots, 5$ | 1 | 2 | 3 | 4 | 5 | Convergence times |
|---|---|---|---|---|---|---|
| Proposed method ($\times 10^{-5}$ kg) | 2.7606 | 3.7437 | 21.199 | 4.0551 | 38.110 | $T \leqslant 9.70$s |
| Existing fractional fixed-time control method ($\times 10^{-5}$ kg) | 14.334 | 76.052 | 140.00 | 10.071 | 58.710 | $T \leqslant 14.44$s |
| Existing practical finite-time control method ($\times 10^{-5}$ kg) | 13.884 | 90.836 | 160.00 | 10.071 | 51.969 | $T \leqslant 14.48$s |

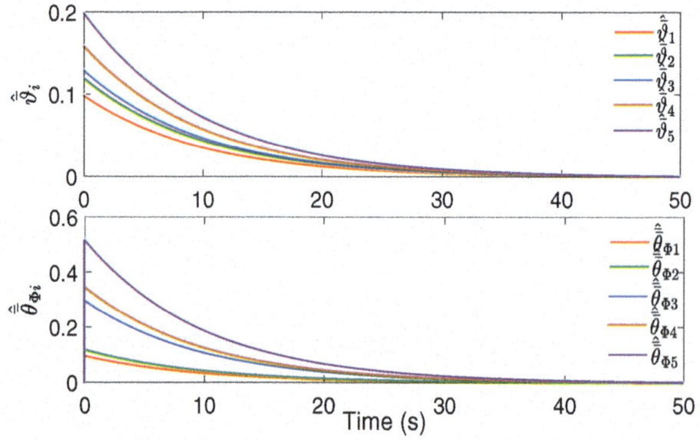


**Fig. 6.9** Adaptive compensation under the proposed method

the maximum interval of triggered event execution under the traditional method is 0.9 s. Figure 6.8a, b shows that the proposed improved event triggered mechanism can ensure a slower execution rate of the actuator and lower control signal transmission in the HICVPS, thus the proposed mechanism can save more control energy. Based on the above comparative analysis, the proposed distributed adaptive event-triggered fixed-time resilient control architecture can maintain a good balance between fixed-time resilient control performance and efficient data transmission.

Figure 6.9 shows the adaptive compensation and updated laws of the proposed method. Figure 6.10a, b shows the unknown disturbances fixed-time estimation error under the proposed method, where Fig. 6.10b shows the observation error of the fixed-time disturbances observer in fractional order form, and the fractional differential operator is 0.9. As shown in Fig. 6.10, the unknown external disturbances of the follower vehicle can be quickly estimated within a fixed time, and the convergence time in Fig. 6.10a, b is less than 1.403 and 0.55 s, respectively.

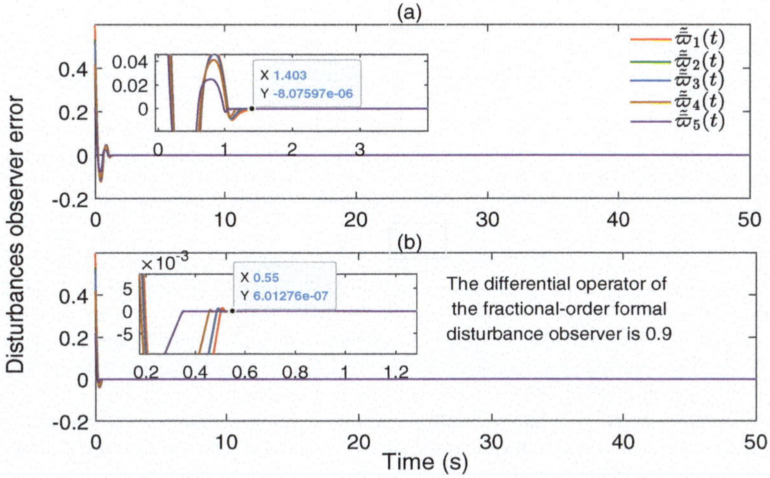


**Fig. 6.10** Fixed time disturbances observer error $\tilde{\bar{\varpi}}_i(t)$ under the proposed method

## 6.5 Summary

In this chapter, a distributed adaptive fixed-time resilient control architecture based on improved event triggered is designed for the HICVPS with unknown communication faults and disturbances. Firstly, an effective sliding mode fixed time disturbance observer is constructed to realize the fixed time fast estimation of unknown external disturbances of the ICVPS under arbitrary initial conditions. Secondly, based on exponential spacing policy and PLF-type information flow topology, the distributed adaptive event-triggered fixed-time resilient control scheme is designed to ensure closed-loop control stability and string stability under the practical fixed-time stability, while the data transmission resources and control energy consumption are reduced. In addition, the fuzzy adaptive fault compensation law is designed to overcome the influence of unknown communication faults. Finally, the effectiveness of the proposed scheme is verified by comparing experimental cases and the SUMO-MATLAB joint simulation platform.

## References


1. Pan C, Chen Y, Chen S, et al (2023) Event-based distributed fixed-time resilient control for heterogeneous vehicular platoon against attack and disturbances. *IEEE Internet of Things Journal*, 10(13): 11858–11868.
2. Wang J, Luo X, Wang L, et al (2020) Integral sliding mode control using a disturbance observer for vehicle platoons. *IEEE Transactions on Industrial Electronics*, 67(8): 6639–6648.
3. Li M, Chen Y, and Liu Y. (2021) Adaptive disturbance observer-based event-triggered fuzzy control for nonlinear system. *Information Sciences*, 575: 485–498.

4. Wen S, Guo G, Chen B, et al (2018) Event-triggered cooperative control of vehicle platoons in vehicular ad hoc networks. *Information Sciences*, 459: 341–354.
5. Xing L, Wen C, et al (2017) Event-triggered adaptive control for a class of uncertain nonlinear systems. *IEEE Transactions on Automatic Control*, 62(4): 2071–2076.
6. Liu Y, and Gao H (2022) Stability, scalability, speedability, and string stability of connected vehicle systems. *IEEE Transactions on Systems, Man, and Cybernetics: Systems*, 52(5): 2819–2832.
7. Y. Xu, G. Guo (2022) Event triggered control of connected vehicles under multiple cyber attacks. *Information Sciences*, 582: 778–796.
8. Schwab A, and Lunze J (2022) Design of platooning controllers that achieve collision avoidance by external positivity. *IEEE Transactions on Intelligent Transportation Systems*, 23(9): 14883–14892.
9. Liu C, Zhao J, and Patton R. J (2023) Distributed antittack fault-tolerant tracking control for vehicle platoon systems under cyber-physical threats. *IEEE Transactions on Industrial Informatics*, 19(6): 7825–7834.
10. Petrillo A, Pescapé A, and Santini S (2021) A secure adaptive control for cooperative driving of autonomous connected vehicles in the presence of heterogeneous communication delays and cyberattacks. *IEEE Transactions on Cybernetics*, 51(3): 1134–1149.
11. Ding L, Li J, Ye M, et al (2022) Fully distributed resilient cooperative control of vehicular platoon systems under Dos Attacks. *IEEE/CAA Journal of Automatica Sinica*, 9(5): 937–940.
12. Li Y, Tang C, Li K, et al (2018) Nonlinear finite-time consensus-based connected vehicle platoon control under fixed and switching communication topologies. *Transportation Research Part C: Emerging Technologies*, 93: 525–543.
13. Du C, Bian Y, Liu H, et al (2021) Cooperative startup control for heterogeneous vehicle platoons: a finite-time output tracking-based approach. *IEEE Transactions on Control of Network Systems*, 8(4): 1767–1777.
14. Sun J, Pu Z, Yi J, et al (2020) Fixed-time control with uncertainty and measurement noise suppression for hypersonic vehicles via augmented sliding mode observers. *IEEE Transactions on Industrial Informatics*, 16(2): 1192–1203.
15. Liu D, Liu Z, Chen C. L. P, et al (2022) Distributed adaptive neural fixed-time tracking control of multiple uncertain mechanical systems with actuation dead zones. *IEEE Transactions on Systems, Man, and Cybernetics: Systems*, 52(6): 3859–3872.
16. Wang C, Wang D, and Peng Z (2022) Distributed output-feedback control of unmanned container transporter platooning with uncertainties and disturbances using event-triggered mechanism. *IEEE Transactions on Vehicular Technology*, 71(1): 162–170.
17. Wu G, Chen G, Zhang H, et al (2021) Fully distributed event-triggered vehicular platooning with actuator uncertainties. *IEEE Transactions on Vehicular Technology*, 70(7): 6601–6612.
18. Yang F, Gu Z, Hua L, et al (2022) A resource-aware control approach to vehicle platoons under false data injection attacks. *ISA Transactions*, 131: 367–376.
19. Liu Y, Yao D, Wang L, et al (2023) Distributed adaptive fixed-time robust platoon control for fully heterogeneous vehicles. *IEEE Transactions on Systems, Man, and Cybernetics: Systems*, 53(1): 264–274.
20. Huang X, Dong J (2020) Learning-based switched reliable control of cyber-physical systems with intermittent communication faults. *IEEE/CAA Journal of Automatica Sinica*, 7(3): 711–724.
21. Gusrialdi A, Qu Z, Simaan M A (2018) Competitive interaction design of cooperative systems against attacks. *IEEE Transactions on Automatic Control*, 63(9): 3159–3166.
22. Zhang D, Liu G, Zhou H, et al (2018) Adaptive sliding mode fault-tolerant coordination control for four-wheel independently driven electric vehicles. *IEEE Transactions on Industrial Electronics*, 65(11): 9090–9100.
23. Guo X, Xu W, Wang J, et al (2021) Distributed neuroadaptive fault-tolerant sliding-mode control for 2-D plane vehicular platoon systems with spacing constraints and unknown direction faults. *Automatica*, 129: 109675.

24. Yu J, Shi P, Lin C, et al (2020) Adaptive neural command filtering control for nonlinear MIMO systems with saturation input and unknown control direction. *IEEE Transactions on Cybernetics*, 50(6): 2536–2545.
25. Cruz Z E, Moreno J A, Fridman L M (2011) Uniform robust exact differentiator. *IEEE Transactions on Automatic Control*, 56(11): 2727–2733.
26. Song X, Sun P, Song S, et al (2022) Event-driven NN adaptive fixed-time control for nonlinear systems with guaranteed performance. *Journal of the Franklin Institute*, 359: 4138–4159.
27. Yu J, Shi P, Zhao L (2018) Finite-time command filtered backstepping control for a class of nonlinear systems. *Automatica*, 92: 173–180.

# Chapter 7
# Distributed Adaptive Prescribed Fixed-Time Resilient Control of Heterogeneous Intelligent Connected Vehicle Platoon Systems Against Unknown Dead-Zone and Faults

In engineering practice, the control problem of HICVPS faces more severe challenges due to physical environment constraints and changing network communication environment. On the one hand, the external disturbances, uncertainties, and unknown input dead zones complicate the control design of HICVPS. On the other hand, the multi-form faults behavior may cause the transient/steady state control performance and string stability of HICVPS to seriously decay, and the control accuracy will be reduced, and the vehicle collision risk will be increased. The existing method of combining prescribed performance control and fixed time control (as we discussed in Chap. 1) can improve the system robustness and control accuracy while ensuring the transient/steady state performance requirements of the system. However, the research on adaptive prescribed performance fixed-time resilient control methods for HICVPS is still limited. It is still a difficult problem for how to maintain the desired inter-vehicle safety space under unknown faults and dead zone constraints, ensure the transient/steady inter-vehicle safety space error within the expected constraints, and improve the rapid recovery ability and convergence accuracy of the system under fault conditions, and ensure fixed time stability and string stability of closed-loop control system.

Based on the above analysis, for the unknown multi-form additive faults and input dead zones, external disturbances, in this chapter, the distributed adaptive platoon resilient control scheme based on the prescribed fixed-time stability. The main contributions of this Chapter are as follows. Firstly, different from the finite time control scheme [1] and the prescribed performance fixed-time tracking control method of a single system [2], in this chapter, a distributed adaptive prescribed fixed-time resilient control scheme for HICVPS to deal with problems such as actuator additive fault, unknown input dead zone nonlinearity and external disturbances. The internal stability and string stability under the prescribed asymmetric constraint performance can also be obtained. Secondly, for realizing the unknown fault and unknown input dead

 
C. Pan and Y. Chen, *Resilient Control of Intelligent Connected Vehicle Platoon Systems*,
https://doi.org/10.1007/978-981-97-8850-7_7

zone compensation, the fixed time compensation strategy is designed to realize the fault and dead zone compensation control of HICVPS based on fuzzy approximation error, unknown fault function and unknown input dead zone parameters. Finally, different from the leader-bilateral information flow topology [3], for reducing the communication burden and save communication information resources, in this chapter, the inter-vehicle communication based on the PF information flow communication topology is carried out. In the sense of fixed-time Lyapunov stability, the spacing error converging to any prescribed asymmetric constraint domain within a fixed time is realized. Meanwhile, the spacing error accuracy and resilience performance of the HICVPS are improved.

## 7.1 Introduction

As an important part of intelligent transportation system, intelligent connected vehicle platoon control technology requires vehicles in the same lane to automatically form a platoon and maintain a given space between vehicles to drive safely, which provides a good feasible idea for solving the urban traffic problems, increasing road throughput, easing traffic congestion and reducing fuel consumption [4]. At present, the basic problems of intelligent connected vehicle platoon control that have been extensively studied in the literature mainly include: longitudinal dynamics of connected and autonomous vehicles, the inter-vehicle spacing policy, information flow topology, closed-loop system stability and string stability [5]. The closed-loop system stability and string stability mean that while the follower vehicle stably tracks the velocity and acceleration of the leader vehicle, the spacing error between neighbor vehicles also converges to zero, and does not transmit and amplify backward along the platoon. However, due to the particularity of the vehicle platoon structure, the intelligent connected vehicle platoon is particularly sensitive to the unknown faults, external disturbances and physical threats inside the vehicle. Therefore, how to design the controller to improve the resilience capability of the vehicle platoon system and maintain the closed-loop stability and the string stability has important application value and practical engineering significance.

For achieving the closed-loop stability and the string stability of the intelligent connected vehicle platoon system, and improving the resilience and robustness of the system, a large number of literatures have been studied. By combining a new quadratic spacing policy, in [1], a distributed finite-time cooperative control method was proposed to deal with the problems of finite-time stability and string stability in the vehicle platoon under parameter uncertainties and unknown disturbances. In [6], a collaborative adaptive control scheme based on V2V communication is proposed to ensure the closed-loop asymptotic stability and string stability of the platoon system under parameter uncertainties and unknown external disturbances. The actuator fault of the vehicular drive system is easy to destroy its output performance and threat the safety and stability of the platoon system. Further, in [7], a coupled sliding mode control algorithm based on disturbance observer is proposed to ensure the finite-time stability and string stability performance of the platoon system under the influence of actuator fault and saturation. On the other hand, due to the high-frequency switching between the braking system and the drive system of the connected and autonomous vehicle, it is easy to cause the dead zone nonlinear inside the actuator, which weakens the transient performance of the system and destroys the stability of the system. In [8], a distributed sliding mode control method is proposed to ensure the string stability performance of HICVPS with unknown actuator input dead zone. However, there is still a lack of research on resilient control of intelligent connected vehicle platoon under actuator faults, unknown input dead zone nonlinearity and external disturbances.

In the problem of intelligent connected vehicle platoon control, the convergence velocity and robustness are important indicators to obtain good closed-loop stability and the string stability. However, in the existing literatures, most of the research results of the vehicle platoon systems focus on asymptotic stability and finite time stability. Compared with asymptotic stability of closed-loop system, the finite-time stability can achieve faster convergence velocity and stronger robustness, but the finite-time stability depends heavily on the initial state value of the system. When the initial state value of the vehicle platoon system is too large, the finite-time convergence rate and the robustness under faults and input dead zone conditions may not be guaranteed. In the design process, the fixed-time stability can get rid of the dependence on the initial state value of the vehicle platoon system and obtain stronger robust performance. However, the research on the fixed-time stability in the vehicle platoon control problem is still scarce. Therefore, how to design a distributed fixed-time resilient control method to ensure the closed-loop stability and the string stability of intelligent connected vehicle platoon systems under the actuator faults, unknown dead zone nonlinearity and external disturbances is still an open research hotspot.

It is worth noting that the string stability and the closed-loop stability of the vehicle platoon system are particularly sensitive to faults and disturbances in the process of high-speed movement due to the physical structure constraints of the intelligent connected vehicle platoon and the limitations of the communication range of the ad hoc network of the mobile vehicle. The higher demands on the error accuracy performance of the vehicle platoon system in the transient and steady-state process are put forward. Therefore, the research of prescribed constraint performance is of great practical engineering significance to meet the performance requirements of transient and steady-state spacing errors or overshoot. In [2], an observer-based adaptive prescribed performance control strategy is proposed to meet the requirement that the tracking error of pure feedback nonlinear systems can converge to the prescribed error constraint range. In [3], a fixed-time prescribed performance resilient control scheme was proposed to address the unknown steering mechanism fault and motor traction loss fault of ground unmanned vehicles. In [9], an adaptive tracking control method with fixed time prescribed constraint performance is proposed to realize that the position and attitude tracking errors of a single four-rotor UAV system can converge to the prescribed constraint domain within a fixed time. In [10], a finite time control scheme based on prescribed performance is proposed to address the uncertainties and the disturbances of the vehicle platoon system, and the spacing error can converges to the prescribed constraint domain within a finite time. However, there is still a lack of research results on the design of fixed time resilient control for distributed prescribed performance of HICVPS with actuator additive faults, unknown input dead zones and external disturbances.

## 7.2 Problem Statement

In this chapter, the problems such as actuator additive faults, input dead zone nonlinearity and external disturbances suffered by HICVPS are focused on as shown in Fig. 7.1. Firstly, the constant time spacing policy and prescribed asymmetric performance function are used to ensure the steady/transient performance of system spacing error. Then, a distributed adaptive prescribed performance fixed-time resilient control scheme is designed to ensure the closed-loop stability and the string stability of the system, and realize that the spacing errors between neighbors can converge to the prescribed asymmetric constraint region within a fixed time and evolve strictly in it. At the same time, the adaptive fixed time compensation laws are designed to eliminate the influence of actuator faults, unknown input dead zone and the external disturbances on the system stability.

### *7.2.1 System Modeling and Information Flow Topology Design*

In this chapter, the HICVPS is considered to be composed of $n$ follower vehicles and one leader vehicle, and its driving environment is considered to be a straight highway, as shown in Fig. 7.1. During the process of movement of the vehicle platoon, the vehicle platoon accompanied by the complex external traffic environment disturbances, such as potholes, wind gusts and strong sand, and it is assumed that no overtaking occurs during the control process.

Based on the above conditions, the third-order nonlinear longitudinal dynamics intelligent connected vehicle platoon system model considered in this chapter is as follows [5, 8]

$$\begin{aligned}\dot{\bar{p}}_i(t) &= \bar{v}_i(t)\\ \dot{\bar{v}}_i(t) &= \bar{a}_i(t) + w_{i,v}(t)\\ \dot{\bar{a}}_i(t) &= \bar{F}_{i,a}(\bar{v}_i(t), \bar{a}_i(t)) + \bar{b}_{i,a}\bar{u}_i(t) + w_{i,a}(t)\end{aligned} \tag{7.1}$$

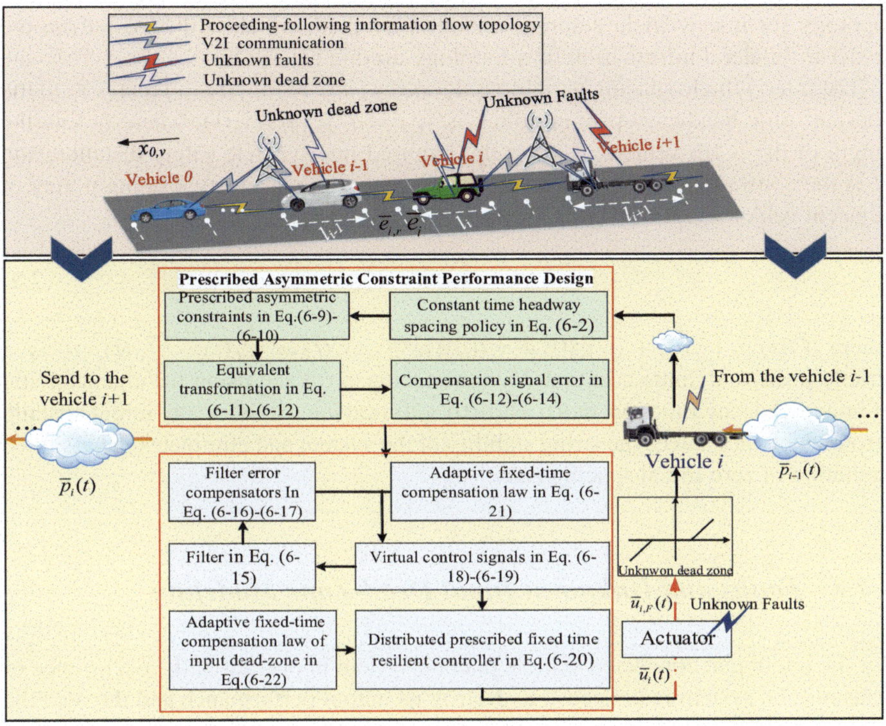


**Fig. 7.1** The control architecture for HICVPS

where $\bar{F}_{i,a}(\bar{v}_i(t), \bar{a}_i(t)) = -\frac{\bar{\varphi}^* A_i K_{ai}}{2\tau_i \bar{m}_i}(2\tau_i \bar{v}_i \bar{a}_i + \bar{v}_i^2) - \frac{1}{\tau_i}(\bar{a}_i + \frac{D_{mi}}{\bar{m}_i})$, $\bar{p}_i(t)$, $\bar{v}_i(t)$, and $\bar{a}_i(t)$ respectively represent the position, velocity, acceleration of the vehicle platoon system state information, $i = 1, 2, \cdots, n$. $\bar{\varphi}^*$ represents the air quality, $A_i$ is the cross-sectional area of the vehicle, $D_{mi}$ is the mechanical resistance, and $K_{ai}$ represents the resistance coefficient of the vehicle. $w_{i,v}(t)$ and $w_{i,a}(t)$ indicate the external environmental disturbances suffered by autonomous vehicles on the road, such as potholes, gusts and strong sand, etc. It should be noted that in the actual situation, they can always meet: $|w_{i,r}(t)| \leq \bar{w}_{i,r}$, $r = v, a$. $\bar{F}_{i,a}(\bar{v}_i(t), \bar{a}_i(t))$ is the nonlinear function term of the vehicle system. $\bar{b}_{i,a} = \tau_i^{-1}$,$\tau_i$ is the time constant of the engine. $\bar{u}_i(t) = \bar{m}_i^{-1} \bar{F}_i$ is for the control input of the vehicle system. $\bar{m}_i$ is the mass of the vehicle, $\bar{F}_i$ is the driving/braking torque of the vehicle. For the leader vehicle $i = 0$, all state information such as position, velocity and acceleration can be obtained by the follower vehicles, and $w_{0,r}(t)$ is not considered in the leader vehicle system.

The intelligent connected vehicle platoon system considered in this chapter realizes V2V and V2I communication and interaction by means of ad hoc network equipments. In order to reduce communication burden and save communication information resources, in this chapter, the PF information flow communication topology for inter-vehicle communication is adopted, which is different from the leader-to-bilateral information flow topology used in [3].

The inter-vehicle spacing variable is defined as $\bar{y}_{i,r} = \bar{p}_{i-1}(t) - \bar{p}_i(t) - l_{i-1}$, the constant time headway spacing policy is $\bar{y}_{i,d} = \bar{d}_{i,des} + H\bar{v}_i(t)$, where $l_{i-1}$ is the length of the $i$−1th vehicle, $\bar{d}_{i,des}$ is the required inter-vehicle safety distance, and $H$ is the constant headway constant. The constant time headway spacing policy of adjacent vehicles is defined as follows:

$$\bar{\sigma}_i(t) = \bar{\bar{y}}_i(t) - \bar{\Gamma}_i(t) \tag{7.2}$$

where $\bar{\Gamma}_i(t) = \{\bar{\bar{y}}_i(0) + (\iota_i \bar{\bar{y}}_i(0) + \dot{\bar{\bar{y}}}_i(0))t\}e^{-\iota_i t}$, $\bar{\bar{y}}_i(t) = \bar{y}_{i,r}(t) - \bar{y}_{i,d}(t)$, $\iota_i$ is a strictly positive definite constant. By introducing $\bar{\Gamma}_i(t)$ and using the characteristic of bounded decay function, $\bar{\sigma}_i(0) = \dot{\bar{\sigma}}_i(0) = 0$ is always true for any initial spacing error, which guarantees the string stability of the system and eliminates the assumed limitation of zero initial spacing error.

### 7.2.2 *Faults and Unknown Input Dead-Zone Modeling*

For the intelligent connected vehicle platoon system in operation, the occurrence of transmission system fault is easy to destroy its output performance and threaten the safety and stability of the vehicle platoon system. In this chapter, several kinds of actuator additive fault that are difficult to detect and adjust in practice are mainly considered, such as abrupt fault, periodic fault and intermittent fault. Therefore, for the th connected and autonomous vehicle, the fault signal of the actuator unit can be modeled as:

$$\bar{u}_{i,F}(t) = \bar{u}_i(t) + \bar{\Lambda}_i(t), t \geq t_{i,F} \tag{7.3}$$

where $\bar{\Lambda}_i(t)$ represents the time-varying fault function, $t_{i,F}$ represents the unknown time when the fault occurs. The above fault model can represent the following situations [11, 12]: 1. Abrupt fault. Due to the leakage hole inside the hydraulic cylinder, or the constant torque caused by the mechanical transmission lock fault, it will cause abrupt fault. At this time, the control input changes from the nominal value to the unknown fault constant value, the form is $\bar{\Lambda}_i(t) = \begin{cases} 0, t < t_{i,F} \\ \lambda_i, t \geq t_{i,F} \end{cases}$, where, $\lambda_i$ is the unknown non-zero constant. 2. Periodic fault. the insulated-gate bipolar transistor fault can result in the periodic faults of the known frequency in the drive system. The form can be $\bar{\Lambda}_i(t) = \begin{cases} 0, t < t_{i,F} \\ \bar{\lambda}_i \sin(\omega t), t \geq t_{i,F} \end{cases}$, where $\bar{\lambda}_i$ is the unknown non-zero constant, and $\omega$ is the known frequency. 3.Intermittent fault. Intermittent faults usually occur intermittently in an unforeseeable form, and the same fault form may occur repeatedly in different time periods. The unpredictability of intermittent faults makes it extremely difficult to deal with. In this chapter, it is assumed that intermittent faults occur periodically and are bounded. Its form is $\bar{\Lambda}_i(t) = -0.5\ \widehat{\lambda}_{i,1}\ sign(\sin(2\pi(t - t_{i,F})T^{-1})) + 0.5\ \widehat{\lambda}_{i,2}$, where $\widehat{\lambda}_{i,1}$ and $\widehat{\lambda}_{i,2}$ are the unknown fault constant, $T$ represents the fault period.

On the other hand, based on the fault of the actuator unit, due to the high-frequency switching between the braking system and the drive system of the connected and autonomous vehicles, the internal dead zone of the actuator is easily nonlinear, which will weaken the transient performance of the system and destroy the stability of the system [8]. The input dead zone nonlinearity of the vehicle is defined as:

$$\bar{u}_{i,dz}(\bar{u}_{i,F}, t) = \begin{cases} \bar{b}_{i,R}(t)(\bar{u}_{i,F}(t) - \bar{\varphi}_{i,R}), \bar{u}_{i,F}(t) \geq \bar{\varphi}_{i,R} \\ 0, -\bar{\varphi}_{i,L} < \bar{u}_{i,F}(t) < \bar{\varphi}_{i,R} \\ \bar{b}_{i,L}(t)(\bar{u}_{i,F}(t) + \bar{\varphi}_{i,L}), \bar{u}_{i,F}(t) \leq -\bar{\varphi}_{i,L} \end{cases} \tag{7.4}$$

where $\bar{\varphi}_{i,R} \geq 0$ and $\bar{\varphi}_{i,L} \geq 0$ respectively represent the unknown dead zone parameters. $\bar{b}_{i,R}(t) > 0$ and $\bar{b}_{i,L}(t) > 0$ respectively represent the unknown dead zone slope function. the input dead zone nonlinearity can be rewritten as [13]:

$$\bar{u}_{i,dz}(\bar{u}_{i,F}, t) = \bar{b}_i(t)\bar{u}_{i,F}(t) + \bar{u}_{i,\varphi}(t) \tag{7.5}$$

where $\bar{u}_{i,\varphi}(t) = \begin{cases} -\bar{b}_{i,R}(t)\bar{\varphi}_{i,R}, \bar{u}_{i,F}(t) \geq \bar{\varphi}_{i,R} \\ -\bar{b}_i(t)\bar{u}_{i,F}, -\bar{\varphi}_{i,L} < \bar{u}_{i,F}(t) < \bar{\varphi}_{i,R} \\ \bar{b}_{i,L}(t)\bar{\varphi}_{i,L}, \bar{u}_{i,F}(t) \leq -\bar{\varphi}_{i,L} \end{cases}, \bar{b}_i(t) = \begin{cases} \bar{b}_{i,R}(t), \bar{u}_{i,F} > 0 \\ \bar{b}_{i,L}(t), \bar{u}_{i,F} \leq 0 \end{cases}$.

### 7.2.3 Prescribed Asymmetric Constraint Performance Design

For ensuring the transient and steady-state performance of the spacing error, the spacing error between neighbor vehicles can converge to the prescribed asymmetric constraint region and evolve strictly in it. The prescribed asymmetric constraints can be expressed by the following inequalities,

$$-\bar{\sigma}_{i,L}\bar{\chi}_i(t) < \bar{\sigma}_i(t) < \bar{\sigma}_{i,H}\bar{\chi}_i(t) \tag{7.6}$$

where $\bar{\sigma}_{i,L} > 0$ and $\bar{\sigma}_{i,H} > 0$ are adjustable upper and lower bound parameters. $\bar{\chi}_i(t)$ is the strictly monotone decreasing function,

$$\bar{\chi}_i(t) = \begin{cases} (\bar{\chi}_i(0) - \bar{\chi}_i(\infty))(\frac{\sin(2\pi t/T_\chi)}{2\pi} - \frac{t}{T_\chi}) + \bar{\chi}_i(0), 0 \leq t < T_\chi \\ \bar{\chi}_i(\infty), t \geq T_\chi \end{cases} \tag{7.7}$$

where $\bar{\chi}_i(\infty) = \lim_{t \to 0} \bar{\chi}_i(t)$, $\bar{\chi}_i(0)$, $\bar{\chi}_i(\infty)$ are strictly positive real numbers, and satisfy $\bar{\chi}_i(0) > \bar{\chi}_i(\infty)$. For arbitrary $0 \leq t < T_\chi$, $\bar{\chi}_i(t)$ is strictly monotone decreasing, and satisfy $\dot{\bar{\chi}}_i(t) \leq 0$. When $t = T_\chi$, $\bar{\chi}_i(\infty)$ is the minimum value of the function $\bar{\chi}_i(t)$, and $\bar{\chi}_i(\infty)$ is true for any $t \geq T_\chi$. In this chapter, since formula (7.2) is always

$\bar{\sigma}_i(0) = \dot{\bar{\sigma}}_i(0) = 0$ is true for any initial spacing error. It is obvious that formula $-\bar{\sigma}_{i,L}\bar{\chi}_i(0) < \bar{\sigma}_i(0) < \bar{\sigma}_{i,H}\bar{\chi}_i(0)$ is always true by (7.6).

**Control Objective** The control objective in this chapter is to design a distributed adaptive prescribed performance fixed-time resilient control method architecture based on the consideration of actuator additive faults and unknown input dead-zone nonlinearities. The architecture mainly meets the following two conditions:

(1) Closed-loop stability of the vehicle platoon system. Under the conditions of fault, disturbances and input dead-zone, each connected and autonomous follower vehicle can track the velocity and acceleration of the leader in a fixed time, and the spacing error is maintained within the prescribed constraint range to avoid collision.

(2) String stability. After the time, based on the definition of string stability in Chap. 3, the spacing error does not get larger as the platoon passes backwards.

In order to facilitate the subsequent control design and stability analysis, the following assumptions, lemmas and definitions are given.

**Assumption 7.1** Since $\bar{\varphi}_{i,R}(t), \bar{\varphi}_{i,L}(t), \bar{b}_{i,R}(t)$ and $\bar{b}_{i,L}(t)$ are both positively definite and bounded, it is reasonable to assume that there is an extremely small unknown positive definite real number $\bar{\bar{b}}_{i,m}$ that makes $0 < \bar{\bar{b}}_{i,m} \leq \bar{b}_i(t) \leq \bar{\bar{b}}_{i,\max}$ true, where $\bar{\bar{b}}_{i,\max} = \max\{\bar{b}_{i,R}, \bar{b}_{i,L}\}$. The above hypothesis can satisfy that the connected and autonomous vehicle can still be controllable while suffering from the input dead zone non-linearities.

**Lemma 7.1** ([13]) *For arbitrary* $\bar{\phi}_s \in \mathrm{R}$, $s = 1, \cdots, n$, *there exists* $(\sum_{s=1}^{n} |\bar{\phi}_s|)^{a_1} \leq \sum_{s=1}^{n} |\bar{\phi}_s|^{a_1}$, *and* $n^{1-a_2}(\sum_{s=1}^{n} |\bar{\phi}_s|)^{a_2} \leq \sum_{s=1}^{n} |\bar{\phi}_s|^{a_2}$, *where* $0 < a_1 \leq 1$, $a_2 > 1$.

**Lemma 7.2** ([9, 13]) *For arbitrary* $\bar{\kappa}_1, \bar{\kappa}_2 \in \mathrm{R}$, *there exists* $|\bar{\kappa}_1|^c |\bar{\kappa}_2|^d \leq \frac{c}{c+d}\lambda |\bar{\kappa}_1|^{c+d} + \frac{d}{c+d}\lambda^{-\frac{c}{d}} |\bar{\kappa}_1|^{c+d}$, *where* $\lambda > 0$, $c > 0$, *and* $d > 0$.

**Lemma 7.3** ([9, 10]) *Consider a dynamic system of the following form:* $\dot{\bar{X}}(t) = \bar{F}(\bar{X}(t))$, $\bar{X}(0) = 0$, *and* $\bar{X} \in \mathrm{R}^N$ *is the state variable of the system,* $\bar{F}(\cdot)$ *is a continuous function. For the above system, if there exists a positive definite function* $\bar{V}(\bar{X})$ *such that* $\dot{\bar{V}}(\bar{X}) \leq -a_1\bar{V}^{\mu}(\bar{X}) - a_2\bar{V}^{\lambda}(\bar{X}) + \bar{\Phi}$, *where* $0 < \mu < 1$, $\lambda > 1$, *and* $0 < \bar{\Phi} < \infty$, $a_1$ *and* $a_2$ *are positive real numbers, then the system is practically fixed-time stable at the equilibrium point, and the convergence time is* $T_{\mathrm{s}} \leq \frac{1}{a_1\eta(1-\mu)} + \frac{1}{a_2\eta(\lambda-1)}$, *where* $0 < \eta < 1$. *The residual set of the system solution can be obtained as* $\bar{X} \in \{\bar{V}(\bar{X}) \leq \min\{(\frac{\bar{\Phi}}{(1-\eta)a_1})^{\frac{1}{\mu}}, (\frac{\bar{\Phi}}{(1-\eta)a_2})^{\frac{1}{\lambda}}\}\}$.

**Lemma 7.4** ([2]) *For any continuous function* $\bar{F}(\bar{X})$ *on a bounded set* $T_X$, *there is a fuzzy logic system* $\gamma^T R(\bar{X})$ *satisfied that* $\sup\left|F(\bar{X}) - \gamma^{*T} R(\bar{X})\right| \leq \mu$, *where* $\gamma^*$ *is the ideal fuzzy weight vector,* $\mu > 0$ *is the normal number.*

**Lemma 7.5** ([7]) *There exists* $0 \leq \left|\bar{X}\right| < \gamma + (\bar{X}^2 + \gamma^2)^{-\frac{1}{2}}\bar{X}^2$ *for any positive real numbers* $\bar{X} \in \mathrm{R}$ *and* $\gamma > 0$.

## 7.3 Design of Distributed Fixed Time Resilient Control Scheme with Prescribed Performance

It is difficult to design the controller directly according to the prescribed asymmetric constraints of inequality (7.6), so the unconstrained equivalent transformation is performed for the constrained spacing error $\bar{\sigma}_i(t)$ and the prescribed constraint performance decay function $\bar{\chi}_i(t)$. It defined as

$$\bar{\sigma}_i(t) = \bar{\chi}_i(t)\bar{\phi}_i(\bar{\alpha}_i(t)) \tag{7.8}$$

where $\bar{\phi}_i(\bar{\alpha}_i(t))$ is strictly monotonically increasing smooth function on the $(-\infty, +\infty) \to (-\bar{\sigma}_{i,L}, \bar{\sigma}_{i,H})$, $\bar{\phi}_i(\bar{\alpha}_i(t))=\frac{\bar{\sigma}_{i,H}e^{\alpha_i}-\bar{\sigma}_{i,L}e^{-\alpha_i}}{e^{\alpha_i}+e^{-\alpha_i}}$, where $\bar{\alpha}_i(t)$ is the conversion error. Further, $\bar{\alpha}_i(t)=\bar{\phi}_i^{-1}(\frac{\bar{\sigma}_i}{\bar{\chi}_i}) = \frac{1}{2}\ln\frac{\bar{\chi}_i\bar{\sigma}_{i,L}+\bar{\sigma}_i}{\bar{\chi}_i\bar{\sigma}_{i,H}-\bar{\sigma}_i}$, the spacing error of unconstrained conversion variable $\bar{\alpha}_i(t)$ can be defined as

$$\bar{\partial}_{i,1} = \bar{\alpha}_i(t) - \bar{\alpha}_{i,c}(t) \tag{7.9}$$

where $\bar{\alpha}_{i,c}(t) = \frac{1}{2}\ln\frac{\bar{\sigma}_{i,L}}{\bar{\sigma}_{i,H}}$ is the reference signal of conversion error $\bar{\alpha}_i(t)$. According to Eq. (7.9), $\dot{\bar{\partial}}_{i,1}$ can be calculated as

$$\dot{\bar{\partial}}_{i,1}(t) = \bar{\bar{\chi}}_i(\dot{\bar{\sigma}}_i - \dot{\bar{\chi}}_i\bar{\chi}_i^{-1}\bar{\sigma}_i) \tag{7.10}$$

where $\bar{\bar{\chi}}_i=\frac{1}{2}(\frac{1}{\bar{\chi}_i\bar{\sigma}_{i,L}+\bar{\sigma}_i} - \frac{1}{-\bar{\chi}_i\bar{\sigma}_{i,H}+\bar{\sigma}_i})$, and $0 < \bar{\bar{\chi}}_i \leq \widehat{\bar{\chi}}_i= \frac{\bar{\sigma}_{i,L}+\bar{\sigma}_{i,H}}{\bar{\chi}_i(\infty)\bar{\sigma}_{i,L}\bar{\sigma}_{i,H}}$. The compensation signal error is defined as

$$\bar{\partial}_{i,2}(t) = \bar{v}_{i-1}(t) - \bar{v}_i(t) - H\bar{a}_i(t) - \bar{\zeta}_{i,2}(t) \tag{7.11}$$

where $\bar{\zeta}_{i,2}(t)$ represents the output signal of the designed filter. First, for reducing the computational amount caused by repeated differentiation in the subsequent design process, the filter (7.12) is introduced, in which the virtual controller $\bar{\eta}_{i,1}(t)$ is the input signal of the filter and $\bar{\zeta}_{i,2}(t)$ is the output signal of the filter.

$$\begin{aligned} \dot{\bar{\zeta}}_{i,2}(t) &= \kappa_{1,i}\bar{\zeta}_{i,3}(t) \\ \dot{\bar{\zeta}}_{i,3}(t) &= -2o_i\kappa_{1,i}\bar{\zeta}_{i,3}(t) - \kappa_{1,i}(\bar{\zeta}_{i,2}(t) - \bar{\eta}_{i,1}(t)) \end{aligned} \tag{7.12}$$

where $\kappa_{1,i} > 0, o_i \in (0, 1]$ are positive constants, $\bar{\zeta}_{i,2}(0) = \bar{\eta}_{i,1}(0)$, and $\bar{\zeta}_{i,3}(0) = 0$, $\left|\bar{\zeta}_{i,2}(t) - \bar{\eta}_{i,1}(t)\right| \leq \Delta\bar{\zeta}_i$, $\Delta\bar{\zeta}_i$ is the unknown positive constant.

Secondly, the compensation error signal is defined as $\bar{\upsilon}_{i,1}(t) = \bar{\partial}_{i,1}(t) - \bar{\psi}_{i,1}(t)$, $\bar{\upsilon}_{i,2}(t) = \bar{\partial}_{i,2}(t) - \bar{\psi}_{i,2}(t)$. Meanwhile, for eliminating the error $(\bar{\zeta}_{i,2}(t) - \bar{\eta}_{i,1}(t))$ caused by the introduced filter, the compensators $\bar{\psi}_{i,1}(t)$ and $\bar{\psi}_{i,2}(t)$ are designed as

$$\dot{\bar{\psi}}_{i,1}(t) = -\varepsilon_{i,1}\bar{\psi}_{i,1}^{\frac{1}{2}}(t) - \varepsilon_{i,2}\bar{\psi}_{i,1}^{3}(t) + \bar{\chi}_i(\bar{\psi}_{i,2}(t) + \bar{\zeta}_{i,2}(t) - \bar{\eta}_{i,1}(t)) \tag{7.13}$$

$$\dot{\bar{\psi}}_{i,2}(t) = -\varepsilon_{i,3}\bar{\psi}_{i,2}^{\frac{1}{2}}(t) - \varepsilon_{i,4}\bar{\psi}_{i,2}^{3}(t) - \bar{\bar{\chi}}_i\bar{\psi}_{i,1}(t) \tag{7.14}$$

where $\varepsilon_{i,1}$, $\varepsilon_{i,2}$, $\varepsilon_{i,3}$, and $\varepsilon_{i,4}$ are designed positive constants. $\varepsilon_{i,1} > \frac{3}{4}$, $\varepsilon_{i,2} > 2$, $\varepsilon_{i,3} > 0$, and $\varepsilon_{i,4} > 0$, $\bar{\psi}_{i,1}(0) = \bar{\psi}_{i,2}(0) = 0$.

The designed virtual control signals $\bar{\eta}_{i,1}(t)$, $\bar{\eta}_{i,2}(t)$ are, respectively

$$\bar{\eta}_{i,1}(t) = -\frac{\bar{\upsilon}_{i,1}(t)\bar{\bar{\chi}}_i(t)\bar{\bar{\eta}}_{i,1}^{*2}(t)}{\sqrt{\bar{\upsilon}_{i,1}^{2}(t)\bar{\bar{\chi}}_i^{2}(t)\bar{\bar{\eta}}_{i,1}^{*2}(t) + \mu_{i,1}^{2}}} \tag{7.15a}$$

$$\begin{aligned}\bar{\bar{\eta}}_{i,1}(t) = {} & K_{i,1}(\bar{\upsilon}_{i,1}^{2}(t))^{\frac{1}{4}} + K_{i,2}\bar{\upsilon}_{i,1}^{3}(t) + \frac{\bar{\upsilon}_{i,1}(t)\bar{\bar{\chi}}_i^{2}(t)H^2}{2\bar{\bar{\tau}}_{i,1}^{2}} \\ & - \bar{\bar{\chi}}_i(t)\dot{\bar{\varGamma}}_i(t) - \dot{\bar{\chi}}_i(t)\bar{\chi}_i^{-1}(t)\bar{\sigma}_i(t)\bar{\bar{\chi}}_i(t)\end{aligned} \tag{7.15b}$$

where $\bar{\bar{\eta}}_{i,1}^{*}(t) = \bar{\bar{\chi}}_i^{-1}(t)\bar{\bar{\eta}}_{i,1}(t)$, $K_{i,1}$, $K_{i,2}$, $\mu_{i,1}$ and $\bar{\bar{\tau}}_{i,1} > 0$ are positive real constants. $\bar{K}_{i,1} = K_{i,1}2^{\frac{3}{4}} - \frac{2}{3}\varepsilon_{i,1}2^{\frac{3}{4}} > 0$, $\bar{K}_{i,2} = K_{i,2}2^2 - \frac{1}{4}\varepsilon_{i,2}2^2 > 0$.

$$\bar{\eta}_{i,2}(t) = -\frac{\bar{\upsilon}_{i,2}(t)\bar{\bar{\eta}}_{i,2}^{2}(t)}{\sqrt{\bar{\upsilon}_{i,2}^{2}(t)\bar{\bar{\eta}}_{i,2}^{2}(t) + \mu_{i,2}^{2}}} \tag{7.16a}$$

$$\begin{aligned}\bar{\bar{\eta}}_{i,2}(t) = {} & K_{i,3}(\bar{\upsilon}_{i,2}^{2}(t))^{\frac{1}{4}} + K_{i,4}\bar{\upsilon}_{i,2}^{3}(t) + \bar{a}_{i-1}(t) - \bar{a}_i(t) \\ & - \dot{\bar{\zeta}}_{i,2}(t) + \frac{1}{2\bar{\bar{\tau}}_{i,2}^{2}}\bar{\upsilon}_{i,2}(t)\hat{\bar{\varOmega}}_{i,1}(t)R_{i,1}^{T}R_{i,1} + \bar{\bar{\chi}}_i(t)\bar{\partial}_{i,1}(t) \\ & + 1.5\bar{\upsilon}_{i,2}(t) + 0.5\bar{\upsilon}_{i,2}(t)H^2\end{aligned} \tag{7.16b}$$

where $K_{i,3}$, $K_{i,4}$, $\mu_{i,2}$, $\bar{\bar{\tau}}_{i,2} > 0$ are positive real constants. $\bar{K}_{i,3} = K_{i,3}2^{\frac{3}{4}} - \frac{2}{3}\varepsilon_{i,3}2^{\frac{3}{4}} > 0$, $\bar{K}_{i,4} = K_{i,4}2^2 - \frac{1}{4}\varepsilon_{i,4}2^2 > 0$. The distributed prescribed fixed time resilient controller $\bar{u}_i(t)$ is designed as

$$\bar{u}_i(t) = \frac{H\bar{b}_{i,a}\bar{\upsilon}_{i,2}(t)\hat{\bar{\bar{b}}}_{i,m}^{*2}(t)\bar{u}_i^{*2}(t)}{\sqrt{(H\bar{b}_{i,a})^2\bar{\upsilon}_{i,2}^{2}(t)\hat{\bar{\bar{b}}}_{i,m}^{*2}(t)\bar{u}_i^{*2}(t) + \mu_{i,3}^{2}}} \tag{7.17a}$$

$$\bar{u}_i^{*}(t) = \frac{1}{H\bar{b}_{i,a}}\bar{\eta}_{i,2}(t)\tanh\left(\frac{\bar{\eta}_{i,2}(t)\bar{\upsilon}_{i,2}(t)}{\bar{\mu}_i}\right) \tag{7.17b}$$

where $\bar{\bar{b}}_{i,m}^{*}(t) = \bar{\bar{b}}_{i,m}^{-1}(t)$, $\hat{\bar{\bar{b}}}_{i,m}^{*}(t)$ is the estimation of $\bar{\bar{b}}_{i,m}^{*}(t)$, $\mu_{i,3}$ and $\bar{\mu}_i$ are positive constants. The designed compensation law of unknown fault and nonlinearity is

$$\dot{\hat{\bar{\Omega}}}_{i,1}(t) = \frac{r_{i,1}}{2\bar{\bar{\tau}}_{i,2}^2}\bar{\upsilon}_{i,2}^2(t)R_{i,1}^T R_{i,1} - \bar{\varepsilon}_i \hat{\bar{\Omega}}_{i,1}^{\frac{1}{2}}(t) - \frac{\bar{o}_{i,1}}{r_{i,1}}\hat{\bar{\Omega}}_{i,1}^3(t) \tag{7.18}$$

where $\tilde{\bar{\Omega}}_{i,1}(t) = \bar{\Omega}_{i,1}(t) - \hat{\bar{\Omega}}_{i,1}(t)$ is the estimation errors, $\hat{\bar{\Omega}}_{i,1}(t)$ is an estimation of unknown parameter $\bar{\varepsilon}_i$, $\bar{o}_{i,1}$ and $r_{i,1}$ are the given positive constants. The nonlinearity compensation law of input dead-zone is

$$\dot{\hat{\bar{\bar{b}}}}_{i,m}^*(t)=r_{i,2}H\bar{b}_{i,a}\bar{\upsilon}_{i,2}(t)\bar{u}_i^*(t) - g_{i,1}\hat{\bar{\bar{b}}}_{i,m}^{*\frac{1}{2}}(t) - \frac{\bar{o}_{i,2}}{r_{i,2}}\hat{\bar{\bar{b}}}_{i,m}^{*3}(t) \tag{7.19}$$

where $g_{i,1}$, $\bar{o}_{i,2}$ and $r_{i,2}$ are given positive constants.

**Theorem 7.1** *Considering the HICVPS (7.1) that suffers from actuator fault (7.3) and unknown input dead zone nonlinearity (7.4), with the conditions of the constant time headway spacing policy (7.2) based on the PF information flow communication topology, and the prescribed asymmetric constraint performance conditions (7.6), the error compensators (7.13)–(7.14) and the virtual control signals (7.15a), (7.15b)–(7.15a), (7.16b) are designed. Combining with the distributed prescribed performance fixed time resilient controller (7.17a), (7.17b), and the unknown faults and nonlinear compensation laws (7.18) and the compensation law of input dead zone nonlinearity (7.19), the spacing error between vehicles can ultimately converge to the set* $\bar{\sigma}_i(t) \in \{|\bar{\sigma}_i(t)| \le \bar{\chi}_i(t)\phi_i(\bar{\alpha}_i)\}$ *after a fixed time* $T_s$*, and the string stability of the system can be guaranteed.*

For facilitating the calculation, the time subscript of the function is omitted in the subsequent process, for example $\bar{\psi}_i(t) = \bar{\psi}_i$. Based on recursive algorithm, the design steps of distributed prescribed performance fixed time control are as follows:

First of all, based on the formula (7.1), (7.2), (7.9) and (7.11), taking derivation of $\bar{\upsilon}_{i,1} = \bar{\partial}_{i,1} - \bar{\psi}_{i,1}$, we get

$$\dot{\bar{\upsilon}}_{i,1} = \bar{\bar{\chi}}_i(\bar{\partial}_{i,2} + (\bar{\zeta}_{i,2} - \bar{\eta}_{i,1}) + \bar{\eta}_{i,1} - \dot{\bar{\Gamma}}_i \tag{7.20}$$

The Lyapunov candidate function is defined as $\bar{V}_{1,i} = 0.5\bar{\upsilon}_{i,1}^2 + 0.5\bar{\psi}_{i,1}^2$, based on the formula (7.17a), (7.17b) and (7.13), $\dot{\bar{V}}_{1,i}$ can be obtained as

$$\dot{\bar{V}}_{1,i} = \bar{\upsilon}_{i,1}\bar{\bar{\chi}}_i(\bar{\eta}_{i,1} - Hw_{i,v} - \dot{\bar{\Gamma}}_i - \dot{\bar{\chi}}_i\bar{\chi}_i^{-1}\bar{\sigma}_i) + \bar{\bar{\chi}}_i\bar{\upsilon}_{i,1}\bar{\upsilon}_{i,2} \tag{7.21}$$

Using the perfect square formula, we can get

$$-\bar{\upsilon}_{i,1}\bar{\bar{\chi}}_i Hw_{i,v} = -\left(\frac{\bar{\upsilon}_{i,1}\bar{\bar{\chi}}_i H}{\sqrt{2}\bar{\bar{\tau}}_{i,1}} + \frac{\bar{\bar{\tau}}_{i,1}w_{i,v}}{\sqrt{2}}\right)^2 + \frac{\bar{\upsilon}_{i,1}^2\bar{\bar{\chi}}_i^2 H^2}{2\bar{\bar{\tau}}_{i,1}^2} + \frac{\bar{\bar{\tau}}_{i,1}^2\bar{w}_{i,v}^2}{2} \tag{7.22}$$

By substituting the formula (7.22) into (7.21), we have

$$\begin{aligned}\dot{\bar{V}}_{1,i} \leq \bar{\upsilon}_{i,1}\bar{\bar{\chi}}_i\left(\eta_{i,1} + \frac{\bar{\upsilon}_{i,1}\bar{\bar{\chi}}_i H^2}{2\bar{\tau}_{i,1}^2} - \dot{\bar{\Gamma}}_i - \dot{\bar{\chi}}_i\bar{\bar{\chi}}_i^{-1}\dot{\bar{\sigma}}_i\right) + \bar{\bar{\chi}}_i\bar{\upsilon}_{i,1}\bar{\upsilon}_{i,2} \\ + \varepsilon_{i,1}\bar{\upsilon}_{i,1}\bar{\psi}_{i,1}^{\frac{1}{2}} + \varepsilon_{i,2}\bar{\upsilon}_{i,1}\bar{\psi}_{i,1}^3 - \varepsilon_{i,1}\bar{\psi}_{i,1}^{\frac{3}{2}} - \varepsilon_{i,2}\bar{\psi}_{i,1}^4 \\ + \bar{\bar{\chi}}_i\bar{\psi}_{i,1}\bar{\psi}_{i,2} + \bar{\psi}_{i,1}\bar{\bar{\chi}}_i(\bar{\zeta}_{i,2} - \bar{\eta}_{i,1}) + 0.5\bar{\bar{\tau}}_{i,1}^2\bar{w}_{i,v}^2\end{aligned} \tag{7.23}$$

By Lemma 7.5, we know that

$$\begin{aligned}\bar{\upsilon}_{i,1}\bar{\bar{\chi}}_i\bar{\eta}_{i,1} &= -\frac{\bar{\upsilon}_{i,1}^2\bar{\bar{\chi}}_i^2\bar{\bar{\eta}}_{i,1}^{*2}}{\sqrt{\bar{\upsilon}_{i,1}^2\bar{\bar{\chi}}_i^2\bar{\bar{\eta}}_{i,1}^{*2} + \mu_{i,1}^2}} \\ &\leq \mu_{i,1} - \bar{\upsilon}_{i,1}\bar{\bar{\chi}}_i\bar{\bar{\eta}}_{i,1}^* \\ &= \mu_{i,1} - \bar{\upsilon}_{i,1}\bar{\bar{\eta}}_{i,1}\end{aligned} \tag{7.24}$$

Substituting equations (7.15a), (7.15b), and (7.24) into (7.23), one has

$$\begin{aligned}\dot{\bar{V}}_{1,i} \leq -K_{i,1}(\bar{\upsilon}_{i,1}^2)^{\frac{3}{4}} - K_{i,2}\bar{\upsilon}_{i,1}^4 + \bar{\bar{\chi}}_i\bar{\upsilon}_{i,1}\bar{\upsilon}_{i,2} + \varepsilon_{i,1}\bar{\upsilon}_{i,1}\bar{\psi}_{i,1}^{\frac{1}{2}} \\ + \varepsilon_{i,2}\bar{\upsilon}_{i,1}\bar{\psi}_{i,1}^3 - \varepsilon_{i,1}\bar{\psi}_{i,1}^{\frac{3}{2}} - \varepsilon_{i,2}\bar{\psi}_{i,1}^4 + \bar{\bar{\chi}}_i\bar{\psi}_{i,1}\bar{\psi}_{i,2} \\ + \bar{\psi}_{i,1}\bar{\bar{\chi}}_i(\bar{\zeta}_{i,2} - \bar{\eta}_{i,1}) + 0.5\bar{\bar{\tau}}_{i,1}^2\bar{w}_{i,v}^2 + \mu_{i,1}\end{aligned} \tag{7.25}$$

Further, taking the derivation of $\bar{\upsilon}_{i,2}(t) = \bar{\partial}_{i,2}(t) - \bar{\psi}_{i,2}(t)$, and based on formula (7.1), (7.3) and (7.5), we can be obtain that

$$\begin{aligned}\bar{\upsilon}_{i,2}\dot{\bar{\upsilon}}_{i,2} = \bar{\upsilon}_{i,2}(\bar{a}_{i-1} - \bar{a}_i - \dot{\bar{\zeta}}_{i,2} - H\bar{F}_{i,a}(\bar{v}_i, \bar{a}_i)) \\ + \bar{\upsilon}_{i,2}(-Hw_{i,a} - H\bar{b}_{i,a}(\bar{b}_i\bar{u}_i + \bar{b}_i\bar{\Lambda}_i + \bar{u}_{i,\varphi})) \\ + \bar{\upsilon}_{i,2}(w_{i-1,v} - w_{i,v} - \dot{\bar{\psi}}_{i,2})\end{aligned} \tag{7.26}$$

The same as formula (7.24), one can be obtain

$$\bar{\upsilon}_{i,2}(-Hw_{i,a} + w_{i-1,v} - w_{i,v}) \leq \bar{\upsilon}_{i,2}^2 + 0.5\bar{\upsilon}_{i,2}^2H^2 + 0.5(\bar{w}_{i-1,v}^2 + \bar{w}_{i,a}^2 + \bar{w}_{i,v}^2) \tag{7.27}$$

According to formula (7.27) and (7.26) can be further derived as

$$\begin{aligned}\bar{\upsilon}_{i,2}\dot{\bar{\upsilon}}_{i,2} \leq \bar{\upsilon}_{i,2}(\bar{a}_{i-1} - \bar{a}_i - \dot{\bar{\zeta}}_{i,2} + \bar{\Omega}_{i,f}(\bar{X}_i)) \\ + \bar{\upsilon}_{i,2}(\bar{\upsilon}_{i,2} + 0.5\bar{\upsilon}_{i,2}H^2 - H\bar{b}_{i,a}\bar{b}_i\bar{u}_i - \dot{\bar{\psi}}_{i,2}) \\ + 0.5(\bar{w}_{i-1,v}^2 + \bar{w}_{i,a}^2 + \bar{w}_{i,v}^2)\end{aligned} \tag{7.28}$$

where $\bar{\Omega}_{i,f}(\bar{X}_i) = -H\bar{F}_{i,a}(\bar{v}_i, \bar{a}_i) - H\bar{b}_{i,a}(\bar{b}_i\bar{\Lambda}_i + \bar{u}_{i,\varphi})$.

Using the fuzzy system approximation principle, $\bar{\Omega}_{i,f}(\bar{X}_i)$ can be approximated by the fuzzy system $W_{i,1}^T R_{i,1}(\bar{X}_i)$, and the approximation error is $\nu_{i,1}(\bar{X}_i)$. Thus, from Lemma 7.4 we can obtain that $\bar{\Omega}_{i,f}(\bar{X}_i) = W_{i,1}^T R_{i,1}(\bar{X}_i) + \nu_{i,1}(\bar{X}_i)$, where $\left|\nu_{i,1}(\bar{X}_i)\right| \leq \bar{\nu}_{i,1}$, $\bar{\nu}_{i,1}$ is an arbitrarily small positive constant. Based on the above,

$$\begin{aligned} \bar{\upsilon}_{i,2}\bar{\Omega}_{i,f}(\bar{X}_i) &= \bar{\upsilon}_{i,2} W_{i,1}^T R_{i,1}(\bar{X}_i) + \bar{\upsilon}_{i,2}\nu_{i,1}(\bar{X}_i) \\ &\leq \frac{1}{2\bar{\bar{\tau}}_{i,2}^2}\bar{\upsilon}_{i,2}^2\bar{\Omega}_{i,1} R_{i,1}^T R_{i,1} + 0.5\bar{\upsilon}_{i,2}^2 + 0.5(\bar{\bar{\tau}}_{i,2}^2 + \bar{\nu}_{i,1}^2) \end{aligned} \tag{7.29}$$

where $\bar{\Omega}_{i,1} = \max\{\left\| W_{i,1} \right\|^2\}$.

$$\begin{aligned} \bar{\upsilon}_{i,2}\dot{\bar{\upsilon}}_{i,2} \leq{}& \dot{\upsilon}_{i,2}\left(\bar{a}_{i-1} - \bar{a}_i - \dot{\bar{\zeta}}_{i,2} + \frac{1}{2\bar{\bar{\tau}}_{i,2}^2}\bar{\upsilon}_{i,2}\bar{\Omega}_{i,1} R_{i,1}^T R_{i,1}\right) \\ &+ \dot{\upsilon}_{i,2}(1.5\bar{\upsilon}_{i,2} + 0.5\bar{\upsilon}_{i,2}H^2 - H\bar{b}_{i,a}\bar{b}_i\bar{u}_i - \dot{\bar{\psi}}_{i,2}) + h_i^* \end{aligned} \tag{7.30}$$

where $h_i^* = 0.5(\bar{w}_{i-1,v}^2 + \bar{w}_{i,a}^2 + \bar{w}_{i,v}^2 + \bar{\bar{\tau}}_{i,2}^2 + \bar{\nu}_{i,1}^2)$.

Further, the Lyapunov candidate function $\bar{V}_{2,i} = \bar{V}_{1,i} + 0.5\bar{\upsilon}_{i,2}^2 + 0.5\bar{\psi}_{i,2}^2 + \tilde{\bar{\Omega}}_{i,1}^2(2r_{i,1})^{-1}$ is considered, using Eqs. (7.25) and (7.30), one has

$$\begin{aligned} \dot{\bar{V}}_{2,i} \leq{}& -K_{i,1}(\bar{\upsilon}_{i,1}^2)^{\frac{3}{4}} - K_{i,2}\bar{\upsilon}_{i,1}^4 - \varepsilon_{i,2}\bar{\psi}_{i,1}^4 - \varepsilon_{i,1}\bar{\psi}_{i,1}^{\frac{3}{2}} + \varepsilon_{i,1}\bar{\upsilon}_{i,1}\bar{\psi}_{i,1}^{\frac{1}{2}} \\ &+ \varepsilon_{i,2}\bar{\upsilon}_{i,1}\bar{\psi}_{i,1}^3 + 0.5\bar{\bar{\tau}}_{i,1}^2\bar{w}_{i,v}^2 + \mu_{i,1} + \bar{\upsilon}_{i,2}\bar{\eta}_{i,2} + \bar{\psi}_{i,1}\bar{\bar{\chi}}_i(\bar{\zeta}_{i,2} - \bar{\eta}_{i,1}) \\ &+ \bar{\upsilon}_{i,2}(-\bar{\eta}_{i,2} + \bar{\bar{\chi}}_i\bar{\partial}_{i,1} + \bar{a}_{i-1} - \bar{a}_i - \dot{\bar{\zeta}}_{i,2} + \frac{1}{2\bar{\bar{\tau}}_{i,2}^2}\bar{\upsilon}_{i,2}\bar{\Omega}_{i,1} R_{i,1}^T R_{i,1} \\ &+ 1.5\bar{\upsilon}_{i,2} + 0.5\bar{\upsilon}_{i,2}H^2 - H\bar{b}_{i,a}\bar{b}_i\bar{u}_i) + h_i^* - \varepsilon_{i,3}\bar{\psi}_{i,2}^{\frac{3}{2}} \\ &- \varepsilon_{i,4}\bar{\psi}_{i,2}^4 - \tilde{\bar{\Omega}}_{i,1}\dot{\hat{\bar{\Omega}}}_{i,1}r_{i,1}^{-1} + \varepsilon_{i,3}\bar{\upsilon}_{i,2}\bar{\psi}_{i,2}^{\frac{1}{2}} + \varepsilon_{i,4}\bar{\upsilon}_{i,2}\bar{\psi}_{i,2}^3 \end{aligned} \tag{7.31}$$

The same as the steps in formula (7.22) and (7.23), we substitute formula (7.16a), (7.16b), and (7.18) into the above equation, and gets

$$\begin{aligned} \dot{\bar{V}}_{2,i} \leq{}& -K_{i,1}(\bar{\upsilon}_{i,1}^2)^{\frac{3}{4}} - K_{i,2}\bar{\upsilon}_{i,1}^4 - K_{i,3}(\bar{\upsilon}_{i,2}^2)^{\frac{3}{4}} - K_{i,4}\bar{\upsilon}_{i,2}^4 \\ &- \varepsilon_{i,2}\bar{\psi}_{i,1}^4 - \varepsilon_{i,1}\bar{\psi}_{i,1}^{\frac{3}{2}} + \varepsilon_{i,1}\bar{\upsilon}_{i,1}\bar{\psi}_{i,1}^{\frac{1}{2}} + 0.5(\bar{\psi}_{i,1}^{\frac{3}{2}} + \bar{\psi}_{i,1}^4) \\ &+ \varepsilon_{i,2}\bar{\upsilon}_{i,1}\bar{\psi}_{i,1}^3 + 0.5\bar{\bar{\tau}}_{i,1}^2\bar{w}_{i,v}^2 + \mu_{i,1} + \mu_{i,2} + 0.5\bar{\bar{\chi}}_i\Delta\bar{\zeta}_i \\ &+ \bar{\upsilon}_{i,2}(-\bar{\eta}_{i,2} - H\bar{b}_{i,a}\bar{b}_i\bar{u}_i) + h_i^* - \varepsilon_{i,3}\bar{\psi}_{i,2}^{\frac{3}{2}} - \varepsilon_{i,4}\bar{\psi}_{i,2}^4 \\ &+ \varepsilon_{i,3}\bar{\upsilon}_{i,2}\bar{\psi}_{i,2}^{\frac{1}{2}} + \varepsilon_{i,4}\bar{\upsilon}_{i,2}\bar{\psi}_{i,2}^3 + r_{i,1}^{-1}\bar{\varepsilon}_i\hat{\bar{\Omega}}_{i,1}^{\frac{1}{2}}\tilde{\bar{\Omega}}_{i,1} + \bar{o}_{i,1}r_{i,1}^{-2}\tilde{\bar{\Omega}}_{i,1}\hat{\bar{\Omega}}_{i,1}^3 \end{aligned} \tag{7.32}$$

where $\bar{\psi}_{i,1}\bar{\bar{\chi}}_i(\bar{\zeta}_{i,2} - \bar{\eta}_{i,1}) \leq 0.5(\bar{\psi}_{i,1}^{\frac{3}{2}} + \bar{\psi}_{i,1}^4) + 0.5\bar{\bar{\chi}}_i\Delta\bar{\zeta}_i$.

Considering the Lyapunov candidate function as $\bar{V}_i = \bar{V}_{2,i} + \bar{\bar{\tilde{b}}}^{*2}_{i,m}(2r_{i,2})^{-1}\bar{\bar{b}}_{i,m}$, and combine with formula (7.17a), (7.17b), (7.19), and (7.32), $\dot{\bar{V}}_i$ can be

$$\begin{aligned}\dot{\bar{V}}_i \le & -K_{i,1}(\bar{\upsilon}^2_{i,1})^{\frac{3}{4}} - K_{i,2}\bar{\upsilon}^4_{i,1} - K_{i,3}(\bar{\upsilon}^2_{i,2})^{\frac{3}{4}} - K_{i,4}\bar{\upsilon}^4_{i,2} \\ & - \varepsilon_{i,2}\bar{\psi}^4_{i,1} - \varepsilon_{i,1}\bar{\psi}^{\frac{3}{2}}_{i,1} + \varepsilon_{i,1}\bar{\upsilon}_{i,1}\bar{\psi}^{\frac{1}{2}}_{i,1} + 0.5(\bar{\psi}^{\frac{3}{2}}_{i,1} + \bar{\psi}^4_{i,1}) \\ & + \varepsilon_{i,2}\bar{\upsilon}_{i,1}\bar{\psi}^3_{i,1} + 0.5\bar{\bar{\tau}}^2_{i,1}\bar{w}^2_{i,v} + \mu_{i,1} + \mu_{i,2} + 0.5\bar{\bar{\chi}}_i\Delta\bar{\zeta}_i \\ & - \bar{\upsilon}_{i,2}\bar{\eta}_{i,2} + \bar{\bar{b}}_{i,m}\mu_{i,3} - H\bar{b}_{i,a}\bar{\upsilon}_{i,2}\bar{u}^*_i + h_i{}^* - \varepsilon_{i,3}\bar{\psi}^{\frac{3}{2}}_{i,2} \\ & - \varepsilon_{i,4}\bar{\psi}^4_{i,2} + \varepsilon_{i,3}\bar{\upsilon}_{i,2}\bar{\psi}^{\frac{1}{2}}_{i,2} + \varepsilon_{i,4}\bar{\upsilon}_{i,2}\bar{\psi}^3_{i,2} + r^{-1}_{i,1}\bar{\varepsilon}_i\hat{\bar{\Omega}}^{\frac{1}{2}}_{i,1}\tilde{\bar{\Omega}}_{i,1} \\ & + \bar{o}_{i,1}r^{-2}_{i,1}\tilde{\bar{\Omega}}_{i,1}\hat{\bar{\Omega}}^3_{i,1} + r^{-1}_{i,2}g_{i,1}\bar{\bar{b}}_{i,m}\tilde{\bar{\bar{b}}}^*_{i,m}\hat{\bar{\bar{b}}}^{*\frac{1}{2}}_{i,m} + r^{-2}_{i,2}\bar{o}_{i,2}\bar{\bar{b}}_{i,m}\tilde{\bar{\bar{b}}}^*_{i,m}\hat{\bar{\bar{b}}}^{*3}_{i,m}\end{aligned} \tag{7.33}$$

where $-\bar{\bar{b}}_{i,m}(H\bar{b}_{i,a})\bar{\upsilon}_{i,2}\hat{\bar{\bar{b}}}^*_{i,m}\bar{u}^*_i + (H\bar{b}_{i,a})\bar{\upsilon}_{i,2}\bar{u}^*_i = \tilde{\bar{\bar{b}}}^*_{i,m}\bar{\bar{b}}_{i,m}H\bar{b}_{i,a}\bar{\upsilon}_{i,2}\bar{u}^*_i$, $0 < \bar{\bar{b}}_{i,m} \le \bar{b}_i(t) \le \bar{\bar{b}}_{i,\max}$.

By substituting (7.33) from the formula (7.17b), and by the facts that: The equation $0 \le |\varphi| - \varphi\tanh(\varphi\eta^{-1}) \le 0.2785\eta$ holds for all $\varphi \in R$ and $\eta > 0$, one obtains

$$\begin{aligned}\dot{\bar{V}}_i \le & -K_{i,1}(\bar{\upsilon}^2_{i,1})^{\frac{3}{4}} - K_{i,2}\bar{\upsilon}^4_{i,1} - K_{i,3}(\bar{\upsilon}^2_{i,2})^{\frac{3}{4}} - K_{i,4}\bar{\upsilon}^4_{i,2} \\ & - (\varepsilon_{i,2} - 0.5)\bar{\psi}^4_{i,1} - (\varepsilon_{i,1} - 0.5)\bar{\psi}^{\frac{3}{2}}_{i,1} + \varepsilon_{i,1}\bar{\upsilon}_{i,1}\bar{\psi}^{\frac{1}{2}}_{i,1} \\ & + \varepsilon_{i,2}\bar{\upsilon}_{i,1}\bar{\psi}^3_{i,1} - \varepsilon_{i,3}\bar{\psi}^{\frac{3}{2}}_{i,2} - \varepsilon_{i,4}\bar{\psi}^4_{i,2} + \varepsilon_{i,3}\bar{\upsilon}_{i,2}\bar{\psi}^{\frac{1}{2}}_{i,2} \\ & + \varepsilon_{i,4}\bar{\upsilon}_{i,2}\bar{\psi}^3_{i,2} + r^{-1}_{i,1}\bar{\varepsilon}_i\hat{\bar{\Omega}}^{\frac{1}{2}}_{i,1}\tilde{\bar{\Omega}}_{i,1} + \bar{o}_{i,1}r^{-2}_{i,1}\tilde{\bar{\Omega}}_{i,1}\hat{\bar{\Omega}}^3_{i,1} \\ & + r^{-1}_{i,2}g_{i,1}\bar{\bar{b}}_{i,m}\tilde{\bar{\bar{b}}}^*_{i,m}\hat{\bar{\bar{b}}}^{*\frac{1}{2}}_{i,m} + r^{-2}_{i,2}\bar{o}_{i,2}\bar{\bar{b}}_{i,m}\tilde{\bar{\bar{b}}}^*_{i,m}\hat{\bar{\bar{b}}}^{*3}_{i,m} + \mu^*_i\end{aligned} \tag{7.34}$$

where $\mu^*_i = 0.5\bar{\bar{\tau}}^2_{i,1}\bar{w}^2_{i,v} + \mu_{i,1} + \mu_{i,2} + 0.5\bar{\bar{\chi}}_i\Delta\bar{\zeta}_i + \bar{\bar{b}}_{i,m}\mu_{i,3} + 0.2785\bar{\mu}_i + h_i{}^*$, $|\bar{\upsilon}_{i,2}\bar{\eta}_{i,2}| - \bar{\upsilon}_{i,2}\bar{\eta}_{i,2}\tanh(\frac{\bar{\eta}_{i,2}\bar{\upsilon}_{i,2}}{\bar{\mu}_i}) \le 0.2785\bar{\mu}_i$.

## 7.4 Analysis of System Stability

***Proof*** For proving the closed-loop stability and string stability of the whole vehicle platoon system under the prescribed asymmetric constraint performance, the Lyapunov candidate function is defined as $\bar{V} = \sum_{i=1}^{n} \bar{V}_i$. Based on Lemma 7.2, $\bar{\upsilon}_{i,j}\bar{\psi}^{\rho}_{i,j} \le \frac{1}{1+\rho}\left|\bar{\upsilon}_{i,j}\right|^{1+\rho} + \frac{\rho}{1+\rho}\left|\bar{\psi}_{i,j}\right|^{1+\rho}$, $\rho = \frac{1}{2}, 3$, $j = 1, 2$. Taking the derivation of $\bar{V}$, and with formula (7.34),

$$\dot{V} \leq \sum_{i=1}^{n}\left\{-\bar{K}_{i,1}\left(\frac{\bar{\upsilon}_{i,1}^{2}}{2}\right)^{\frac{3}{4}}-\bar{K}_{i,2}\left(\frac{\bar{\upsilon}_{i,1}^{2}}{2}\right)^{2}-\bar{K}_{i,3}\left(\frac{\bar{\upsilon}_{i,2}^{2}}{2}\right)^{\frac{3}{4}}\right\}$$
$$+\sum_{i=1}^{n}\left\{-\bar{K}_{i,4}\left(\frac{\bar{\upsilon}_{i,2}^{2}}{2}\right)^{2}-\widehat{\bar{\varepsilon}}_{i,1}\left(\frac{\bar{\psi}_{i,1}^{2}}{2}\right)^{\frac{3}{4}}-\widehat{\bar{\varepsilon}}_{i,3}\left(\frac{\bar{\psi}_{i,2}^{2}}{2}\right)^{\frac{3}{4}}-\widehat{\bar{\varepsilon}}_{i,2}\left(\frac{\bar{\psi}_{i,1}^{2}}{2}\right)^{2}\right\}$$
$$+\sum_{i=1}^{n}\left\{-\widehat{\bar{\varepsilon}}_{i,4}\left(\frac{\bar{\psi}_{i,2}^{2}}{2}\right)^{2}+r_{i,1}^{-1}\bar{\varepsilon}_{i}\tilde{\bar{\Omega}}_{i,1}\hat{\bar{\Omega}}_{i,1}^{\frac{1}{2}}+\bar{o}_{i,1}r_{i,1}^{-2}\tilde{\bar{\Omega}}_{i,1}\hat{\bar{\Omega}}_{i,1}^{3}\right\}$$
$$+\sum_{i=1}^{n}\left\{r_{i,2}^{-1}g_{i,1}\bar{\bar{b}}_{i,m}\tilde{\bar{\bar{b}}}_{i,m}^{*}\hat{\bar{\bar{b}}}_{i,m}^{*\frac{1}{2}}+r_{i,2}^{-2}\bar{o}_{i,2}\bar{\bar{b}}_{i,m}\tilde{\bar{\bar{b}}}_{i,m}^{*}\hat{\bar{\bar{b}}}_{i,m}^{*3}+\mu_{i}^{*}\right\} \tag{7.35}$$

where $\widehat{\bar{\varepsilon}}_{i,1}=(\varepsilon_{i,1}-0.5)2^{\frac{3}{4}}-\frac{1}{3}\varepsilon_{i,1}2^{\frac{3}{4}}>0$, $\widehat{\bar{\varepsilon}}_{i,3}=\varepsilon_{i,3}2^{\frac{3}{4}}-\frac{1}{3}\varepsilon_{i,3}2^{\frac{3}{4}}>0$, $\widehat{\bar{\varepsilon}}_{i,2}=(\varepsilon_{i,2}-0.5)2^{2}-\frac{3}{4}\varepsilon_{i,2}2^{2}>0$, and $\widehat{\bar{\varepsilon}}_{i,4}=\varepsilon_{i,4}2^{2}-\frac{3}{4}\varepsilon_{i,4}2^{2}>0$. Secondly, based on Lemma 7.2 and Young's inequality, we can get

$$\tilde{\bar{\Omega}}_{i,1}\hat{\bar{\Omega}}_{i,1}^{\frac{1}{2}} \leq -s_{1}\left|\tilde{\bar{\Omega}}_{i,1}\right|^{1+\frac{1}{2}}+s_{2}\left|\bar{\Omega}_{i,1}\right|^{1+\frac{1}{2}} \tag{7.36}$$

$$\tilde{\bar{\bar{b}}}_{i,m}^{*}\hat{\bar{\bar{b}}}_{i,m}^{*\frac{1}{2}} \leq -s_{1}\left|\tilde{\bar{\bar{b}}}_{i,m}^{*}\right|^{1+\frac{1}{2}}+s_{2}\left|\bar{\bar{b}}_{i,m}^{*}\right|^{1+\frac{1}{2}} \tag{7.37}$$

where $\varsigma=\frac{1}{2}$,$s_{1}=\frac{1}{1+\varsigma}(2^{\varsigma-1}-2^{\varsigma^{2}-1})$, and $s_{2}=\frac{1}{1+\varsigma}(1-2^{\varsigma-1}+\frac{\varsigma}{1+\varsigma}+\frac{2^{-(\varsigma-1)^{2}(\varsigma+1)}}{1+\varsigma})$. by the same token

$$\tilde{\bar{\Omega}}_{i,1}\hat{\bar{\Omega}}_{i,1}^{3} \leq -\bar{\gamma}_{1}\left|\tilde{\bar{\Omega}}_{i,1}\right|^{1+3}+\bar{\gamma}_{2}\left|\bar{\Omega}_{i,1}\right|^{1+3} \tag{7.38}$$

$$\tilde{\bar{\bar{b}}}_{i,m}^{*}\hat{\bar{\bar{b}}}_{i,m}^{*3} \leq -\bar{\gamma}_{1}\left|\tilde{\bar{\bar{b}}}_{i,m}^{*}\right|^{1+3}+\bar{\gamma}_{2}\left|\bar{\bar{b}}_{i,m}^{*}\right|^{1+3} \tag{7.39}$$

$\bar{\gamma}_{1}=\frac{\xi}{1+\xi}(2^{\xi}-1)$, $\bar{\gamma}_{2}=\frac{\xi 2^{\xi}}{1+\xi}$, and $\xi=3$. Substituting (7.36)–(7.39) into (7.35), also based on Lemma 7.1, yields

$$\dot{V} \leq -\bar{h}_{1}\bar{V}^{\frac{3}{4}}-\bar{h}_{2}\bar{V}^{2}+W_{i}^{*} \tag{7.40}$$

where $\bar{h}_{1}=\min\{\bar{K}_{i,1},\bar{K}_{i,3},\widehat{\bar{\varepsilon}}_{i,1},\widehat{\bar{\varepsilon}}_{i,3},r_{i,2}^{-1}g_{i,1}\bar{\bar{b}}_{i,m}s_{1}2^{\frac{3}{4}},r_{i,1}^{-1}\bar{\varepsilon}_{i}s_{1}2^{\frac{3}{4}}\}$,

$\bar{h}_{2}=\frac{1}{6n^{2}}\min\{\bar{K}_{i,2},\bar{K}_{i,4},\widehat{\bar{\varepsilon}}_{i,2},\widehat{\bar{\varepsilon}}_{i,4},\bar{o}_{i,1}r_{i,1}^{-2}\bar{\gamma}_{1}2^{2},r_{i,2}^{-2}\bar{o}_{i,2}\bar{\bar{b}}_{i,m}\bar{\gamma}_{1}2^{2}\}$, and

$$W_i^* = \sum_{i=1}^{n} \{\mu_i^* + r_{i,1}^{-1} \bar{\varepsilon}_i s_2 |\bar{\Omega}_{i,1}|^{1+\frac{1}{2}} + \bar{o}_{i,1} r_{i,1}^{-2} \bar{\gamma}_2 |\bar{\Omega}_{i,1}|^{1+3}\} + \sum_{i=1}^{n} \{r_{i,2}^{-1} g_{i,1} \bar{b}_{i,m} s_2$$
$$|\bar{b}_{i,m}^*|^{1+\frac{1}{2}} + r_{i,2}^{-2} \bar{o}_{i,2} \bar{b}_{i,m} \bar{\gamma}_2 |\bar{b}_{i,m}^*|^{1+3}\}.$$

Under Lemma 7.3, $\bar{\upsilon}_{i,1}, \bar{\upsilon}_{i,2}, \bar{\psi}_{i,1}, \bar{\psi}_{i,2}, \tilde{\bar{\Omega}}_{i,1}$, and $\tilde{\bar{\bar{b}}}_{i,m}^*$ can converge in a fixed time $T_s \leq \frac{4}{\bar{h}_1 \eta} + \frac{1}{\bar{h}_2 \eta}$ to the set $(\bar{\upsilon}_{i,1}, \bar{\upsilon}_{i,2}, \bar{\psi}_{i,1}, \bar{\psi}_{i,2}, \tilde{\bar{\Omega}}_{i,1}, \tilde{\bar{\bar{b}}}_{i,m}^*) \in \{\bar{V} \leq \mathrm{M}^*\}$, where $0 < \eta < 1, \mathrm{M}^* = \min\{\mathrm{M}_1^*, \mathrm{M}_2^*\}, \mathrm{M}_1^* = (\frac{W_i^*}{(1-\eta)\bar{h}_1})^{\frac{4}{3}}, \mathrm{M}_2^* = (\frac{W_i^*}{(1-\eta)\bar{h}_2})^{\frac{1}{2}}$. Therefore $\bar{\upsilon}_{i,1}, \bar{\upsilon}_{i,2}, \bar{\psi}_{i,1}, \bar{\psi}_{i,2}, \tilde{\bar{\Omega}}_{i,1}$, and $\tilde{\bar{\bar{b}}}_{i,m}^*$ are uniformly bounded at a fixed time. According to the formula $\bar{\upsilon}_{i,j}(t) = \bar{\partial}_{i,j}(t) - \bar{\psi}_{i,j}(t)$, as well as formula (7.9), for time, $\bar{\alpha}_i(t)$ is eventually possible to converge to the set $|\bar{\alpha}_i(t)| \leq \min\{2\sqrt{2M_1^*} + \bar{\alpha}_{i,c}, 2\sqrt{2M_2^*} + \bar{\alpha}_{i,c}\} \triangleq \bar{\alpha}_i$. Therefore, the spacing error $\bar{\sigma}_i(t)$ between two neighboring vehicles can eventually converge to the set $\bar{\sigma}_i(t) \in \{|\bar{\sigma}_i(t)| \leq \bar{\chi}_i(t)\bar{\phi}_i(\bar{\alpha}_i)\}$, which can satisfy the prescribed constraint conditions in formula (7.6). Therefore, the system is practically stable in fixed time under the nonlinear conditions of fault and input dead zone, that is, the closed-loop stability of the system has been proved.

Based on the above analysis, and according to formula (7.6) and (7.7), for, the spacing error $\bar{\sigma}_i(0)=0$ is constant, and the spacing error $\bar{\sigma}_i(t)$ is practically stable at fixed time under the prescribed asymmetric constraint performance (7.6). By the definition of string stability in Chap. 4, for $\exists \xi > 0$, and any $\theta > 0$, we can get that $\|\bar{\delta}_i(0)\|_\infty < \xi \Rightarrow \sup_i \|\bar{\delta}_i(\cdot)\| < \theta$ is true. Therefore, the string stability of the considered vehicle platoon system is guaranteed in the sense of distributed prescribed performance fixed time stability. □

## 7.5 Method Verification and Analysis

In this section, we further verify the effectiveness and superiority of the proposed scheme through comparative experiments. In the experimental analysis, the HICVPS considered is composed of three follower vehicles and one leader vehicle. The system parameters of the vehicle are given as $\bar{\varphi}^* = 1.2\,\mathrm{kg/m^3}$, $A_i = 2.2\,\mathrm{m^2}$, $D_{mi} = 5\,\mathrm{N}$ and $K_{ai} = 0.35$.

The heterogeneous dynamic characteristic parameters of each vehicle are as follows: for the vehicle's mass, $\bar{m}_1 = 1200$ kg, $\bar{m}_2 = 1100$ kg, and $\bar{m}_3 = 1350$ kg. For the time lag of engine, $\tau_1 = 0.12$, $\tau_2 = 0.3$, and $\tau_3 = 0.25$. For the length of the follower vehicle, $l_0 = 3$ m, $l_1 = 4.5$ m, $l_2 = 2$ m. For the initial states of the vehicles $[\bar{p}_j(t), \bar{v}_j(t), \bar{a}_j(t)]^T, j = 0, 1, 2, 3,$ there are $[128, 1.8, 0]^T, [118, 1.6, 1.1]^T, [103, 1.0, 3.4]^T, [88, 0.9, 2.5]^T$. The requirements of inter-vehicle safety distance $\bar{d}_{i,des} = 5$ m, and the constant time headway constant $H = 0.8$.

The external disturbances suffered by the system is considered as $w_{i,v}(t) = 0.02 \times \sin(2\pi t)$, $w_{i,a}(t) = 0.01 \cos(0.5\pi t)$, respectively. Considering the input dead zone nonlinear parameters as $\bar{\varphi}_{i,R} = 0.02$ and $\bar{\varphi}_{i,L} = 0.04$, $\bar{b}_{i,R}(t) = 0.9 + 0.01e^{-t}$ and $\bar{b}_{i,L}(t) = 1.2 + 0.002e^{-t}$. Simultaneously, the parameters of the upper and lower bounds of the prescribed asymmetric constraint function are $\bar{\sigma}_{i,L} = 1$ and $\bar{\sigma}_{i,H} = 1.6$, respectively. For the performance decay function $\bar{\chi}_i(t)$, we take that $\bar{\chi}_i(0) = 0.3$, $\bar{\chi}_i(\infty) = 0.02$, $T_\chi = 3.2$.

For simulating the real situation and verify the effectiveness and resilience of the proposed method, the second vehicle and the third vehicle were respectively considered to suffer from various types of faults at different times. Considering the case that the second vehicle suffered periodic faults, the fault function is as follows: $\bar{\Lambda}_2(t) = \begin{cases} 0, t < 35s \\ 18\sin(0.5t), t \geq 35s \end{cases}$. At the same time, the case of the third vehicle suffering from intermittent fault is considered, and its fault function form is as follows: $\bar{\Lambda}_3(t) = -0.5 \overset{\frown}{\lambda}_{i,1}\ sign(\sin(2\pi(t - t_{i,F})T^{-1})) + 0.5 \overset{\frown}{\lambda}_{i,2}$, where $t_{3,F} \in [15, 25)$, $\overset{\frown}{\lambda}_{i,1} = 20$, $\overset{\frown}{\lambda}_{i,2} = 5$, $T = 10s$.

In this chapter, the experimental results of the proposed scheme applied under fault free conditions are shown in Figs. 7.2 and 7.3. Figure 7.2 shows that with the proposed scheme, under the conditions of external disturbances and input dead zone nonlinearities, the spacing error of the vehicle platoon system can rapidly converge to zero within the prescribed constraint range, and its convergence time is only 2.85 s, and the system can maintain good closed-loop stability and the string stability. Figure 7.3 shows the position, velocity, acceleration and other state information of the vehicle platoon system, as well as the control input. The results show that the follower vehicle can effectively track the velocity and acceleration movement of the leader vehicle under prescribed constraints, and the position information shows that no collision occurs during the control process. The simulation experiment further considers the distributed fixed-time resilient control with prescribed constraint performance under multiple faults to further verify the superiority of the proposed scheme. And we take the existing distributed prescribed performance fixed-time resilient control method [9] and the method without prescribed performance constraints as a comparison reference. Figure 7.4 shows the spacing error performance of the proposed method under dead zones and multiple faults compared with existing distributed prescribed

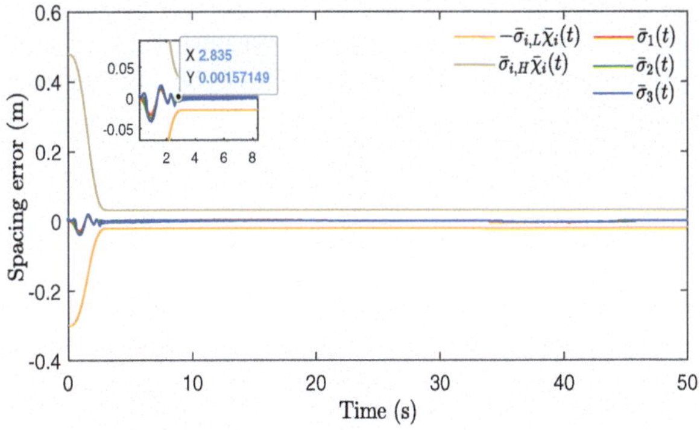


**Fig. 7.2** The inter-vehicle spacing errors $\bar{\delta}_i(t)$ under proposed method with dead-zone and no faults

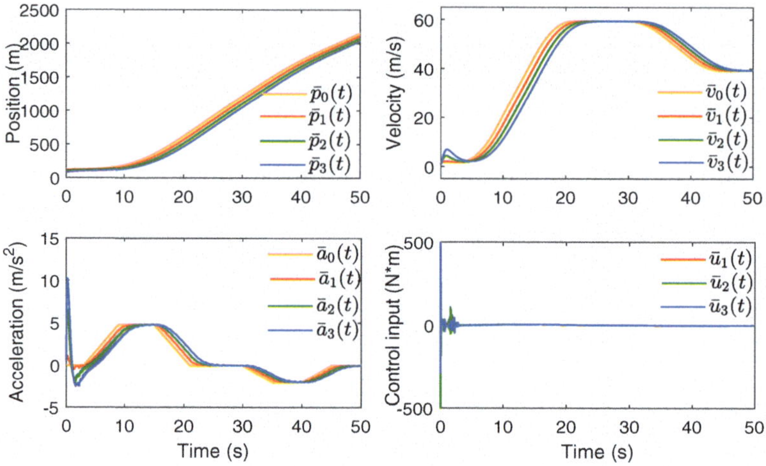


**Fig. 7.3** The system states and control input under proposed method with dead-zone and no faults

performance fixed time resilient control methods and methods without prescribed performance constraints.

As can be seen from Fig. 7.4, compared with the existing method in Fig. 7.4b and the method without prescribed performance constraints in Fig. 7.4c. The proposed method in Fig. 7.4a not only has faster convergence speed and higher convergence accuracy under prescribed constraints, but also maintains better string stability performance when different vehicles suffer periodic faults and intermittent faults respectively. Its convergence accuracy is always kept within the given 0.02 m. It also has high resilience for a variety of faults, and the spacing error accuracy can always be maintained within 0.01 m before and after the fault occurs. The existing method in Fig. 7.4b has a large error fluctuation after the fault occurs, and the error accuracy is greater than 0.01 m. The method in Fig. 7.4c without prescribed performance constraints cannot maintain the spacing error accuracy of the prescribed constraints after the fault occurs. Figures 7.5, 7.7, and 7.8 respectively show the system state and control inputs under input dead zones and multiple faults for the proposed method, the existing distributed prescribed performance fixed time resilient control method, and the method without prescribed performance constraints. Figure 7.6 shows the compensation law designed for the proposed method under input dead zones and multiple faults.

Based on the above analysis, the effectiveness and superiority of the proposed scheme are verified when the HICVPS suffers from dead zone nonlinearity, external disturbances and multiple fault threats. It is also proved that the distributed adaptive fixed-time resilient control scheme proposed in this chapter under prescribed asymmetric constraint performance can effectively improve the resilience performance, transient and steady-state performance of the system. The desired spacing error accuracy can be maintained under the proposed method, and the closed-loop stability and string stability of the system are improved.

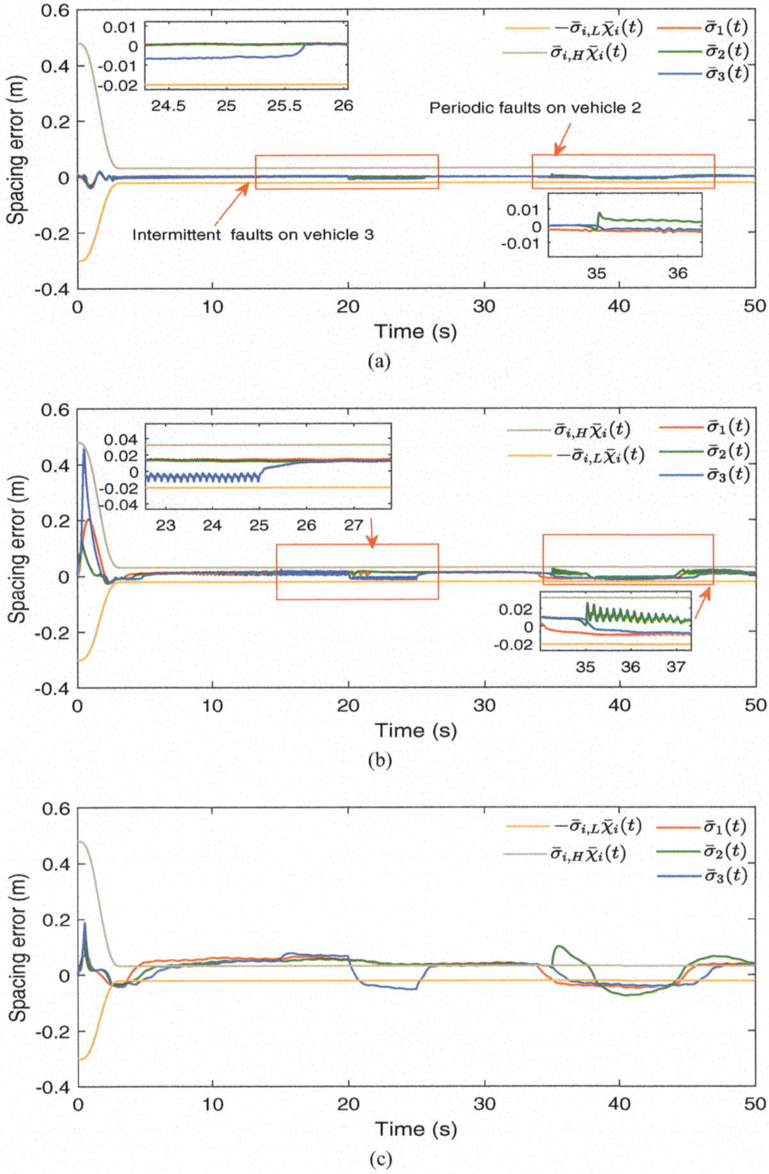


**Fig. 7.4** The spacing errors under proposed method and compared methods with dead-zone and faults. **a** $\bar{\delta}_i(t)$ under the proposed method in Theorem 1, **b** $\bar{\delta}_i(t)$ under the existing distributed prescribed performance fixed-time resilient control method, **c** $\bar{\delta}_i(t)$ under the method without prescribed performance constraint control

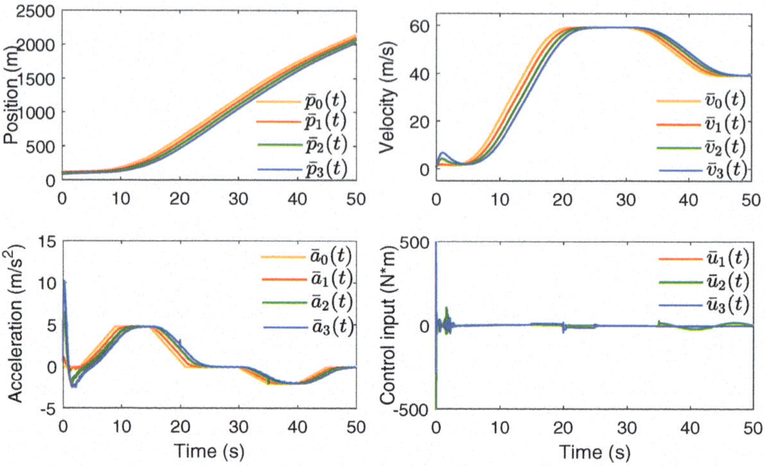


**Fig. 7.5** The system states and control input under proposed method with dead-zone and faults

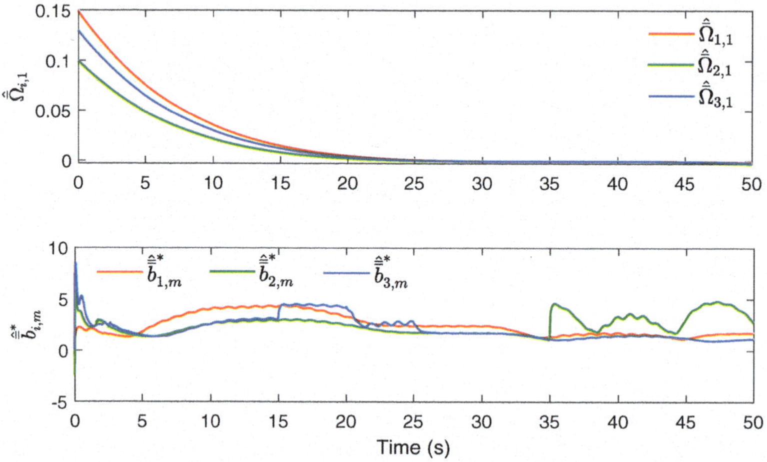


**Fig. 7.6** The adaptive compensation laws under proposed method with dead-zone and faults

## 7.6 Summary

The distributed resilient control problem of HICVPS subjected to unknown actuator additive fault, unknown input dead zone nonlinearity and external disturbances is studied. Firstly, a distributed adaptive fixed-time resilient control scheme with prescribed asymmetric constraint performance is designed to ensure the transient and steady-state performance of the vehicle platoon system. In the sense of fixed-time Lyapunov stability, the spacing errors are achieved to converge to the prescribed

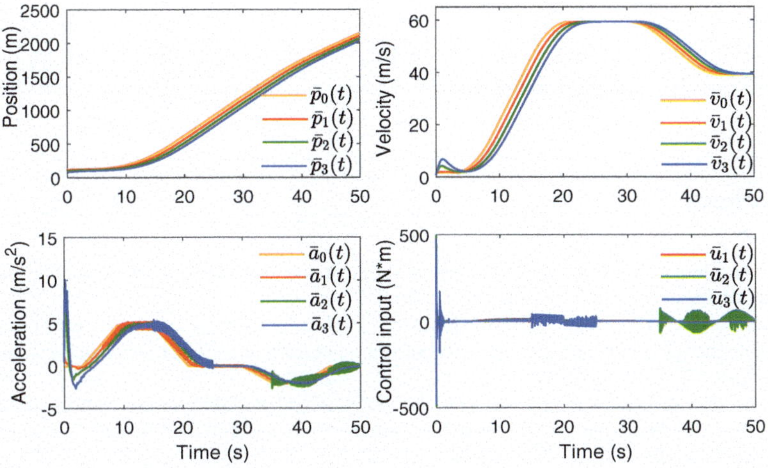


**Fig. 7.7** The system states and control input under the existing distributed prescribed performance fixed-time resilient control method with dead-zone and faults

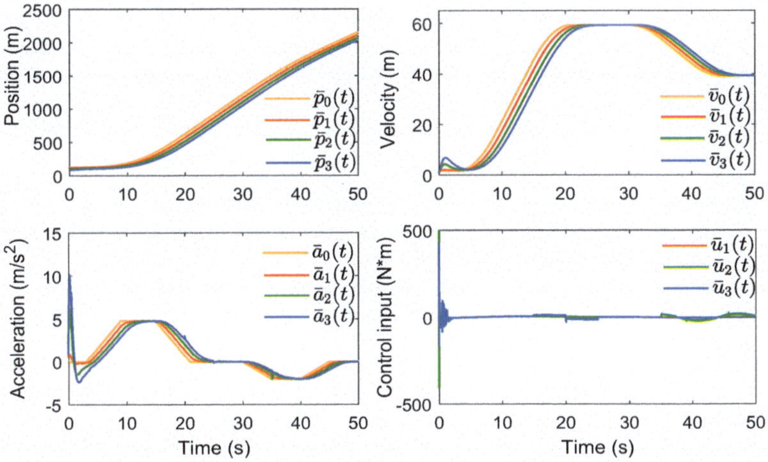


**Fig. 7.8** The system states and control input under the method without prescribed performance constraint control with dead-zone and faults

asymmetric constraint domain within a fixed time. The stability of closed-loop stability and string stability are also ensured. The spacing error accuracy and resilience performance of the intelligent connected vehicle platoon system can be improved. At the same time, the adaptive fixed time compensation laws are designed to eliminate the influence of fault, input dead zone and external disturbances on the stability of the HICVPS. In this chapter, the longitudinal control strategy of HICVPS is mainly considered. However, the design that comprehensively considers both longitudinal control and lateral control is more suitable for practical application. In the future research work, the prescribed time control scheme of HICVPS under longitudinal and lateral conditions will be further considered to improve the engineering application capability of the designed controller.

## References


1. Guo G, Zhao Z (2023) Finite-time terminal sliding mode control of connected vehicle platoons. *Control Theory & Applications*, 40(1): 149.
2. Qiu J, Sun K, Wang T, et al (2019) Observer-based fuzzy adaptive event-triggered control for pure-feedback nonlinear systems with prescribed performance. *IEEE Transactions on Fuzzy Systems*, 27(11): 2152–2162.
3. Guo X, Wang J, Liao F, et al (2017) Distributed adaptive sliding mode control strategy for vehicle-following systems with nonlinear acceleration uncertainties. *IEEE Transactions on Vehicular Technology*, 66(2):981–991.
4. Chen C, Xiao T, Qiu T, et al (2020) Smart-contract-based economical platooning in blockchain-enabled urban internet of vehicles. *IEEE Transactions on Industrial Informatics*, 16(6): 4122–4133.
5. Guo G, and Li D (2019) Adaptive sliding mode control of vehicular platoons with prescribed tracking performance. *IEEE Transactions on Vehicular Technology*, 68(8): 7511–7520.
6. Zhu Y, Wu J, and Su H (2022) V2V-based cooperative control of uncertain, disturbed and constrained nonlinear CAVs platoon. *IEEE Transactions on Intelligent Transportation Systems*, 23(3):1796–1806.
7. Shen H, Yue W (2022) Coupled sliding mode control with actuator faults and saturation of autonomous platoon. *Control Engineering of China*, https://doi.org/10.14107/j.cnki.kzgc.20210677.
8. Guo X, Wang J, Liao F, et al (2019) Neuroadaptive quantized PID sliding-mode control for heterogeneous vehicular platoon with unknown actuator dead-zone. *International Journal of Robust and Nonlinear Control*, 29(1): 188–208.
9. Cui G, Yang W, Yu J, et al (2022) Fixed-time prescribed performance adaptive trajectory tracking control for a QUAV. *IEEE Transactions on Circuits and Systems II: Express Briefs*, 69(2):494–498.
10. Guo G, Zhang X, Gao Z (2022) Finite-time Fixed configuration formation control of intelligent vehicles with prescribed transient and steady-state performance. *China Journal of Highway and Transport*, 35(03):28–42.
11. Yu M, and Wang D (2014) Model-based health monitoring for a vehicle steering system with multiple faults of unknown types. *IEEE Transactions on Industrial Electronics*, 61(7): 3574–3586.
12. Liu D, Liu Z, Chen C L P, et al (2022) Distributed adaptive neural fixed-time tracking control of multiple uncertain mechanical systems with actuation dead zones. *IEEE Transactions on Systems, Man, and Cybernetics: Systems*, 52(6): 3859–3872.
13. Sun X, Guo G, Zhang P (2021) Fixed-time consensus tracking of multi-agent systems under unmatched disturbances. *Acta Automatica Sinica*, 47(6): 1368–1376.

# Chapter 8
# Summarization and Prospect

## 8.1 Summary of Full Book

In this book, based on the problems of physical faults, communication faults under malicious cyber attack and the external disturbances, the adaptive resilient control of intelligent connected vehicle platoon system under fault conditions is studied from the perspective of resilient control. Under the theory of finite-time and fixed-time stability, the adaptive resilient controller design, the inter-vehicle spacing policy design and the disturbance compensation method of intelligent connected vehicle platoon system are studied, and the closed-loop control stability and string stability of the system are analyzed. The research of this book plays an important role in ensuring the safety of intelligent connected vehicle platoon system longitudinal driving and improving the rapid recovery ability and resilience ability of the system under fault conditions. The specific research carried out in this book is summarized as follows

(a) In Chap. 2, several resilient control methods of the intelligent connected vehicle platoon system, such as coupled sliding mode control, command filtered back-stepping control are introduced. Then, the basic structures and design processes of the distributed longitudinal platoon resilient control methods are also discussed and analyzed.
(b) In Chap. 3, a distributed adaptive resilient control method for intelligent connected vehicle platoon system with actuator faults and input saturation is studied. Firstly, the quadratic spacing policy is introduced to ensure the strict string stability of fault and disturbance decreasing layer by layer along the vehicle platoon system length. Then, the adaptive updated and compensation laws are designed to compensate for input saturation and environmental disturbance. Then, a finite time adaptive coupled sliding mode robust resilient controller is designed to ensure the resilient performance and fast stability of the closed-loop control system. The strict string stability of the system is proved by analysis, and its sufficient conditions are given. Based on Lyapunov stability theory, it is proved

 
C. Pan and Y. Chen, *Resilient Control of Intelligent Connected Vehicle Platoon Systems*,
https://doi.org/10.1007/978-981-97-8850-7_8

that the closed-loop system is globally finite-time stable. Finally, the feasibility of the proposed scheme is verified through comparative cases analysis.

(c) A distributed adaptive finite time resilient control method for intelligent connected vehicle platoon systems with improved exponential spacing policy is studied in Chap. 4. Firstly, based on the Chap. 3, by introducing the upper bound of fault factors and the velocity information of the leader vehicle, the improved exponential spacing policy is designed to improve the traffic flow capacity and ensure the string stability of the system. Secondly, based on the error compensator, a nonlinear observer is designed to compensate for the bias fault and the external disturbances. Then, based on the adaptive updated laws of unknown parameters, the distributed adaptive finite time resilient control scheme is designed to compensate the influence of faults. The global finite-time stability, string stability and traffic flow stability of the closed-loop system are proved by theoretical analysis. Finally, the effectiveness of the proposed scheme is verified by the comparative cases.

(d) A distributed adaptive prescribed performance fast finite time resilient control method for the intelligent connected vehicle platoon system under compound cyber attacks is studied in Chap. 5. Firstly, a nonlinear state compensator is constructed to ensure the boundness of system state signals under communication faults caused by compound cyber attacks. Secondly, the adaptive updated laws are designed to realize the effective estimation of the unknown fuzzy weight and the upper bound of the unknown communication faults. Based on the improved exponential spacing policy in Chap. 4 and the asymmetric Barrier Lyapunov function, the distributed adaptive fast finite time resilient controller is designed to ensure the string stability under prescribed performance, the fast finite time stability of the closed-loop system, and the transient/steady state performance of the system. Finally, the comparative cases analysis prove the effectiveness and superiority of the proposed scheme.

(e) In Chap. 6, an event-triggered based distributed adaptive fixed time resilient control method for intelligent connected vehicle platoon system under communication faults and disturbances caused by unknown deception attacks is studied. Firstly, a sliding mode fixed time disturbances observer is constructed to quickly estimate the unknown external disturbances under arbitrary initial conditions. Secondly, based on the exponential spacing policy and proceeding-leader-following information flow topology, a distributed adaptive event-triggered fixed time resilient control scheme is designed to ensure the closed-loop control stability and string stability in the sense of practical fixed time of the system, while reducing data transmission resources and control energy consumption. The fuzzy adaptive fault compensation laws are designed to overcome the influence of communication faults under unknown cyber attacks. Finally, the effectiveness of the proposed scheme is verified by the comparative cases and the SUMO-Matlab joint simulation platform.

(f) A distributed adaptive prescribed performance fixed-time resilient control method for intelligent connected vehicle platoon system with unknown input dead zones, disturbances and actuator faults is studied in Chap. 7. Firstly, by introducing the exponential spacing policy based on the topology of the proceeding-following information flow, the inter-vehicle information interaction is reduced on the basis of the resilient control in Chap. 6. Secondly, a distributed adaptive resilient controller is designed by combining the prescribed fixed-time decay function and the asymmetric spacing error constraint function. The theoretical analysis shows that the practical fixed-time stability and string stability of the closed-loop system can be guaranteed under the prescribed constraint conditions. The fixed time adaptive compensation laws are designed to realize the effective compensation of unknown dead zone parameters and faults. Finally, the effectiveness of the proposed scheme is verified through comparative cases analysis.

## 8.2 Future Work Prospects

Due to the physical and communication faults continue to threaten the stable operation of the intelligent connected vehicle platoon system and seriously hinder the application of the intelligent connected and automated vehicle industry. Multi-dimensional vehicle platoon system modeling, the distributed control and economic optimization problems in this field still have a lot of research space. Based on the existing work progress in this book, further research needs to be carried out in the future as follows:

(a) Distributed multi-task control model issue of two-dimensional intelligent connected vehicle platoon system under the cyber attacks and physical faults, as shown in Fig. 8.1. Considering the characteristics of cooperative control and

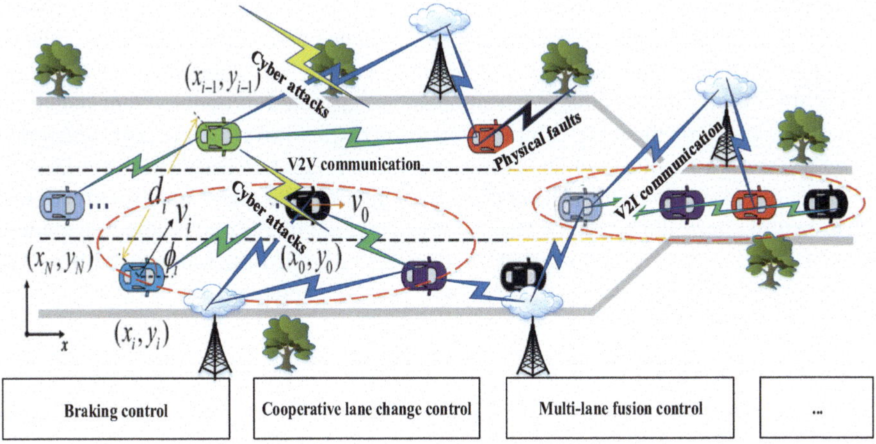


**Fig. 8.1** Distributed multi-task control model of two-dimensional intelligent connected vehicle platoon system under the cyber attacks and physical faults

network communication under braking control, cooperative lane change and multi-lane fusion, the two-dimensional distributed control model of intelligent networked vehicle system under multiple tasks can be constructed. Firstly, based on the longitudinal dynamics and lateral dynamics of the intelligent connected and automated vehicle model, the multi-task characteristics of braking, collaborative lane change and multi-lane fusion, and the communication structure characteristics of the vehicle platoon system, the two-dimensional distributed multi-task mathematical model of the intelligent connected vehicle system based on vehicle-road-network can be constructed. Secondly, considering denial of service attacks, false data injection attacks, man-in-the-middle attacks and other threats, based on the resilient control technology, the distributed control architecture of two-dimensional intelligent connected vehicle platoon system under multi-task conditions can be constructed to realize the establishment of distributed multi-task control model and ensure the multi-task execution and stable operation ability of the vehicle platoon system under fault and attacks conditions.

(b) Active/passive based distributed cooperative secure control of two-dimensional intelligent connected vehicle platoon system based on multiple attacks, as shown in Fig. 8.2. On the one hand, considering the false signals influence induced by malicious communication attacks in the sensors and actuators of intelligent connected vehicle platoon system, a distributed cooperative secure controller is designed to improve the security and stability ability of the system in the sense of passive security control. On the other hand, for solving the information security problem of interactive transmission of high-speed mobile V2V communication in the vehicle platoon system under faults induced by cyber attacks, the distributed cooperative secure control scheme based on privacy preservation can be studied. In the sense of active security defense, the security of VANETs communication interaction and system string stability control can be improved.

(c) Distributed resilient optimal control of two-dimensional intelligent connected vehicle platoon system based on dynamic event-triggered strategy. As shown in Fig. 8.3, considering the dynamic performance and control economic cost of closed-loop system under the condition of the physical faults and the communication faults induced by cyber attacks, a distributed dynamic event-triggered scheme based on the inter-vehicle interaction history information can be designed to better save the inter-vehicle communication interaction resources. At the same time, based on the optimization objective function, the distributed resilient optimization controller based on the two-dimensional intelligent connected vehicle platoon system can be designed to ensure the closed-loop control stability and string stability, and ensure the economy of the system control.

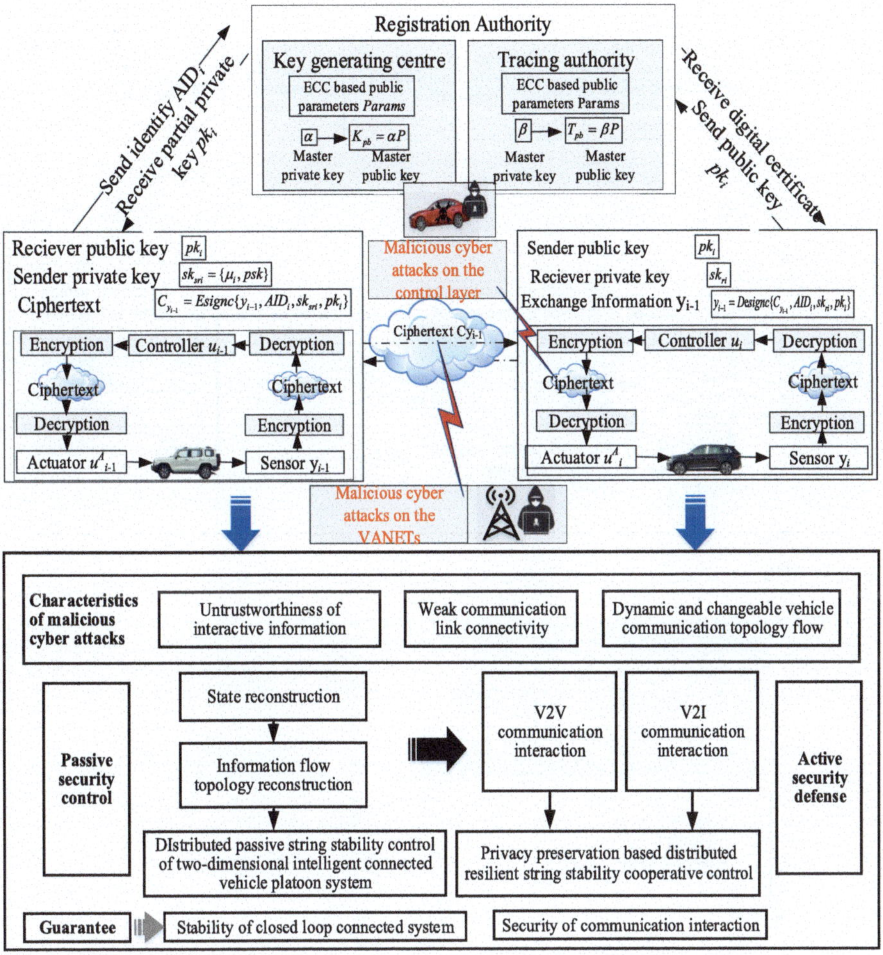


**Fig. 8.2** Active/passive based distributed cooperative secure control of two-dimensional intelligent connected vehicle platoon system based on multiple attacks

(d) As shown in Fig. 8.4, considering the physical faults and the communication faults induced by the deception attack and denial of service attacks that VANETs and feedback loop communication interaction may suffered from under actual intelligent transportation environment, the algorithm proposed can be applied in the Webots-Matlab joint simulation platform and SUMO-MATLAB co-simulation platform. The influence of faults on system stability can be analyzed and the effectiveness of control scheme can be verified. Combined with Omnet++-Veins-SUMO joint vehicular network simulation platform and dedicated short-range communication technology, it can simulate the high-speed

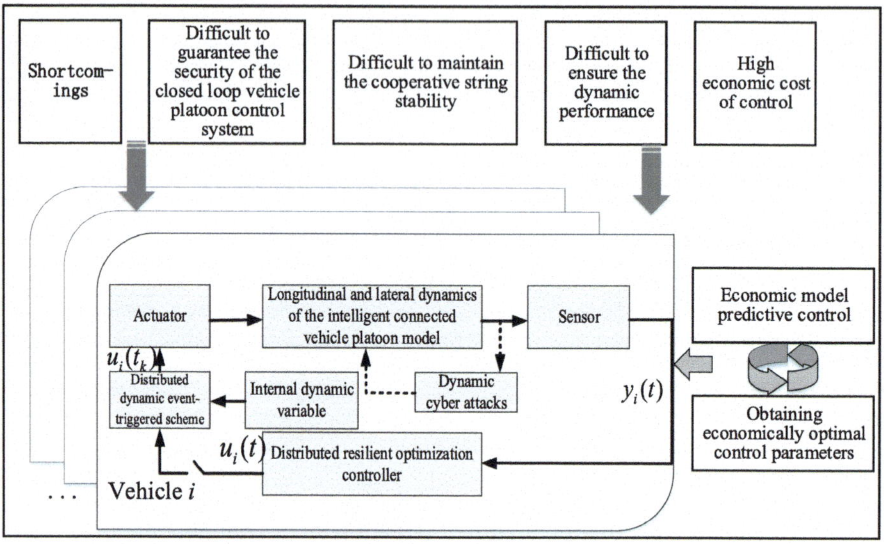


**Fig. 8.3** Distributed resilient optimal control based on dynamic event-triggered strategy

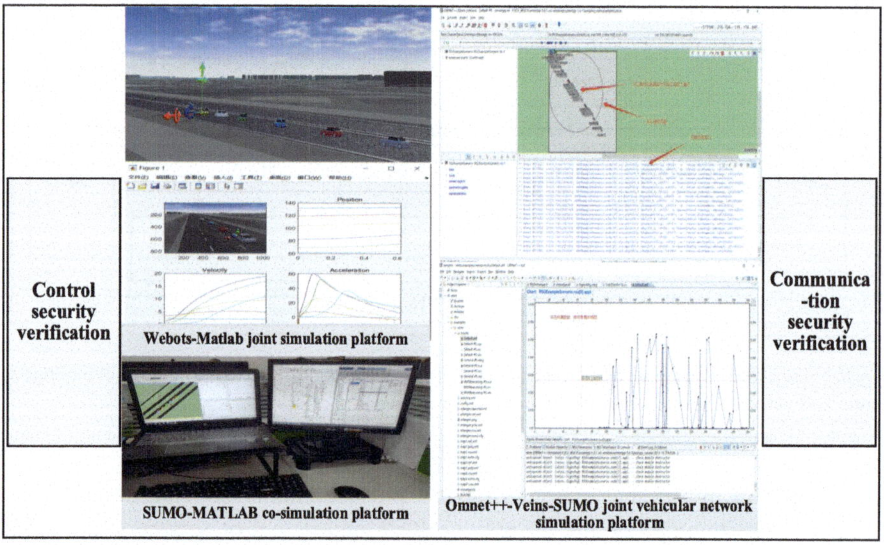


**Fig. 8.4** Experiment platform that can be used to analyze and verify the security of the control scheme

mobile communication in real road vehicular network, analyze the real impact of faults on the communication layer of the vehicle platoon system, and further the effectiveness of the control algorithm proposed in this book can be verified.

The manufacturer's authorised representative in the EU is Springer Nature Customer Service Centre GmbH, Europaplatz 3, 69115 Heidelberg, Germany. If you have any concerns regarding our products, please contact ProductSafety@springernature.com

Printed and bound by CPI Group (UK) Ltd, Croydon, CR0 4YY
01/10/2026
02225980-0018